695
X

RIVERSIDE STUDIES IN BIOLOGY

Editors: William C. Steere
*The New York Botanical Garden*
H. Bentley Glass
*State University of New York
at Stony Brook*

# THE EVOLUTION AND CLASSIFICATION OF FLOWERING PLANTS

# THE EVOLUTION AND CLASSIFICATION

HOUGHTON MIFFLIN COMPANY · BOSTON

CHARLES E. BESSEY (1845–1915), author
of a phylogenetic system and a set of
dicta which profoundly influenced
subsequent taxonomic thought.

*Arthur Cronquist*

SENIOR CURATOR, THE NEW YORK BOTANICAL GARDEN

# OF FLOWERING PLANTS

NEW YORK   ATLANTA   GENEVA, ILL.   DALLAS   PALO ALTO

*Note*: This study was supported in part by grants from the National Science Foundation to the New York Botanical Garden, and also in part by a grant from the Bache Fund of the National Academy of Sciences. The diagrams showing evolutionary relationships are specifically released from copyright.

# Editor's Introduction

WHEN AN editor is called upon to introduce a unique book presenting major new concepts, his sense of pleasure is equalled only by his sense of novelty or even of incredulity. So many "new" texts and reference works today are only abridgments or extensions of earlier works, or updated versions of standard works, that it is an especially pleasant surprise to encounter a book that has not been written again and again, or that, in fact, did not exist until now.

*The Evolution and Classification of Flowering Plants* defines taxonomy as "a study aimed at producing a system of classification of organisms which best reflects the totality of their similarities and differences." Other definitions are possible, and the author's definition could be elaborated upon. In any event, plant taxonomy, or systematic botany, is a transcendental discipline which comprehends and uses information from all other disciplines and sub-disciplines of science; it is an information system containing all pertinent knowledge relating to all plants.

This book now provides one of the few, if not the only, published works in which a student or professional botanist can find a clear and orderly exposition of the basic principles of modern broad-scale taxonomy, as well as the philosophic concepts behind them. It is also the only comprehensive work in which one can find in English a character-by-character treatment of probable evolutionary trends in the angiosperms, based on present-day information and conceptual thought. Furthermore, it throws much new light on questions of evolutionary parallelism, on the relationship between phylogeny and taxonomy, and on the possible evolutionary significance of the nature of the supply of mutations. Students will find especially useful and stimulating the major chapter concerned with the evolution of characters in flowering plants, based on evidence derived from structure, physiology, and ecology.

The greater part of this book is devoted to the presentation of a new and comprehensive system for the angiosperms, dealing particularly with the grouping of families into orders and subclasses on a world-wide basis, something not undertaken by an American botanist since the inspired work of Charles Bessey in the latter part of the nineteenth century. However, far from being parochial or chauvinistic, this new system reflects not only the author's own ideas and research over the years, but also his professional association and collaboration with other outstanding botanists of other countries who have turned their intellectual efforts to finding the most accurate system of higher plants, especially the Russian evolutionist, Armen Takhtajan, and the German, Walter Zimmermann. Among systems, the one presented here is most nearly like that of Takhtajan, published in Russian in 1966, but with many significant differences. Of many systems of angiosperms published in many countries during many eras, the present system is unique in presenting first a comprehensive statement of the philosophical concepts as well as the body of factual information upon which it is based.

This is the only book in any language in which the essential information concerning the arrangement of the orders and families of angiosperms is summarized in the form of clear synoptical keys. The author makes clear that these keys are intended as conceptual aids and not as a precisely accurate means for identification, since they must necessarily ignore or minimize many of the exceptions to the characters listed. These keys summarize for the reader what sorts of things are included in the various groups, and what are the most nearly constant differences between groups, without forcing him to seek this information by laborious line-by-line comparison of detailed descriptions which also provide for all the exceptions.

Finally, the critical synthesis of differing points of view, different systems, and an enormous amount of contemporary information will make this book an indispensable reference work for a long time to come.

William C. Steere, Editor

*Bronx, New York*

# Introduction

THIS BOOK reflects an attempt to develop a general system of angiosperms compatible with presently available information, and to put that system in a proper theoretical frame of reference It is directed to students who have at least a nodding acquaintance with plant taxonomy, and to professional botanists. The glossary is intended mainly as an aid to memory; the student must be reasonably familiar with the taxonomic vocabulary in order to understand the text.

The treatment here presented is a synthesis based on the literature, on discussions with other botanists, and on my own observations. With regard to the dicotyledons, it is an elaboration and modification of the scheme which I presented in 1957.[1] I have used the herbarium and living collections of the New York Botanical Garden to check some points from the literature and to extend my familiarity with the kinds of plants, but in general I have accepted published factual observations as correct. The formal morphology of the families of angiosperms is well covered in the 12th edition of the Engler Syllabus, of which I have made extensive use. The taxonomic and phylogenetic interpretations are my own, by birth or adoption.

The spelling of family names herein follows the list of conserved families in the current (1966) edition of the International Code of Botanical Nomenclature. Although I am not convinced that the authors of the list always reached the right conclusion, I am not prepared to sustain an argument to the contrary in any individual case. The divisional and class names follow the recent paper of Cronquist, Takhtajan, and Zimmermann.[2]

It will be observed that I have accepted a number of small families that have more commonly been submerged in larger ones, while rejecting others. I am not fond of monotypes, and I prefer to submerge them in larger groups when this can be done without undue violence to phyletic or phenetic unity. Segregates which represent intermediates between two larger families are of doubtful value, and I have generally rejected them. The Penthoraceae (linking the Saxifragaceae and Crassulaceae) and Hype-

[1] Outline of a new system of families and orders of dicotyledons. Bull. Jard. Bot. Brux. 27: 13–40. 1957.

[2] Cronquist, A., A. Takhtajan, and W. Zimmermann. On the higher taxa of Embryobionta. Taxon 15: 129–134. 1966.

coaceae (linking the Papaveraceae and Fumariacee) are examples of this type. On the other hand, it is becoming increasingly clear that a number of small groups do not fit properly into any of the traditional families, and that to retain them in their customary havens is more misleading than useful. The three genera here grouped (following Airy-Shaw) as the Alseuosmiaceae, for example, simply do not belong with the Caprifoliaceae, to which they have generally been assigned, nor can they properly be squeezed into any other established family. It is to be expected that other small groups now quietly hiding in large families may have to be segregated as they are reconsidered or as more information becomes available.

It will not escape notice by the critical reader that much of the interpretation here presented is remarkably similar to that given by Takhtajan in 1959 and 1966. This similarity reflects a community of interest and outlook, bolstered by a mutual exchange of views over the past ten years. Our ideas were initially developed independently, but since 1957 we have been in frequent correspondence, and I spent a month consulting with him in Leningrad in 1965. An outline of his newest system, and later the book itself, were available to me during the closing stages of the preparation of this work. We venture to hope that the similarity of our views reflects the necessities of the prsent state of knowledge and heralds the approach of a generally acceptable scheme to replace the highly useful but now moribund Englerian arrangement of families and orders of angiosperms.

Many people have helped me in one way or another during the preparation of this manuscript and while my thoughts were developing. A complete list would probably have to include nearly all of my professional associates, beginning with Ray J. Davis, from whom I took my first botany course. I can not now recall to what extent some of my ideas are original, and to what extent they spring from seeds planted in my mind by others. It should be understood that not all of the people to whom I acknowledge help will agree with the use to which I have put their thoughts or information; some will, I am sure, take vigorous exception.

With these caveats, I present a partial list of people who, in conversation or correspondence, have helped shape my ideas or my presentation of the subject of this book, or who have provided specific information used in the preparation of the manuscript. I am grateful to all of them, and also to those whose contribution I can not now isolate in my memory. Ernst Abbe, Caroline Allen, Edgar Anderson, Chester Arnold, Irving Bailey, Margarita Baranova, Elso Barghoorn, F. A. Barkley, T. M. Barkley, Rupert Barneby, James Brewbaker, Franz Buxbaum, W. H. Camp, James Canright, Sherwin Carlquist, Vernon Cheadle, R. T. Clausen, Herbert Copeland, John Cronquist, Calaway Dodson, James Doyle, John Ebinger, Richard Eyde, David Fairbrothers, D. Frohne, R. Darnley Gibbs, George Gillett, Verne Grant, Wayne Handlos, Charles Heiser, C. L. Hitchcock,

Noel Holmgren, Richard Howard, Hugh Iltis, Howard Irwin, Richard M. Klein, Tetsuo Koyama, Robert Kral, Harlan Lewis, Walter Lewis, Thomas Mabry, Bassett Maguire, Ernst Mayr, Rogers McVaugh, C. R. Metcalfe, H. E. Moore, Maynard Moseley, L. I. Nevling, Dan Nicolson, Marion Ownbey, Barbara Palser, Willard Payne, Ghillean Prance, Peter Raven, John Reeder, D. J. Rogers, Reed Rollins, Rudolf Schuster, Albert C. Smith, C. Earle Smith, Lyman B. Smith, William L. Stern, Armen Takhtajan, Sophia Tamamschian, Robert Thorne, B. L. Turner, Warren Wagner, Thomas Wilson, Carroll Wood, Walter Zimmermann, George Bunting, Ulrich Hamann, L. J. Harms, F. P. Jonker, and Paul Richardson.

The references which are inserted here and there are intended to lead the student into the pertinent literature, and in some cases to document a factual statement or acknowledge the source of an interpretation. I have not hesitated to cite a secondary rather than a primary reference when it suited my purposes to do so; review papers have their uses. When similar information is available both in English and in some other language, I have usually cited the English-language reference, for the convenience of the English-speaking readers who presumably make up the majority of my audience. No attempt has been made to cite all of the pertinent papers. More comprehensive bibliographies can be found in the 12th edition of the Engler Syllabus (1964) and in Takhtajan's 1959 and 1966 books, cited below.

There are several books and papers which I have used so extensively that I prefer to list them here instead of repetitively throughout the text. These are:

Brewbaker, J., The distribution and phylogenetic significance of binucleate and trinucleate pollen grains in angiosperms. Am. Jour. Bot. 54: 1069–1083. 1967. (Manuscript available in advance of publication)

Brown, W. H., The bearing of nectaries on the phylogeny of flowering plants. Proc. Am. Phil. Soc. 79: 549–595. 1938.

Eames, A., *Morphology of the Angiosperms* New York: McGraw-Hill, 1961.

Erdtman, G., *Pollen Morphology and Plant Taxonomy.* Angiosperms. Stockholm: Almqvist & Wiksell, 1952.

Hegnauer, R., Chemotaxonomie der Pflanzen. Band 2. Monocotyledoneae. 1963. Band 3. Dicotyledoneae: Acanthaceae — Cyrillaceae. 1964. Band 4. Dicotyledoneae: Daphniphyllaceae — Lythraceae. 1966. Basel: Birkhauser Verlag.

Hutchinson, J., *The Families of Flowering Plants,* Two Vols., 2nd ed. New York: Oxford University Press. 1959.

Maheshwari, P., An *Introduction to the Embryology of the Angiosperms.* New York: McGraw-Hill. 1950.

Martin, A. C., The comparative internal morphology of seeds. Am. Midl. Nat. 36: 513–560. 1946.

Mauritzon, J., Die Bedeutung der embryologischen Forschung für das natürliche System der Pflanzen. Lunds Univ. Arsskr. 1939. N.F. II. 25 (15): 1–70. 1939.

Melchior, H., ed., A. Engler's Syllabus der Pflanzenfamilien, 12th ed. II. Band. Angiosperm. Gebrüder Borntraeger. Berlin. 1964.

Metcalfe, C. R., and L. Chalk, *Anatomy of the Dicotyledons.* Two Vols. Oxford: The Clarendon Press, 1950.

Takhtajan, A. D., *Die Evolution der Angiospermen.* Jena: Gustav Fischer Verlag, 1959.

———, *Systema et Phylogenia Magnoliophytorum* (in Russian). Moscow and Leningrad: Soviet Publishing Institution "Nauka," 1966.

Wunderlich, R., Zur Frage der Phylogenie der Endospermtypen bei den Angiospermen. Osterr. Bot. Zeits. 106: 203–293. 1959.

# Contents

# THE EVOLUTION AND
# CLASSIFICATION OF
# FLOWERING PLANTS

## Chapter 1

Staminate inflorescences, young leaves, and (inset) fruit of Quercus coccinea, *the scarlet oak. In spite of the clear-cut technical distinctions among the three sections (Cyclobalanus, Erythrobalanus, and Lepidobalanus) of Quercus, it is customary to keep them all in a single genus, because botanists and laymen alike intuitively sense the unity of the group. An oak is an oak is an oak. U. S. Forest Service photos by W. D. Brush.*

# Taxonomic Principles

## Introduction

Taxonomy might be variously defined, according to the definer, but I believe the best definition is "a study aimed at producing a system of classification of organisms which best reflects the totality of their similarities and differences." This is obviously a large task. We don't know everything there is to be known about any one individual organism, let alone a species, or genus, or the more than 300,000 species of organisms that are generally considered to be plants. We know perhaps more about man than about any other kind of organism, but every student of human physiology or biochemistry is impressed by the amount we do not know. We do not know, for example, which individuals can safely smoke large quantities of cigarettes over a period of years and which ones will succumb to lung cancer under the same conditions.

It may be argued that susceptibility to lung cancer is not an important taxonomic character. Indeed it is not. But we know that it is not, only because it does not correlate strongly with other characters found to be useful in human taxonomy because they are correlated among themselves.

Several important principles can be illustrated by this one example: (1) Taxa are properly established on the basis of multiple correlations of characters. (2) The taxonomic importance of a character is determined by how well it correlates with other characters; i.e., the taxonomic value of a character is determined a posteriori rather than a priori. (3) An important feature of taxonomy is its predictive value. If studies had shown that susceptibility to tobacco-induced lung cancer were confined to one race, then (1) this character would be correlated with a whole set of other characters; (2) it would thereby become a significant taxonomic character; and (3) we could accurately predict that an untested individual, known on

3

the basis of other characters to belong to a nonsusceptible race, could smoke without fear of developing lung cancer as a result. In actuality susceptibility to lung cancer is not obviously correlated with race and has no taxonomic significance.

It is perfectly clear that diversity among angiosperms, and among organisms in general, is not completely helterskelter. Patterns do exist. Some combinations of characters occur repeatedly, with minor variations on a major theme. Many other combinations of characters which are theoretically possible simply do not exist. There are gaps of all sizes, and cluster patterns of all sizes and degrees of density and complexity, in the distribution of character combinations. The job of the taxonomist is to recognize these cluster patterns and organize them into a formal hierarchy.

## The Natural System

The concept that there is, or ought to be, a natural system which best reflects the essential nature of different sorts of organisms, is an ancient one, going back at least as far as Aristotle. In *De Partibus* he wrote, "the method then that we must adopt is to attempt to recognize the natural groups — each of which combines a multitude of differentiae, and is not defined by a single one as in dichotomy."

Little progress toward a natural system, above the level of the genus, was made for more than 2000 years after Aristotle. In the middle of the 18th century Linnaeus, the revered "father" of taxonomy, arranged the genera of angiosperms according to his Sexual System, which depended largely on the number of stamens and carpels in the flower. Linnaeus admitted that his system was artificial, and that it would eventually be superseded by a natural system, but he felt that information and understanding were yet inadequate for the production of such a system. He did provide a separate "fragment of a natural system" in which he listed certain groups of genera that he thought belonged together. We now agree with many, but not all, of the suggestions he made.

Linnaeus' successors immediately set to work to devise a natural system. Michel Adanson, Bernard and Antoine Laurent de Jussieu, Robert Brown, and especially Augustin Pyramus de Candolle made signal contributions, and by the middle of the 19th century most of the families of angiosperms which we now recognize had been perceived as natural groups. The meaning of "natural" was nebulous, but the groups were recognized all the same.

The concept of organic evolution, first made scientifically respectable in 1859 by Charles Darwin in his monumental *Origin of Species*, provided the missing rationale: that the natural system reflects evolutionary relationships. Darwin was well aware of the significance of the evolutionary concept to taxonomy. In Chapter 14 he wrote

. . . the Natural System is founded on descent with modification . . . .
the characters which naturalists consider as showing true affinity between
any two or more species, are those which have been inherited from a common parent, all true classification being genealogical . . . . community of
descent is the hidden bond which naturalists have unconsciously been
seeking . . . .

Taxonomists quickly adopted the concept of organic evolution and made
it their own. A natural system must reflect evolutionary relationships, and
if it could be shown that any particular taxonomic group was not monophyletic, but came instead from diverse ancestors, then that group must be
abandoned as polyphyletic and therefore unnatural. For example, as soon
as taxonomists came to believe that the Sympetalae were not a monophyletic group, but rather a group of diverse origins which had independently
achieved the sympetalous condition, then the group Sympetalae had to be
abandoned and its members reassigned according to their probable evolutionary relationships.

The reasons why an evolutionary classification is preferred to one which
cuts across evolutionary relationships are simple. Only if our taxa represent
truly evolutionary groups will new information, from characters as yet unstudied, fall into the pattern which has been established on a relatively
limited amount of information. If the system is to have predictive value, if
it is to reflect the totality of similarities and differences in addition to the
formal critical taxonomic characters, it must have an evolutionary foundation. Artificial classifications, using a few arbitrarily selected characters, are
easily devised, but they do not have the predictive value of a natural
classification; new information will not tend to fall into line.

A perceptive reader may note that we are here using the term predictive
value in a slightly different sense than on the first page of the chapter. A
proper taxonomic system has predictive value not only for previously
studied characters on unstudied individuals, but also for characters which
have been studied only in related taxa, or which are indeed wholly unstudied or unknown. The predictions will not always be correct, but they
have a much better than random chance of being so. The more closely
related the taxa, the more likely they are to be similar in characters as yet
unstudied.

A closely related, complementary reason why a sound taxonomic system
must be evolutionary is that the gaps in the distribution of diversity reflect
evolutionary history. In a taxonomic system an attempt is made to draw
the lines between groups through the gaps in the pattern of diversity. The
detection of these gaps, the unraveling of evolutionary history, and the
establishment of a taxonomic system are closely interrelated processes.

As we shall see, the gaps in the pattern of diversity are often relative
rather than absolute. Sometimes it is useful to think of a distributional

curve with humps and hollows of various sizes. The hollows may not all go down to the base line, but it is still useful to draw the lines in the hollows instead of through the humps. The fewer the units lying near or astraddle the boundary, the easier the system is to comprehend and use, and the greater the likelihood that the system is truly natural. In setting up our system we try to define our groups so as to have few or no intermediates.

One thing which should be kept clearly in mind in trying to match the taxonomic scheme to the evolutionary history is that it is the *amount* of evolutionary divergence which is important, not the antiquity of the split. *Selaginella* has been distinct from *Lycopodium* for much longer than the angiosperms have been in existence, but that has no bearing on the rank at which the groups should be recognized, or how closely related they are to each other. Likewise the iron bacteria have been distinct from the photosynthetic bacteria since precambrian times, long before the origin of the subkingdom Embryobionta, but that is no barrier to holding the various groups of bacteria within a single taxonomic class.

The evolutionary relationships among certain groups of organisms are attested by the fossil record. Our concepts of the phylogeny and taxonomy of the vertebrates, for example, have been profoundly influenced by the evidence from fossils. In many instances the fossils not only provide connecting links between distinct modern groups, but they also furnish a closely graded series of steps between the ancestral and the modern forms. The series of fossils connecting the modern horse with the Eocene *Hyracotherium* (eohippus), a 3-toed mammal the size of a small dog, is a familiar example.

In the angiosperms, on the other hand, the fossil record gives but little help to the taxonomist. The most generally useful characters for recognizing higher taxa of angiosperms lie in the flowers, and these are seldom fossilized. Such fossil flowers as do exist are mostly mere impressions which tell little about the actual structure. Identification of angiosperm fossils is difficult at best, and depends largely on matching fruits or vegetative fragments with those of modern species or genera. Fossil pollen can also frequently be identified by matching it with that of modern plants. Such identifications cannot provide new or independent information on the evolutionary diversification of a group, or on the transitions between groups; they merely document the existence of a modern group at a particular time in the past. We shall not ignore such evidence from fossils as does exist, but this evidence obviously cannot be a mainstay of the system of angiosperms.

Because the fossil record gives us so little help, we are forced to establish our concepts of the evolutionary origin and diversification of the angiosperms largely by comparison of living members of the group. The development of a taxonomic system and a phylogenetic scheme thus go hand in hand, each influencing and depending on the other. A phylogenetic scheme which provides for all the available information and hangs together without serious internal contradictions is regarded as not only satisfactory but also something of a triumph. When a reasonable amount of evidence on the characteristics of a group of related taxa has been assembled, the number of phyletic schemes which will adequately account for that evidence is always very small, typically only one.

All gaps between taxonomic groups are ultimately bridged if the ancestors as well as the modern members of the groups are taken into account. It often turns out, however, that fossil connecting forms between modern groups are absent, or relatively few and confined to a short span of geologic time. This scarcity of fossil intermediates may reflect the fact that major new taxa often result from the achievement of a new adaptive level or the penetration of a selective barrier. The phylad leading to the new type may be only a thin line, poorly preserved in the fossil record, but when it reaches a new adaptive plateau it increases greatly in numbers and radiates explosively into a host of slightly differing new types. Thus even when the fossil record is taken into account, there are important clusters separated by relatively empty spaces in the pattern of diversity.

The major structure of the overall taxonomic scheme is designed primarily to provide for modern organisms. Fossil groups are intercalated as necessary, with due attention to those which are abundant and varied. Fossil connecting links between distinct modern groups are regarded more as an evidence of relationship than as a disturbance to the system.

The scarcity of recognized fossil intermediates connecting modern families and orders of angiosperms may depend on a special set of circumstances. We shall see that there is good reason to doubt the selective significance of many of the families and orders of angiosperms, and the breakthrough theory to explain the scarcity of intermediates may not apply very well. On the other hand, fossil intermediates between the modern families and orders of angiosperms would in most cases not be recognized as such, because the critical characters are seldom adequately preserved. The net result of all these factors is that the system of angiosperms is to a large degree independent of the fossil record.

## Typological Concepts

Much has been written in recent years about the dangers and faults of typological thinking in taxonomy. Although serving as a useful corrective to unnecessary rigidity, such commentary often overlooks the fact that a certain amount of typology is necessary not only in taxonomy but in other kinds of thought as well.

The endless array of natural and artificial objects must somehow be organized in our minds into a set of named categories, in order that we may think about them and communicate our thoughts to others. Consciously or unconsciously we seek repetitive patterns so that we may not be lost in the jungle of diversity. The small child who points to something and asks his mother, "What?," is trying to establish categories and names, and the one who calls a stranger "daddy" is at least making progress. He has recognized the category which other people call man, and has only to establish the right name for it.

We all have type concepts for chair, bed, table, window, house, road, mountain, river, lake, and countless other objects, as well as for plant, animal, tree, snake, parasite, and human. We have a mental picture, in greater or lesser detail, of what something ought to be like to fit into each of these categories. The fact that a chaise longue is somewhat intermediate between a chair and a bed does not interfere with the utility of the type concepts for chair and bed. A person unfamiliar with a chaise longue will simply consider it an unusual type of chair, or an unusual type of bed, or something between a chair and a bed, and the person who sees a great many chaises longues will develop a type concept and learn the name for this category as well.

Plant taxonomists necessarily have type concepts for taxonomic groups, such as moss, fern, gymnosperm, monocot, dicot, Magnoliales, Compositae, *Helianthus*, maple, sugar maple, etc. We cannot think effectively without them. The fact that it is possible to develop a type concept for a group does not prove that it is a natural, acceptable taxonomic group, but on the other hand any proposed group which does not permit the establishment of a type concept is of doubtful value.

It should of course be understood that the taxonomic types we refer to here are biological rather than nomenclatural. The nomenclatural type may or may not be biologically typical; all that is necessary is that it fall within the confines of the group.

## The Value of Characters

Ideally, the comparative studies on which taxonomic conclusions are based should include all characteristics of all species of a group, and of many individuals throughout the geographic range of each species. Morphological comparisons should include not only the obvious floral and vegetative structure, but also the various kinds of micromorphology, as observed both with the light microscope and with the electron microscope. Equally, attention should be given to all phases of physiology in its broadest sense, from habitat requirements and genetic compatibilities to serology and chemical analysis.

Practically, this ideal is not yet attainable or even in sight. Acquisition of the requisite information in physiology and micromorphology is a time-consuming procedure which as yet has been hardly more than begun. It is clear on the basis of experience as well as from a priori theory that information from these sources is going to be increasingly useful in the continuing development of our taxonomic and evolutionary scheme, but at present it cannot be applied with any consistency because the gaps in our information bulk so much larger than the actual data. It is always tempting to assume that the more recondite characters are inherently more stable and thus more important than the ones which are easily observed, but the assumption is unwarranted and gives rise to unfortunate and unnecessary error.

We have by now become accustomed to the idea that obvious morphological characters of any sort do not have a fixed, inherent taxonomic significance. A character which distinguishes families in one order may distinguish genera in another family, or species in another genus, or it may vary on a single individual of another species. For example, the Boraginaceae may generally be distinguished from the Verbenaceae by having alternate leaves, in contrast to the opposite leaves of the Verbenaceae. (There are, to be sure, other differences between the two families.) Within the family Scrophulariaceae the arrangement of leaves is usually a good

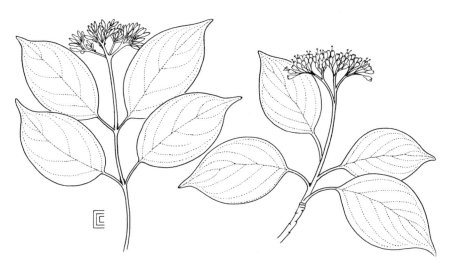

FIGURE 1.1

Cornus amomum (left) and Cornus alternifolia (right). These two species, although closely related, differ in a character (leaf arrangement) which usually marks groups of generic or higher rank.

generic character, some genera having the leaves opposite, others alternate. Within the genus *Cornus*, the species *Cornus alternifolia* has alternate leaves, whereas most of the other species have opposite leaves. In the common sunflower, *Helianthus annuus*, the leaves near the base of the stem are opposite, but the others are alternate. In each case the taxonomic importance of the character is determined after the groups have been perceived.

How, then, may taxonomic groups be recognized, and taxonomic systems set up, if particular characters do not have an inherent importance? The answer is that sound taxonomy proceeds by the use of multiple correlations. We are beginning to use digital computers to help recognize or at least to demonstrate these correlations, but in the past they have been made almost entirely by inspection. The groups are perceived on the basis of all the available information, and the critical characters are selected thereafter. Thus it can be stated as a general principle that individual characters do not have a fixed, a priori importance; a character is only as important as it proves to be in each individual instance in marking a group which has been recognized on the basis of all the available evidence.

We have found by experience that certain characters and kinds of characters are more likely to be important than others. Floral characters are more likely to be important than vegetative characters; and the arrangement of parts, either floral or vegetative, is likely to be more important than their size and shape. But in all cases the likelihood is only a likelihood rather than a certainty. The proof of the pudding remains in the eating. The position of the ovary in the flower commonly marks families or even orders (e.g., Campanulales, Cornales, Dipsacales, and Umbellales, all with inferior ovaries), but in the Rosaceae it marks only subfamilies (e.g., the Pomoideae with inferior ovary, the other subfamilies with superior ovary), and in the genus *Saxifraga* some species have the ovary wholly superior, others have it partly inferior, and a few have it more or less completely inferior.

One of the interesting things which has turned up in evaluating the significance of taxonomic characters is that the presence of a structure or substance is likely to be more important than its absence.[1] We can now give a genetic explanation of why this should be so, and at the same time strengthen our conviction that it really is so. Studies in genetics and developmental morphology have amply demonstrated that individual phenotypic characters are commonly governed by a complex system of genes, rather than by a single gene unaffected by others. The appearance of a particular characteristic in the phenotype requires the whole genetic

[1] This concept is elucidated by Marion Ownbey in Res. Stud. State Coll. Wash. Monog. Suppl. 1. 1956.

a

b

c

FIGURE 1.2

Peloric (Fig. 1.2a, 1.2b) and normal (Fig. 1.2c) flower types in the
common snapdragon, *Antirrhinum majus*. Courtesy of the George
W. Park Seed Co., Greenwood, S.C. Peloric snapdragons differ
from the ordinary type in only a single pair of genes.

system, and if two individuals or two allied taxa share the same structure then they probably also share the same genetic system governing its development. If any one of the essential genes in the system is lost, then the structure fails to develop, and a plant which lacks a particular structure may differ by only a single gene from one which has it. It has been demonstrated that peloric (regular-flowered) forms of the common snapdragon differ from the ordinary bilabiate form in only a single pair of genes, but this does not mean that the complex, irregular corolla of the snapdragon arose from a regular-flowered ancestry by a single mutation; a complex system of genes is certainly required instead.

Even an individual chemical component in the cell often represents the end product of a biosynthetic chain which can be broken at any link. Anthocyanin is a familiar example. The production of purple-flowered sweet peas by crossing two different white-flowered strains is often studied in beginning genetics under the heading of factor interaction. Each of the white-flowered strains in this cross lacks one of the essential links in the biosynthetic chain leading to the formation of anthocyanin. The $F_1$ hybrid, although it is heterozygous for genes governing two of these links, has all the links intact and produces purple flowers.

The betacyanins are a unique group of flower pigments which are known only in the order Caryophyllales. The restriction of betacyanins to this group of families is one of several characters which leads us to believe that the group really does hang together. The presence of betacyanins in the Cactaceae is one of a number of characters which leads modern taxonomists to the conviction that this family, whose affinities were once obscure, is properly associated with the Caryophyllales. On the other hand, the apparent absence of betacyanins from the family Caryophyllaceae is not necessarily of any great moment. It does suggest the possibility that this family may not belong with the others in the order to which it gives its name, but the suggestion is not borne out by further study. The similarities in several other features combine to support the traditional definition of the group.

It is a well known fact that apetalous genera, species, or individuals occur in various groups of plants which usually have petals. It is now generally believed that the apetalous condition in angiosperms usually reflects the loss of petals rather than an original absence, and that this loss has occurred repeatedly in various groups only distantly related to each other.

It is of course also clear that the absence of a character may reflect the complete absence of the genetic system which would be required to produce it. There is no reason to believe that the prokaryotic organization of the protoplast of bacteria and blue-green algae could be transformed into the eukaryotic organization typical of other plants by merely supplying one or two genes missing from an otherwise complete system, nor is there any reason to think that the gymnosperms could be induced to form an angiospermous ovary merely by supplying a missing gene or so.

Thus we are led on both theoretical and practical grounds to the conclusion that the absence of a character is a less reliable guide to taxonomic affinities than its presence.

## The Phylogenetic Concept in Taxonomy

### THE MONOPHYLETIC REQUIREMENT

We have, thus far, developed the principles that taxonomy is based on multiple correlations, that the value of characters is determined a posteriori, that the presence of a character is likely to be more important than its absence, and that a proper taxonomic system must reflect evolutionary relationships. This last item, the reflection of evolutionary relationships, requires further consideration.

It has become practically axiomatic that a taxonomic group must be monophyletic, and that a polyphyletic group is per se unnatural and must be abandoned. Too rigid an adherence to this rule, however, would wreak havoc with our system, without providing anything useful to replace it. Thus, on the basis of comparative morphology and physiology it is perfectly clear that the protozoa, and through them eventually the whole animal kingdom, arose in several more or less parallel lines from different groups of algal flagellates. There was no original protozoan, from which all other protozoans evolved, and no original animal that was ancestral to all others.

Must the protozoan phylum and the animal kingdom therefore be abandoned as unnatural? Not at all. The assumption of the holozoic mode of nutrition by certain members of several different groups of algal flagellates had such an all-pervasive influence on their subsequent evolutionary development that we are justified in regarding them and their descendants as a coherent natural group. All of the other features which we think of as characterizing animals as opposed to plants were evolved in response to or in association with the holozoic mode of nutrition, and many of these characters persist even in such animals as tapeworms, which have abandoned holozoic nutrition and taken to absorbing their food.

We need not further pursue the still knotty problem of where to draw a line between the plant and animal kingdoms. Our point is that the monophyletic requirement must be considerably stretched in order to permit the retention of the very natural and useful grouping of organisms into plants and animals. As an aside, it might be mentioned that efforts to avoid this problem by recognizing three or more kingdoms create at least as many difficulties as they avoid. And to refer all the unicellular flagellates to the protozoa would deprive the animal kingdom of its essential characters, as well as making the plant kingdom polyphyletic.

The problem of applying the monophyletic criterion to the distinction of the plant and animal kingdoms is not an isolated example. Over and over again it appears, when enough evidence has been assembled, that the common ancestor of a group would have to be excluded from that group and referred to an ancestral one. In the Compositae, for example, it seems likely that the common ancestor to all species of *Baccharis*, if we had it, would be an *Archibaccharis*, the common ancestor to all species of *Archibaccharis* would be a *Conyza*, and the common ancestor to all species of *Conyza* would be an *Erigeron*. The most primitive existing species of *Erigeron*, in turn, would on morphological grounds just as well be referred to *Aster*, and in fact it was first described as *Aster peregrinus* Pursh. (In this case it happens that the primitive item goes with the derived group. But if a few of the right species were to die out, then this near-ancestral species would be more at home in *Aster*.) Carrying things back another step, it seems likely that the nearest common ancestor to all the genera in the tribe Astereae, if we had it, would be better placed in the Heliantheae.

The angiosperms themselves probably do not have a single common ancestor short of the gymnosperms. The characteristic xylem vessels of angiosperms have evidently originated several times among the primitive members of the group. Stages in the development of the closed carpel, usually regarded as an essential feature of angiosperms, can be observed among living members of the primitive order Magnoliales. Within the Magnoliales one can also see all stages in the evolution of the angiosperm stamen from the ancestral sporophyll with embedded microsporangia. Furthermore, it seems clear, again on the comparative morphology of living species, that the development of the closed carpel and the typical angiospermous stamen with filament and anther, rather than being strictly monophyletic, took place in several related evolutionary lines within the Magnoliales. Differentiation of the perianth into calyx and corolla has likewise taken place independently in various families, as has also the origin of petals from staminodes. The ancestry of the angiosperms is further considered in the following chapter.

The foregoing examples of difficulty with the monophyletic requirement are based largely on comparisons of modern organisms and are thus subject to the challenge that the phylogeny may have been incorrectly interpreted. However, the same lack of strict monophylesis also permeates vertebrate taxonomy, where the fossil record is heavily relied on for phylogenetic interpretations. Thus, George Gaylord Simpson concluded in 1945 that

> it is not probable on the basis of present knowledge that all the animals
> here included in the Mammalia arose from the Reptilia as a single

species, genus, or even family, but it is not suggested on this account that some of them should be returned to the Reptilia or that another class should be created for them. They certainly arose from a unified group of reptiles of much smaller scope than a class, perhaps a family or perhaps a super-family, and for practical purposes this is an adequate fulfillment of the requirement of monophyly.[2]

In general, it appears that whenever we do get reasonable evidence on the phylogeny of characters which mark major taxonomic groups, these turn out to have developed through parallel evolution in various closely related but separate lines which collectively make up the ancestral stock of the group.

Should we then divorce taxonomy from phylogeny, returning to the nebulous pre-Darwinian concepts of a natural system? Not at all. The marriage is highly useful and clearly worth saving. It is the evolutionary concept which has given meaning to the whole idea of a natural system, just as Darwin said it would. Furthermore, it is the fact of evolution and evolutionary relationships which permits taxonomy to have predictive value. Because of their common ancestry, the members of a natural group tend to share many more recondite characters beyond the obvious ones which permit us to recognize the group.

If the monophyletic requirement is interpreted loosely rather than strictly, most of the conflict between phylogeny and taxonomy disappears. Monophylesis and polyphylesis are not such utterly distinct things as the terms would suggest. There is a continuous graduation from the strictest monophylesis to the most utter polyphylesis in proposed taxonomic groups. (Species which originate by a single act of genome doubling following hybridization may be taken to represent the strictest monophylesis.) In order to be natural and acceptable, a taxonomic group must fall somewhere toward the monophyletic end of this scale. Simpson has proposed (in the paper previously quoted) the useful rule of thumb that if a taxonomic group of a particular rank is derived wholly from another group of lesser rank, that is a sufficient degree of monophylesis for taxonomic purposes.

It thus appears that a workable taxonomic system cannot provide a perfect reflection of evolution, no matter how abundant the evidence on which it is based. Indeed, the more abundant the evidence, and the better our understanding of phylogeny, the clearer it becomes that the correlation between phylogeny and taxonomy must be general rather than exact. However, the phylogenetic concept still provides the underlying rationale for the natural system. Taxonomy can provide only a somewhat muddy reflection of evolution, but a reflection all the same.

[2] The principles of classification and a classification of mammals. Am. Mus. Nat. Hist. Bull. 86.

PARALLELISM

Once we admit the necessity for a broad interpretation of the monophyletic requirement, we are committed to the position that similarities
due to evolutionary parallelism, as well as those due strictly to inheritance
from a common ancestor, provide some indication of relationship and
should be considered in the formulation of the taxonomic system. As long
ago as 1912 Wernham pointed out that "critical tendencies are no less
important than critical characters." He went on to say that

> the general relation between the significant features of the ancestry and
> those of the descendants is, that in the former the characters in question
> are not constant throughout the group, nor may they be completely
> evolved. In other words, we are dealing with *tendencies* to characters, and
> not with the critical characters themselves, in the case of the ancestry.
> In the progeny, on the other hand, the characters are constant and
> completely evolved; and the line which unites ancestor and descendants
> represents the transition between the tendencies and their realization.[3]

In order to understand why similarities due to parallelism give any indication at all of phyletic relationship, we must consider something about
the mutative process and the mechanism of evolution. We are here concerned with these processes only insofar as they bear on the establishment
of a general system of angiosperms. There is already an abundant literature
dealing with the mechanics of evolution per se.

It is commonly said that mutation is at random, but this is true only
in the sense that we can neither control nor predict the time and place of
individual mutations. It is not random in the sense that one mutation is as
likely as any other. Some of the possible mutations in any given gene are
much more frequent than others, and the mutation rate in opposite directions at the same locus (from $a_1$ to $a_2$ and from $a_2$ back to $a_1$) is likely to
be different. Furthermore, the rate and direction of mutations in a given
gene are governed not only by the intrinsic nature of the gene but also by
the other genes it is associated with — the remainder of the genotype of
the organism.

Looking at things from the opposite end, the chances of producing a
particular mutant gene depend largely on what you start with. Any mutant
gene which can be fitted into the functioning economy of the organism is
likely to differ from its immediate ancestor in only a single nucleotide
sequence. The greater the number of such differences between two functioning genes, the less — astronomically less — the likelihood that the
changes have occurred simultaneously as a single mutation.

[3] New Phytologist 11: 390. 1912.

Thus we are led inescapably to the position that insofar as mutation is a controlling force in evolution, the greater the genetic similarity between two groups, the greater is the likelihood that they will produce similar mutations, have similar evolutionary potentialities, and undergo parallel evolutionary change. Evolutionary progress is like scientific progress in that when the time is ripe the same advance may be achieved independently by different individuals (or taxa) of similar background.

Conversely, groups which have undergone parallel evolutionary change may well have been rather similar to begin with. We shall see that the evolutionary history of angiosperms is so beset with parallelism as to indicate a basic genetic homogeneity of the whole group, with the consequence that the taxonomic and phyletic significance of similarities in individual characters must be interpreted with the greatest of caution.

On phenotypic as well as genotypic and mutative bases, different groups have different evolutionary potentialities, and not all evolutionary channels are open to any one group. At the grosser levels, this is of course immediately obvious. An oak doesn't have much chance of developing into a carnivore, nor is a dog likely to develop photosynthesis. Once evolution has proceeded beyond the unicellular flagellate stage, the evolutionary barrier between the plant and animal modes of life is essentially insuperable.

Aside from the necessity of having the proper mutations, a group undergoing evolutionary change must at all times remain well enough adapted to the environment so that its individuals can compete, survive, and leave offspring. The greater the difference between the old ecologic niche and the new one being moved into, the greater the difficulty of maintaining adaptation to the environment during the change.

Thus, insofar as natural selection is a controlling force in evolution, the greater the phenotypic similarity between two groups, the greater is their potentiality to undergo parallel evolutionary change. Since the heritable features of the phenotype depend on the genotype, the foregoing sentence is equally true if the word genetic is substituted for phenotypic.

We therefore come to the general principle that evolutionary parallelism tends to indicate relationship, and that it should be given due weight, along with other factors, in arriving at taxonomic conclusions. If the similarities resulting from parallelism are numerous and pervasive, and especially if they do not all relate to a single ecological change, then the ancestors were probably very similar to begin with. This is true regardless of whether one believes that all evolutionary trends must be explained in terms of survival value, or whether one believes that some evolutionary trends chiefly reflect the kind and frequency of mutations rather than natural selection.

Most evolutionary geneticists nowadays maintain that the nature of the supply of mutations merely imposes limits on what natural selection can accomplish, rather than having any positive effect in governing evolutionary trends. Not all taxonomists are so convinced, however. If there is any validity to the thought that mutation pressure can itself promote an evolutionary trend, there are certain consequences to taxonomic theory.

To examine these consequences, let us assume that there is every gradation from evolutionary trends which are essentially selective to those which are governed primarily by the kind and frequency of mutations. Then, the greater the selective control, the less the taxonomic significance of the parallelism, and vice versa. If the control is essentially selective, then all the similarity that is required between the different ancestors is the ability to produce the necessary mutations as part of a functioning, ecologically adapted genotype; selection does the rest. If the control is essentially internal (based on the nature of the supply of mutations), then the ancestors must be sufficiently similar so that both of them produce the requisite mutations in considerably greater frequency than other mutations, with the result that the mutation pressure, operating along with other causes of speciation, causes parallel evolutionary developments.

On theoretical grounds, therefore (continuing the assumption from the previous paragraph), characters which are not closely related to survival value and ecologic niches may often prove to be better guides to relationship than characters which are directly related to selection. This is especially true in the angiosperms, in which the evolutionary barriers between different ecologic niches are frequently minimal, and one family may fill highly diverse niches. In groups such as the vertebrates, on the other hand, in which the evolutionary barriers between different major ecologic niches (or adaptive zones) are typically very strong, the taxonomic consequences of a breakthrough may be correspondingly greater.

On purely pragmatic grounds, plant taxonomists over the last two centuries have come to emphasize the importance of seemingly nonadaptive characters in contrast to the obviously adaptive ones — not because of their possible nonadaptiveness, but because they have turned out to be useful in marking groups which have been perceived on the basis of all the available evidence. The characters used to distinguish the families and orders of angiosperms are in large measure things which are difficult to relate to adaptation and survival value: hypogyny, perigyny, and epigyny; polypetaly, sympetaly, and apetaly; apocarpy and syncarpy; placentation; numbers of floral parts of each kind; the presence or absence of endosperm or perisperm in the seed; sequence of development of the stamens; and the like. Some of these characters turn out on close inspection to have at least a modicum of adaptive significance, as we shall note on subsequent pages, yet it hardly seems accidental that there appears to be a sort of inverse rough correlation between the taxonomic importance of a character and the obviousness of its adaptive significance.

Adaptive characters are not automatically excluded from consideration at the higher taxonomic levels, however. For example, the obviously adaptive feature of parasitism has evolved so rarely in the angiosperms that it can sometimes be used, along with other characters, to help mark major groups. Furthermore, we shall see that some characters of obvious adaptive significance turn out to be much more useful among the monocots than they generally are among the dicots.

### Ecologic Niches, Survival Value, and Taxonomic Groups[4]

Each of the obvious ecologic niches for land plants is occupied by species representing diverse families and orders, whether these niches are conceived of in terms of habitat, growth habit, method of pollination, method of seed dispersal, or various combinations of these and other features. Even such specialties as parasitism, entomophagy, mycotrophy, the succulent habit, the megaphytic habit, and the mangrove habit have evolved repeatedly in different families and orders. The principle of competitive exclusion[5] may be valid for vertebrates or for animals in general, but its application to plants is much more limited — or at least less obvious.

Conversely, a single family may fill widely varying ecologic niches. The large and highly natural family Compositae includes trees, shrubs, vines, succulents, megaphytes, and ordinary herbs, the herbs being annual, biennial, or perennial, the perennials sometimes monocarpic (i.e., flowering only once and then dying). The leaves of the woody species are evergreen in some, deciduous in others. The plants occur from tropical to arctic and antarctic regions, from the tidal zone to above timberline in the mountains, in open or forested places, in very dry to very wet or fully aquatic habitats. Some species are adapted to alkaline or salty soils, some to seleniferous soils, some to serpentine soils, some to cliff-crevices, some to shifting sand dunes, and some to disturbed sites around human habitations. Various species are insect-pollinated, wind-pollinated, self-pollinated, or apomictic. The achenes are distributed by wind, or by animals, or have

----

[4] The reader may wish to compare the interpretation here presented with that of G. L. Stebbins (Adaptive radiation and trends of evolution in higher plants. Vol. I, pp. 101–142 in Th. Dobzhansky, M. K. Hecht, and W. C. Steere, eds., *Evolutionary Biology* (New York: Appleton-Century-Crofts, 1967), who makes some of the same points but insists that all evolutionary trends are adaptive.

[5] The principle is usually stated at the specific level, as for example: "in equilibrium communities no two species occupy the same niche." (G. E. Hutchinson, *The Ecological Theater and the Evolutionary Play* [New Haven: Yale University Press, 1965], p. 27.) Another formulation of the principle is given by Garrett Hardin (The competitive exclusion principle. Science 131: 1292–1297. 1960) as "complete competition cannot coexist." Recent concepts about vertebrate taxonomy are permeated by the application of this same thought to higher taxa.

no very obvious means of distribution. Yet this family is so sharply defined that, if it is taken in the customary broad sense, there is not a single species about which there is any doubt as to its inclusion in or exclusion from the group.

It is noteworthy that none of the purportedly phylogenetic systems of angiosperms gives more than scant attention to the possible adaptive significance of the characters which mark the major groups. This is no accident, and it reflects a basic difference between the families and orders of angiosperms, on the one hand, and those of many vertebrates on the other.

Students of mammalian taxonomy, and of vertebrate taxonomy in general, are accustomed to seeing a fairly good correlation between the major taxonomic groups and broad ecologic niches (adaptive zones). Simpson[6] has elevated this correlation to a principle: "The event that leads, forthwith or later, to the development of a higher category is the occupation of a new adaptive zone." This is a natural corollary of the principle of competitive exclusion, and for a large part of the animal kingdom, at least, it appears to be perfectly sound. As we shall see, its application to the plant kingdom is much more limited. The angiosperms as a unit fit the principle, but the classes, subclasses, orders, and families of angiosperms very often do not — at least not in any obvious way.

The whole structure of a higher animal is intimately correlated with the way it makes its living — where it lives, what kind of food it eats, how it captures that food, and how it keeps from being used as food by something else. Most birds (class Aves) are adapted to flight, whereas most other vertebrates are not. Among the mammals, the whales and porpoises (order Cetacea) are adapted to life in the ocean, never coming out on land; the seals and walruses (order Pinnipedia) are adapted to a marine habitat, but return to land to breed; most other mammals are primarily terrestrial. Only one order of mammals, the bats (Chiroptera), has developed true flight. The rodents (order Rodentia) and the rabbits (order Lagomorpha) are gnawing animals, whereas most other mammals are not. The dogs (family Canidae) and the cats (family Felidae), although they are both carnivores (order Carnivora), have different habits of catching, attacking, and devouring their prey, and these differences are reflected in the appearance of the two families.

The overriding problem for animals in general is food. Having given up (in the course of evolution) the ability to make their own food, they depend directly or indirectly on the food-makers, the plants. Motility, a nervous system, a tightly integrated morphology, indeed all of the characters which we think of as distinctly animal, relate eventually to nutrition.

[6] G. G. Simpson, *The Major Features of Evolution* (New York: Columbia University Press, 1953).

These characters, in turn, permit animals to do many things which plants cannot do well, if at all; e.g., to fight, to exercise territoriality, to choose their mates, to move, as individuals, from one habitat or area to another, and to use active rather than passive means to escape being used as food by something else. With all these abilities, animal populations are still faced by an absolute limit of food supply. Directly or indirectly the food supply limits the size (or at least the potential size) of the population. Once a group has adopted a particular sort of food and a way to obtain it, the evolutionary pressure is toward perfecting the adaptation to its particular mode of life. All ecologic niches are filled by one or another group which is so well adapted to its niche that other, less well adapted groups cannot easily invade.

Only on a newly formed, isolated oceanic island is there any unfilled niche not soon invaded by well adapted populations from nearby. When a few kinds of animals are introduced to such islands, there is a rapid adaptive radiation into many of the unfilled niches. Freed from the necessity to compete with better adapted organisms occupying the same niche, individuals which are relatively poorly adapted can still survive and reproduce, giving rise to a new evolutionary line which may become progressively

FIGURE 1.3

*Argyroxiphium sandwicense*, a species of Hawaiian silversword. From *Island Life*, copyright by Sherwin Carlquist, 1965.

better adapted to that niche and eventually block new phylads from entering it. The Hawaiian honey-creepers provide a famous example of the ecologic and morphologic diversity which can burst forth from an undistinguished ancestry in such circumstances.

The same relaxation of competition leads to rapid and extensive evolutionary change in plants of oceanic islands, as witness the famed Hawaiian silverswords (*Argyroxiphium*), which are related to the tarweeds of California. The contrast between the evolutionary behavior of plants on oceanic islands and those on continental land masses is not so striking as that for animals, however, because even on the mainland a given taxonomic group of plants often includes highly diverse ecologic types.

Because of the general correlation of structure, appearance, ecologic niche, and taxonomic affinity, many of the families, orders, and classes of vertebrates are well known to the general public and have well established common names. Such common names as fish, shark, reptile, snake, lizard, bird, hawk, owl, penguin, squirrel, kangaroo, and monkey may serve as examples. These are genuine common names, reflecting a folk classification that does not depend on a formal scientific taxonomy. Many other taxonomic groups, though not necessarily recognized in purely folk classification, are so easily grasped that they have become familiar to a large part of the general public. Marsupials, thrushes, finches, carnivores, and amphibians are examples. It is not here suggested that folk classification of vertebrates conforms in all respects to the formal taxonomy, but only that a great many of the higher taxa are readily recognized not only by taxonomists but also by the general public.

How different is the situation among angiosperms! The thoroughgoing structural differences which mark the higher taxa of vertebrates have no real parallel among the higher taxa of angiosperms. Differences in growth habit, which might perhaps be roughly compared to the evident differences among the higher taxa of vertebrates, occur repeatedly *within* the higher taxa of angiosperms, especially among the dicotyledons, and they are at best useful chiefly in combination with more technical characters of flowers and fruits. The ancient folk classification of land plants into trees, shrubs, and herbs cuts squarely across the natural taxonomic arrangement.

Once the angiospermous condition has been achieved, the obviously adaptive changes which can take place mostly occur so easily and frequently that they tend to mark species and genera rather than larger groups. Or, as in the case of some evolutionary changes in the xylem structure, the same selective forces operate so consistently throughout the group that the sharing of advanced features provides but little indication of relationship. The vast majority of angiosperms make and use essentially the same kinds of foods, using the same raw materials which are obtained in the same way, and they rely on the same source of energy for photosynthesis. As we have seen, the families and orders, by and large, are not restricted to well defined or even approximately mutually exclusive ecologic niches.

Plants do compete with each other for space, light, water, and minerals, but the number of devices available to them in this competition is relatively limited, and most of the changes are genetically not difficult, at least among the dicotyledons. Variation from one habitat to another is often gradual, so that a phylad may move from one to another without ever being deprived of an appropriate habitat. Evolutionary changes which result in better general adaptation, such as those affecting the structure of the conducting tissues, are available to and have been adopted by many different groups. The concept of adaptive peaks, separated by adaptive valleys, is more significant for plants at the specific and generic levels than above them.

Even at the specific level, natural selection often favors greater ecological amplitude for plants than for animals. The plant cannot move from place to place. It must grow where the seed lodges, or not at all. An ability to survive and reproduce in a less than optimal environment is therefore favored. Each of several different species of trees intermingled in a forest may have slightly different optimal requirements, yet some individuals of each species may actually be growing in sites to which one of the other species is a little better suited.

The plant must of course be adapted to its environment if it is to survive and leave offspring, but the plant system is more loosely integrated and more tolerant of casual variation than that of animals. *Verbesina alternifolia*, an eastern American species of the Compositae, usually has the stem evidently winged by the decurrent leaf-bases, but the occasional individuals with wingless stems are under no obvious selective disadvantage. Anyone who has been responsible for laboratory instruction in a beginning taxonomy course will recall how often students happen upon flowers which are unusual in some way and do not fit the formal description for their species or genus.

The separate character complexes for various organs of plants are only rather loosely articulated. The growth habit is certainly important to the plant, but each of several growth habits is compatible with each of several types of flower and fruit, the one exerting relatively little influence on the other. The independence of the various character complexes is only relative rather than absolute, however. Samaras and nuts are certainly better adapted to trees than to low herbs.

Even within such a seemingly well integrated structure as the flower there are two separate sets of adaptations, one relating to pollination, the other to seed dispersal. Again, each of several different methods of pollination is compatible with each of several different methods of seed dispersal. The adaptations for seed dispersal do not reach their full

development until the fruit is mature, but the gynoecium is the precursor to the fruit, and the structure of the gynoecium is an integral part
of the set of characters relating to seed dispersal.

The number of genuine common names of angiosperms which conform
at all closely to major taxonomic groups is very limited. Grass, palm,
orchid, lily, and cactus might be suggested as possibilities, but even for
some of these the conformity is not very good. Some members of the
family Euphorbiaceae occupy in Africa the niche commonly occupied by
cacti in America, and these euphorbiads look so much like cacti that only
a botanist would draw the distinction. On the other hand, only a botanist
would recognize some tropical American epiphytic cacti (such as the
Christmas cactus) as being cacti at all. Most of the Cyperaceae and at least
some of the Juncaceae would be considered as grasses by many people
without taxonomic training. The name lily is often applied to a great
many plants that do not belong to the Liliaceae, and some of these are
not at all closely related to the "true" lilies. The name orchid does conform fairly closely to the Orchidaceae, but this conformity is due in large
part to the interest of horticulturists and the influence of botanists. The
vast majority of Americans would not recognize the vast majority of orchids
as being such, and the situation is no different in the tropics where the
orchids are much more abundant than they are in temperate regions. Only
the palms are left as a widely recognized folk taxon which compares closely
to a scientific one. A few cycads would pass as palms to many people, but
here the confusion may be no greater than in many folk groups of
vertebrates. We may note in passing that the palms and some of the
other taxa mentioned in this paragraph are monocots. As we shall see, the
monocots labor under special handicaps which make habital changes much
more difficult for them than for the dicots.

The principal characters left to mark the families and orders of angiosperms are technical floral characters such as we have already enumerated.
The selective significance of most of these characters is dubious. Some
botanists believe that they have a real (though usually unspecified) survival
value. This view reflects the thought that the logic of the evolutionary
principles which have been worked out primarily from studies on animals
is so compelling that these principles must necessarily apply to plants as
well. Many other botanists avoid the question by ignoring it. A few are
bold enough to assert that the major groups of angiosperms, and the
characters which mark them, more nearly reflect the supply of mutations
than the effect of natural selection. The concept of internally directed
evolutionary trends is out of favor among evolutionary geneticists, and the
dominant school of evolutionary thought now holds that all evolutionary
trends must be explained in terms of survival value. Nevertheless, the
application of this concept (that survival value motivates all evolutionary

trends) to the major taxa of angiosperms has not been notably successful, and the alternative of mutation pressure deserves serious consideration.

The difficulty in recognizing higher taxa of angiosperms afflicts even trained taxonomists. There are some tendencies and some intangibles of aspect that can be learned, but most taxonomists will not even try to identify unfamiliar material which does not have flowers or fruits. Even when flowers, fruits, and vegetative parts are available it may take some time to determine the family of a plant belonging to one of the less familiar groups.

The orders of angiosperm are even more difficult than the families. Except for the problem of whether certain groups, such as the legumes, should be lumped into one large family or split into several smaller ones, there is pretty general agreement as to which genera go into which families. A few genera are booted about from one family to another, and some others are doubtfully attached to particular families for lack of a better place to put them, but the proportion of these is small indeed. The arrangement of families into orders, on the other hand, still arouses active and major disagreement among competent botanists. In all systems, the orders of angiosperms are apt to be nebulous and ill-defined, and their formal characterization studded with exceptions. In many manuals the keys for identification bypass the orders and go directly to the families. Some botanists go so far as to say that the orders can be defined only by the lists of families to be included. Others are more hopeful and believe that natural orders with a reasonable degree of coherence can be characterized, even though the individual characters are subject to exception. I belong to the latter group.

At the level of genera and species, many taxa of both plants and animals are readily recognized and have acquired genuine common names. Alder, beech, birch, dandelion, elm, goldenrod, hickory, manzanita, maple, oak, rose, and violet are a few of the many folk names which conform fairly well to botanical genera, and such names as sugar maple, silver maple, black maple, red maple, striped maple, paper birch, cherry birch, yellow birch, and red oak, white oak, black oak, post oak, and bur oak show that species as well as genera are often recognized. The present generation of city-bred Americans may be unacquainted with most or all of these plants, but our pioneer forebears knew and had names for these and many other genera, sections, and species. The English colonists, familiar with oaks at home, had no difficulty in recognizing the American species of Quercus to be oaks, although the species were different. Nor did it take them long to learn to recognize the wholly new (to them) and unfamiliar genus Carya, for which they adopted an already existing Indian name. (Carya, Hicoria, and hickory are all variants of the same Indian word.)

Suburban gardeners have become acquainted with many genera and species of plants, and the botanical generic or sectional names have often been taken over into common speech. Azalea, ageratum, calendula, chrysanthemum, cosmos, cyclamen, dahlia, petunia, phlox, rhododendron, and verbena are familiar examples.

Here again it is not suggested that common names and folk classification at generic and infrageneric levels conform closely to formal taxonomy (for they do not), but only that many botanical genera, sections, and species, in contrast to the higher taxa such as families and orders, are readily recognizable by their general appearance. Many of the characteristics which enable people to recognize genera and lesser taxa by their appearance are doubtless adaptive. On the other hand, few systematists would deny that related genera and species sometimes differ chiefly in technical characters which are obscure to the layman and have no very obvious ecologic or adaptive significance. Some of these characters are presumably governed by genes which are useful for some other, less obvious effect, but alternative possibilities relating to the supply of mutations should not be ruled out a priori.

## The Size and Delimitation of Taxa

Any useful system of classification must concern itself both with the objects being classified and with the characters used to differentiate the groups. The scheme must permit the units to be assigned to the proper groups, and the groups should be characterized, however imperfectly, by something more than a mere list of the units to be included. The taxonomic system is no exception.

It is obviously desirable that each taxon be sharply delineated and well characterized. The existence of intermediate forms, which could as well be assigned to one group as another, is disturbing to our sense of order. It is even more disturbing to find a unit which on the basis of the critical distinguishing character belongs to one group, but which appears, on the basis of the rest of the evidence, to be more closely related to another group. Obviously the system should, insofar as possible, be devised to avoid such flaws.

From the standpoint of understanding and remembering the scheme of classification, it would be useful if all units of a given rank contained the same number of subordinate units, if differences among units of the same rank were always of the same nature, significance, or importance, and if all characters had a fixed value in the scheme of classification. Thus it would appeal to the orderly sense of our minds if all taxonomic orders contained ten families, all families ten genera, and all genera ten species, or even if all orders contained ten families, all families 20 genera, and all genera 40 species. Likewise it would be appealing if the arrangement of leaves on the stem (alternate, opposite, or whorled) always marked genera within a family, but never species within a genus nor families within an order.

In practice, as we shall see, any such consistency in the taxonomic scheme is impossible. The objectives of uniform size of groups and uniform weight of characters often clash and cannot both be achieved at the same time. It is also, unfortunately, impractical to insist that there never be any intermediates between recognized taxa, however much the existence of these intermediates may disturb our sense of propriety. All of these objectives, however, are properly kept in mind by the taxonomist.

If a small group is so different from others that it cannot be included elsewhere without undue difficulty, it is kept as a separate unit. Thus we have some genera with only one species, some families with only one genus, and some orders with only one family. The genus *Casuarina* contains perhaps 50 species, but it is the only genus of the family Casuarinaceae, which in turn is the only family of the order Casuarinales. Among the gymnosperms, the well known maidenhair tree, *Ginkgo biloba*, is the only living species of its genus, family, and order.

On the other hand, if a large and mentally unwieldy group can be divided by using characters of less evident importance, or by ignoring the existence of a few transitional units, it may be useful to undertake this division. Thus the genera in such large and closely knit families as the Compositae, Cruciferae, and Umbelliferae are notoriously difficult to define. In order to have any genera at all, we must use relatively trivial differentiating characters, and even then be prepared to face the existence of transitional species. In the Compositae we would otherwise confront the prospect of including the whole subfamily Asteroideae (Tubuliflorae), with some 15,000 or more species, in a single genus. At the beginning of the 20th century one disgusted botanist did include all species of the family Cruciferae in the single genus *Crucifera*, but his treatment has been generally ignored. In such a large group, an imperfect organization is better than no organization at all.

The larger the group, the greater the mental pressure to divide it, even if minor characters must be used and some intermediates must be arbitrarily placed. Conversely, the smaller the group, the greater the differences must be between it and its relatives, if the group is to be maintained. The family Bataceae, consisting only of the genus *Batis* with two species, has usually been maintained in a distinct order of uncertain affinities, the *Batales*. In recent years it has been suggested that the affinity of the Bataceae is with the order Caryophyllales, and indeed the differences between *Batis* and some genera of the Chenopodiaceae (of the order Caryophyllales) are perhaps not so formidable as they have seemed. The question thus arises as to whether the family Bataceae might not better be included as a somewhat aberrant member of the Caryophyllales, instead of being kept as a separate order. It still seems necessary, in any case, to maintain the Bataceae as a distinct family, even though the family has only a single genus and two species. We shall have occasion to refer to the Bataceae again in a subsequent chapter.

We are thus led to the principle of seeking the best compromise among the often conflicting objectives of the taxonomic system. The taxonomic system is a general purpose system, and it cannot properly serve any one purpose so faithfully as to exclude some other purpose altogether. There can be as many special systems as there are special purposes, but only the taxonomic system attempts to balance all these needs against one another in a single scheme. It should not be surprising that taxonomists often differ about how best to meet the multitude of conflicting objectives.

The fact that recognizable groups are not always sharply limited should not be surprising. The evolutionary concept provides the basic rationale for the taxonomic system. All groups, no matter how distinctive, are connected through a common ancestry tracing back to the unicellular flagellates and eventually to the precursors of the autotrophic bacteria. If the fossil record were complete, all gaps between groups would disappear. At any particular time level there are gaps, but these gaps developed only through the extinction of intermediate forms after evolutionary divergence. It is only to be expected that extinction of the connecting forms will progress irregularly, and that at any given time there are perceptible groups which are still connected by an evolutionary remnant of transitional forms. Asa Gray once observed that the several segregates of the genus *Gilia* just fail as genera, "by want of a little extinction." Most students of the Polemoniaceae now regard these segregates as distinct genera, in spite of the connecting species.

## The Level or Rank of Taxa

The formal taxonomic hierarchy provides for species, genera, families, orders, classes, and divisions, in ascending sequence, within the plant kingdom. Every species belongs to a genus, every genus to a family, every family to an order, and so on.

The number of levels of relationship which we wish to recognize in any particular group may or may not coincide with the number provided in the required set. If there are too many required categories for the number of levels to be recognized, the categories are telescoped together by having monotypic groups. Thus the order Plumbaginales contains only the family *Plumbaginaceae*, and family *Paeoniaceae* contains only the genus *Paeonia*, and the genus *Linnaea* contains only the species *Linnaea borealis*. *Adoxa moschatellina* is the only species of its family, and we have already noted that *Ginkgo biloba* is the only species of its order.

If there are not enough categories for the number of levels of relationship which we wish to recognize, additional categories may be inserted.

Thus the species of the large genus *Carex* are organized into four sub-genera and numerous sections, and the genera of the Compositae are organized into about 12 tribes. The Compositae provide an especially interesting case, because many botanists consider them to be the only family of an order Asterales. Thus we seem to have at once too many and too few standard categories in dealing with the Compositae. This brings us to the question of the level at which groups should be recognized.

Aside from infraspecific taxa, the only taxonomic category with an inherent rank is the species. An exact definition of the species is impossible, and the more precise one attempts to be, the larger the number of species which do not fit the definition. Still, the basic concept is simple enough. A species is the smallest population which is permanently (in terms of human time) distinct and distinguishable from all others. It is the smallest unit which simply cannot be ignored in the scheme of classification. It is the primary taxonomic unit, and it may also be thought of as the basic evolutionary unit. In sexual populations, gene exchange by hybridization within a species is ordinarily rampant, whereas such gene exchange between different species is restricted or even impossible. Interspecific hybrids are not always wholly sterile, but they are not so fertile and so competitively adapted as to swamp out the parents. Although there are some differences in interpretation, a reasonable degree of reproductive isolation from other species, under natural conditions, is an essential specific quality. Without such isolation, the population would lose its identity through interbreeding.

Many species have varieties which are partially segregated from each other, nearly always with a geographic or ecologic correlation, but connected by numerous intermediates. When parts of what had been a single species have diverged enough so that the distinction between them is reasonably sharp and prospectively permanent, the parts have become distinct species. It is only to be expected that at any particular time some species will be in the process of breaking up into separate species. The fact that polyploidy may introduce a barrier to interbreeding without any other significant difference complicates the problem further. The line between strong varieties and weak species is necessarily an arbitrary one, involving subjective taxonomic judgment. The weak species of one taxonomist may be the strong varieties (or subspecies) of another.

Many pages could profitably be devoted to a discussion of how to decide whether a pair (or group) of taxa would better be treated as distinct species or parts of one species, but we are here concerned primarily with the broader aspects of angiosperm taxonomy and evolution. The difficulties in interpretation do not alter the fact that the species is an inherent taxonomic level, the smallest unit which is permanently distinct and distinguishable.

It is perfectly clear that natural, recognizable groups of species, and groups of groups, exist. The ranks at which these groups should be received are not inherent in the nature of the group, but depend on subjective individual judgment. The criteria on which that judgment should be based are recondite. They come down to a personal evaluation of the importance of the differences and the size and coherence of the group, in the context of the system as a whole. Since we have already pointed out that individual characters cannot reasonably have a fixed, inherent taxonomic value, any evaluation of the importance of the characters marking a group is likely to be difficult and subject to unresolvable differences of opinion.

In spite of these inherent difficulties, a very considerable agreement has grown up regarding the groups to be recognized as genera and families of angiosperms. In part, at least, this agreement involves typological thinking. If the circumstances permit, we try to define genera in such a way that one can recognize a genus from its aspect, without recourse to technical characters not readily visible to the naked eye. Thus a person who is familiar with one species of *Typha* can readily recognize other species of *Typha* when he first sees them; familiarity with a few species of *Acer*, *Aster*, and *Solidago* will permit one to recognize other species of these genera; etc.

In the case of the maples, the typological concept that "a genus ought to look like a genus" has clearly influenced the traditional definition of the group. There are enough well correlated differences in the flowers and inflorescence of various species of *Acer* to permit the recognition of several genera with characters just as strong as those of genera in other families; but the trees all "look so much like maples," because of the opposite, usually palmately lobed leaves and characteristic double samaras, that it is customary to define the genus broadly.

The trouble with such typological concepts is that they often do not provide a clear answer. Although it is possible to learn to recognize an *Aster*, or a *Carex*, or a *Polygonum*, or a *Quercus*, or a *Solidago* by its general appearance, it is equally possible to learn the sections or subgenera of these and many other genera in the same way. There is nothing but custom and individual opinion to determine whether the segregates from these genera should be held as distinct genera or considered as sections or subgenera of more broadly defined genera. Custom, in such matters, is merely the sum of a series of individual opinions, plus inertia.

To at least some degree, the same sort of typological concepts which influence our thinking on genera operate also at higher taxonomic levels. It is customary to maintain the Verbenaceae as a family distinct from the Labiatae in spite of the fact that there is no clear break between the two families. The reason is that the differences are so pronounced that it is much easier to maintain two different but fairly clear concepts than a single rather amorphous one. Exactly the same situation exists regarding the Apocynaceae and Asclepiadaceae.

It should by now be clear that there is good reason to allow custom to enter into the decision on the rank at which a group is received. Since the rank is not inherent in supraspecific groups, it is only by giving some weight to custom that any stability in the taxonomic scheme can ever be achieved. A good operating principle is to maintain the existing classification whenever it can be defended on natural grounds, and to avoid changing the rank of groups if no significant change in the concept of their relationship to each other and to other groups is involved.

On the other hand, if a traditional arrangement runs counter to evolutionary relationships, then it should give way to a more natural arrangement when such can be devised, no matter how heavy the weight of tradition. It has been customary for nearly two centuries to hold *Franseria* and *Ambrosia* as distinct, but the recent work of Willard Payne[7] has made it necessary to combine these two genera. He demonstrated that *Ambrosia* consists of several different groups of *Franseria* that had independently undergone further coalescence of the involucral bracts (thus attaining the morphological character of *Ambrosia*), and that the closest relationships of the various groups of species of *Ambrosia* (as traditionally defined) are not with each other but with diverse species of *Franseria*. Shinners had independently reached the conclusion a few years earlier that the two genera should be combined, but he did not develop the essential argument that the distinction runs counter to natural relationships. As a sidelight we may note that from the standpoint of formal nomenclature (as contrasted to phylogeny), *Ambrosia* is the older name, and that the enlarged genus consisting of traditional *Franseria* and traditional *Ambrosia* must be called *Ambrosia*.

In some cases custom is divided and is likely to remain so. Distinguished precedent can be found for treating the legumes as three closely related families, or as three well marked subfamilies of a single broadly defined family. No one questions the relationships; it is only the rank of the groups which is in dispute. Here is the legitimate, but narrow, ground for difference between taxonomic splitters, who would recognize as many taxa at each level as the evidence permits, and the lumpers, who would recognize as few as possible. The splitter prefers to recognize the agreed units at a higher taxonomic level, the lumper at a lower one. There are no objective criteria to determine who is right. Those who compare the system here presented with the conceptually similar scheme of Takhtajan will note that he often splits where I lump. These two systems do not represent the

[7] A re-evaluation of the genus Ambrosia (Compositae). Jour. Arn. Arb. 45: 401–430. 1964.

limits of possibility, however. Another system which I understand to be in late stages of development reflects an even broader concept of families and orders than my own.

The problem of lumping or splitting also exists at the specific level, as shown by Hooker's advice that botanists should seek "to determine *how few, not how many* species" make up the flora of a region.[8] However there is now, as we have noted, a good measure of agreement on the criteria for recognizing species, and the difference between lumpers and splitters mainly comes down to how liberally we should interpret the requirements in borderline cases. There is no ultimate authority in such matters, as indeed there is not in any scientific endeavor.

## SELECTED REFERENCES

Carlquist, S., *Island Life*. Garden City, New York: Natural History Press, 1965.

Cronquist, A., Orthogenesis in evolution. Res. Stud. State Coll. Wash. 19: 1–18. 1951.

———, The taxonomic significance of evolutionary parallelism. Sida 1: 109–116. 1963.

Dansereau, P., Ecosystems of the world and the play of natural selection. A manuscript prepared for publication in Vol. II of *Evolutionary Biology*, to be published by Appleton-Century-Crofts in 1968.

Davis, P. H., and V. H. Heywood, *Principles of Angiosperm Taxonomy*. Edinburgh: Oliver & Boyd, 1963.

Simpson, G. G., *Principles of Animal Taxonomy*. New York: Columbia University Press, 1961. (The reader may note that this and the foregoing work present rather different points of view, but both are useful.)

Stebbins, G. L., Natural selection and differentiation of angiosperm families. Evolution 5: 299–324. 1951. (A comprehensive but speculative attempt to interpret the evolution of families and orders in terms of survival value.)

———, Adaptive radiation and trends of evolution in higher plants. Vol. I, pp. 101–142 in Th. Dobzhansky, M. K. Hecht, and W. C. Steere, eds., *Evolutionary Biology*. New York: Appleton-Century-Crofts, 1967.

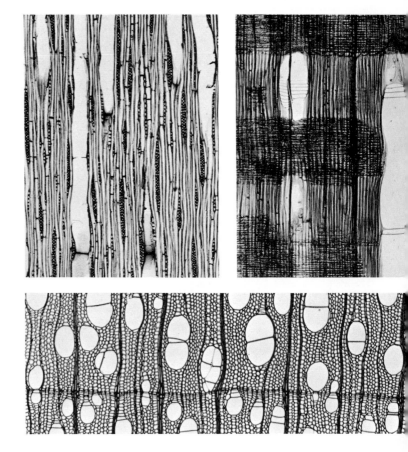

## Chapter 2

*Wood of river birch,* Betula nigra. *Top left, tangential long section; top right, radial long section; bottom, cross section. The xylem vessel confers a major competitive advantage on most angiosperms, as compared with other groups of land plants. Photos courtesy of the U.S. Forest Products Laboratory.*

# The Origin of the
# Angiosperms

## Characteristics of the Group

The origin of the angiosperms was an "abominable mystery" to Charles Darwin, and it remains scarcely less so to modern students of evolution. It is clear that they are vascular plants, related to other vascular plants, that they belong to the pteropsid rather than to the lycopsid or sphenopsid line within the vascular plants, and that their immediate ancestors must have been, by definition, gymnosperms.

The most distinctive features of the angiosperms, as opposed to the gymnosperms, are the enclosure of the ovules in an ovary, the further reduction of the female gametophyte, and the characteristic "double fertilization" which leads to the development of an entirely new food storage tissue in the seed, the endosperm. The term endosperm is sometimes also applied to the food storage tissue of gymnosperm seeds, which consists of the body of the female gametophyte. But this is wholly different from the true endosperm of angiosperm seeds, which arises from the fusion of a sperm nucleus with typically two nuclei (or the product of fusion of two nuclei) of the female gametophyte.

Although these several differences clearly represent evolutionary modifications from the situation in gymnosperms, their survival value is debatable. The enclosure of the ovules in an ovary is often thought to provide additional protection, but this is hard to verify by observation. Furthermore, although the ovules of both conifers and cydads are briefly exposed to the air at the time of pollination, they are during their subsequent development as effectively covered and protected by the cone scales as angiosperm ovules are by the ovary. The significance of the substitution of a triploid nutritive tissue (the endosperm) in the seed for a haploid one is even more obscure. We know only that it has occurred.

Two other characteristic but taxonomically less dependable features of the angiosperms are probably much more important than the morphology of the ovules and ovary in explaining the dominant position of the angiosperms in the vegetation of the earth. One is the exploitation of insects and other animals as agents of pollination. The second, and perhaps more important, is the development of a more complex and efficient conducting system, both xylem and phloem. The xylem vessel, in particular, appears to be of critical importance.

The gymnosperms are largely wind-pollinated. In order for wind pollination to be effective and economical, there must be many individuals of a kind, preferably growing close together. (Anyone who has ever tried to grow just a few hills of corn, well separated from any large corn field, will recognize the problem.) Even so, great quantities of pollen are wasted. A huge amount of metabolically precious protein must be produced, only to be thrown to the winds, never finding its mark.

Pollination by insects and other animals requires a much smaller amount of pollen and does not require such large populations for efficiency, but the difference is not all pure gain. The plant must produce something, such as nectar, that will attract the pollinator, and it must also use some energy in producing petals or other structures which help guide the pollinator to the source of supply. A greater waste is in a sense traded for a lesser one.

In taking up insect pollination, angiosperms have exploited a new evolutionary opportunity, yet without wholly giving up the old. Unlike some other evolutionary opportunities which are exploited by one or another group of organisms, this one does not have insurmountable barriers preventing exit. A considerable number of angiosperms have reverted to wind pollination, even though they doubtless had insect-pollinated ancestors. Thus in a sense the angiosperms have an extra string to their pollinating bow, permitting them to occupy diverse ecological niches.

The xylem vessel confers probably the most important competitive advantage which most angiosperms have over most gymnosperms. True, there are some primitive angiosperms which lack vessels, and some aquatic angiosperms which do not need vessels and have lost them, but the vast majority of angiosperms do have vessels and do enjoy a competitive advantage over other plants as a result. A few gymnosperms (the Gneticae) and even a few pteridophytes have more or less well developed vessels, but, for reasons unknown, they have not been able to exploit the evolutionary opportunity thus seemingly opened to them. Perhaps the Gnetican vessel evolved too late, after the angiosperms had already pre-empted the field. Certainly the vast majority of gymnosperms are wholly dependent on tracheids for the movement of water from the roots, up through the stem, and into the leaves. Tracheids are less efficient conductors than vessels;

given an adequate supply they simply cannot deliver the same rate of flow.

Having a less ample and assured water supply than angiosperms, gymnosperms must perforce have more rigorous means of transpiration control to meet the same degree of evaporative stress, or they must be able to survive greater desiccation. It is no accident that the gymnosperms tend to be sclerophyllous, and to have a more xeromorphic aspect than most angiosperms.

The necessary control of transpiration entails a restriction of photosynthesis as well. Anything which permits the exchange of gases necessary for photosynthesis will necessarily permit the evaporation of water also. Thus the advantage of vessels over tracheids in water conduction leads to an advantage of most angiosperms over most gymnosperms in the rate of photosynthesis. Individual kinds of gymnosperms may have other adaptations which permit them to survive in competition with angiosperms, but it is only in special circumstances that gymnosperms in general have any competitive advantage over angiosperms.

The most extensive of the special habitats in which the gymnosperms (as represented by the conifers) do seem to have a competitive advantage over angiosperms is in the boreal forest region of North America and Eurasia. Here the evergreen habit of the conifers, which is made possible by their sclerophyllous leaf structure and resistance to desiccation, permits them to carry on photosynthesis on any day when conditions are right, in contrast to the angiosperms, which are bare most of the year and which must produce great quantities of new leaves each year, only to drop them at the end of the short summer. Furthermore the sun, being at a lower angle in the sky, does not cause as much evaporative stress here as in lower latitudes, and there is a favorable p/e (precipitation-evaporation) ratio, so that the lack of vessels is not such a great handicap. The boreal forest does have some angiospermous trees, notably species of *Populus*, just as the deciduous forest to the south has some conifers, but the dominance of the conifers reflects a set of environmental conditions more favorable to them than to angiospermous trees in general.

The sieve tube–companion cell structure of phloem, so widely taught in general botany, is a feature of angiosperms alone. The sieve elements of gymnosperms and vascular cryptograms characteristically have lateral sieve areas rather than terminal sieve plates, and they are not united end to end into sieve tubes. Companion cells are lacking in gymnosperms and vascular cryptograms, although there are scattered parenchyma cells in the phloem. The more complex structure and organization of angiosperm phloem must be presumed to be more efficient than that of gymnosperms, although I do not know that this has been experimentally demonstrated.

There is no obvious reason why the gymnosperms should not have developed a complex phloem of angiospermous type. We know only that they did not (except, of course, in the group that became angiosperms). Nor is there any obvious reason why the several features which collectively separate the angiosperms from the gymnosperms should all have evolved concomitantly. The concomitance is not perfect: There are some primitive angiosperms which lack vessels, and others in which the carpels are merely folded and unsealed at anthesis. Development of the pollen tube is obviously correlated with enclosure of the ovules, and double fertilization doubtless results at least in part from the extreme reduction of the female gametophyte; but otherwise there seems to be no inherent reason why these several advances should have occurred in concert. They are not parts of a necessarily interlocking system; each could operate without the others.

On the basis of present information it seems that the origin of the angiosperms as we know them was fortuitous. One may speculate that the genetic revolution which permitted the development of angiospermous reproductive structures also shattered the stability of other characters such as the structure of the xylem and phloem. For the present, at least, that is mere speculation; another time it might have happened differently.

The angiosperms evidently do fit Simpson's dictum, previously quoted, about the relationship of taxonomic groups to adaptive zones. The changes in structure of xylem and phloem, combined with the changes in the reproductive apparatus and pollinating mechanism, certainly do put the angiosperms into a very broad but distinctly new adaptive zone. As a result, they have become the dominant vegetation of the earth. We shall see that the application of Simpson's dictum to the families and orders of angiosperms is much more dubious.

## The Fossil Record

What sort of gymnosperms may have been ancestral to the angiosperms is still to some extent an open question. The fossil record, which as we have pointed out has only a very limited usefulness in clarifying relationships among the various groups of angiosperms, is no more helpful in determining their ancestry. Morphologically, the angiosperms are so sharply limited and so well set off from all known gymnosperms that a series of hypothetical connecting forms must be postulated in order to tie them to anything else at all. Here the "missing link" is indeed still missing — or at least has not been recognized as such.

So far as the megafossil record shows, the angiosperms might have arisen full blown as a number of separate families, highly diverse among themselves, early in the Cretaceous period. In the earliest Cretaceous, the angiosperm fossils are few and scattered, and even where they occur they are much outnumbered by ferns and gymnosperms. They are progressively more abundant in progressively younger beds of the Lower Cretaceous period, and in the Upper Cretaceous they become dominant.

Allowing always for the possibility of error, it appears that at least three modern genera, in as many families, are represented by Lower Cretaceous megafossils. These are *Sassafras, Cercidiphyllum,* and *Populus.* The Lauraceae, to which *Sassafras* belongs, are generally considered to be among the more primitive families of angiosperms, although they are clearly not the most primitive family. The Cercidiphyllaceae are somewhat more advanced, and the Salicaceae are now generally regarded as moderately advanced plants, not at all primitive. Many other Lower Cretaceous fossils have been given names indicating their resemblance to modern genera (e.g., *Celastrophyllum, Ficophyllum, Juglandophyllum, Menispermites, Nelumbites, Quercophyllum, Sapindopsis*), but their actual affinities are uncertain. It does seem likely that some of these Lower Cretaceous fossils may represent the order Nymphaeales, a fairly primitive order of dicots which is sometimes regarded as near the ancestry of the monocots.

By the middle of the Upper Cretaceous, the number of angiosperm fossils which can be referred to modern genera with some confidence becomes rather large. Among these are *Artocarpus, Dalbergia, Ficus, Liquidambar, Magnolia, Platanus,* and *Typha.*

Aside from the fact that the angiosperms did become progressively more abundant after their first appearance in the early Lower Cretaceous, there is very little in the megafossil record to indicate the nature of the evolutionary history and relationships among the more primitive members of the group. On the contrary, the diversity among these early angiosperm megafossils suggests that the group originated well before Cretaceous time.

On the other hand, the diversity of Lower Cretaceous angiosperms may turn out to be more apparent than real. The diversity of leaf form which can exist even within a closely related group of modern angiosperms is a constant source of difficulty to taxonomists, and many a "new" species which has been confidently described and assigned to a particular genus on the basis of sterile material has turned out to belong to a quite different family. Attempts to match Lower Cretaceous leaves with those of modern angiosperms may have led earlier paleobotanists to erroneous conclusions about the age of a number of families.

In contrast to the leaves, the pollen of Lower Cretaceous angiosperms is remarkably uniform. Aside from uniaperturate types which are difficult to distinguish from gymnosperm pollen, Lower Cretaceous angiosperm pollen consists very largely of triaperturate forms which fit in toward the base of the evolutionary series postulated by palynologists on grounds of comparative morphology. Triporate and polyporate types, as well as the various highly specialized types now found in many families, are notable by their absence from the Lower Cretaceous. (Information from an un-published review in an Honors Thesis at Harvard University by James A. Doyle, 1966.)

In spite of the megafossil evidence which has been thought to require a pre-Cretaceous origin of the group, all pre-Cretaceous fossils which have been thought or suspected to be angiospermous are subject to question. Some Jurassic pollen grains have been identified by competent botanists as being not only angiospermous but even referable to the Nymphaeaceae; other equally competent botanists consider these pollen grains, which are monosulcate, to be gymnospermous. Similar controversy surrounds the identity of some Jurassic bits of wood which lack vessels and may or may not be angiospermous.

Perhaps the most interesting pre-Cretaceous fossil that might prove to be an angiosperm is *Sanmiguelia*, a palm-like plant from Triassic deposits in Colorado. The fossils consist only of leaf impressions, without cellular detail. They do indeed look like parts of palm leaves, and if they had been found in Cretaceous or later deposits they would probably pass as palms without serious question. However, they are also something like cycad or cycadeoid leaves, and if there were not such things as palms they would perhaps be referred to the Cycadicae. Palms are not usually con-sidered to be very primitive among the angiosperms; indeed they are one of the more highly specialized groups, at least vegetatively. If *Sanmiguelia* is really a palm, some rethinking of our concepts may be in order, but without flowers, fruit, or wood its status as a palm must rank as an interest-ing possibility rather than as a fact or even a probability. Paleobotany is replete with examples, continuing even until the present time, of drastic reinterpretations of the affinities of fossils consisting of imprints of vege-tative parts.

### Possible Ancestors

In view of the inadequacy and ambiguity of the fossil record, our thoughts on the relationships of the angiosperms must be guided largely by the comparative morphology of modern angiosperms and modern and fossil gymnosperms. Most of the orders of the gymnosperms have at one time or another been suggested as possibly ancestral to the angiosperms. If the gymnosperms (division Pinophyta) are divided into three sub-divisions — the Pinicae, Gneticae, and Cycadicae, as in the system of Cronquist, Takhtajan, and Zimmermann[1] — the Pinicae may quickly be excluded from serious consideration, the Gneticae may be excluded after more careful study, and only the Cycadicae (including the seed ferns) re-main as possible ancestors to the angiosperms.

[1] On the higher taxa of Embryobionta. Taxon 15: 129–134. 1966.

PINICAE

Nothing in the Pinicae offers any reasonable possibility of being ancestral to the angiosperms. The uniquely specialized xylem of the Pinicae and Gneticae is discussed under the latter group. The seed-bearing apparatus of the Pinicae and the angiosperms follows distinctly different lines of evolutionary specialization: The angiosperms are clearly phyllosporous; carpels are modified leaves. The seed-bearing organs of the Pinicae are variously modified but never distinctly leaflike; probably they are modified telomes or telome-systems that have never been vegetative leaves. In the Cordaitales and Ginkgoales the ovules are obviously terminal on slender stalks; in the Cordaitales these ovuliferous stalks are aggregated into an apparently simple cone. In the Pinales the female cones are compound, and the ovules are borne on "ovuliferous scales" which are axillary to the primary bracts of the cone. The ovular apparatus of the Taxales is even more specialized. The possibility of deriving the angiosperm flower from such structures seems most remote, and the more so when we recall that the primitive angiosperm flower is bisexual, whereas the cones of the Pinicae (and Gneticae) are uniformly unisexual. Neither is there anything in the leaves to suggest a relationship to the angiosperms. Nothing in the whole subdivision Pinicae has net-veined leaves, whereas net venation is standard among the angiosperms and occurs also in some of the ferns and seed ferns. Even the monocotyledons, which are said to have parallel-veined leaves, have abundant cross-connections between the main veins.

GNETICAE

The Gneticae resemble the angiosperms in several respects and were at one time taken seriously as their possible ancestors. The Gneticae are dicotyledonous, and they differ from all other gymnosperms in having vessels in the secondary wood. The female gametophyte of the Gneticae is more reduced than that of most gymnosperms. In *Gnetum* there is no suggestion of an archegonium, and fertilization takes place while the female gametophyte is still in free-nuclear condition. The ovules of the Gneticae differ from those of most gymnosperms and resemble those of most angiosperms in having two integuments instead of only one. *Gnetum* has net-veined leaves and is very angiospermous in general appearance; indeed *Gnetum gnemon* could easily pass for *Coffaea arabica* when in sterile condition, and I am told that it has done so in some college greenhouses. The unisexual, scaly-bracted inflorescences of *Gnetum* and other Gneticae might be compared, without too much stretch of the imagination, to the catkins of some amentiferous angiosperms, such as the Salicaceae or Betulaceae. It is therefore not surprising that some credence was once given to the idea that the Gneticae, particularly *Gnetum* itself, might be closely allied to the angiosperms.

Unfortunately, the first appearances are not borne out by further study. The vessels, once thought to provide strong evidence of relationship, are now seen to indicate quite the contrary. As I. W. Bailey[2] has pointed out, the perforations of vessel members in the angiosperms arose initially through the dissolution of the pit-membranes in scalariform-bordered pitting, whereas in the Gneticae they developed through modifications of circular bordered pits of a typically coniferous type. The mature structures look similar, but they get there by different routes; the similarity is due to evolutionary convergence. The angiospermous vessels therefore cannot be derived from the Gnetican vessel; any possible connection between the Gneticae and the angiosperms must antedate the origin of vessels in both groups.

The potentiality to develop vessels exists in the vascular cryptograms as well as in the seed plants. Xylem elements which fit the morphological criteria of vessels occur in *Pteridium*, *Selaginella*, and *Equisetum*, as well as in the Gneticae and the angiosperms. Only the latter two groups have capitalized on the water-conducting potentiality of a vessel system, but evidently not much can be made of this similarity in terms of phylogenetic relationship.

The Pinicae and Gneticae have

> a highly specialized and peculiar type of primary xylem which is entirely unlike that of other known vascular plants with the possible exception of the Ophioglossales. Typical scalariformly pitted tracheids are eliminated from both the primary and the secondary xylem, and circular bordered pits have worked backward in ontogeny from the secondary xylem through the primary xylem. The development of vessels in the Gnetales from such circular pitted tracheids is unique in type and entirely unlike the derivation of vessel members from scalariformly bordered-pitted tracheids [in angiosperms and certain vascular cryptogams].[3]

Changes in our concepts of evolutionary relationships among the angiosperms have placed a different light on the similarity of the Gnetican inflorescence to that of the Amentiferae. As we shall see, the Amentiferae are now believed to have reduced flowers. The primitive angiosperm is thought to have had strobiloid, bisexual flowers, with a well developed perianth, numerous stamens, and numerous separate carpels, each flower representing a complete bisexual strobilus. Such flowers would be most difficult to derive from the unisexual, scaly-bracteate, compound strobili of the Gneticae. The similarity between the Gnetican strobilus and the aments of some angiosperms is now seen as superficial and due only to a

---

[2] The development of vessels in angiosperms and its significance in morphological research. Am. Jour. Bot. 31: 421–428. 1944.

[3] Bailey, op. cit.

degree of convergence. Any phyletic connection between the two groups would have to antedate the development of such similarity in this respect as does exist.

The angiosperms may well be derived from polycotyledonous rather than dicotyledonous ancestors. *Degeneria*, one of the most primitive existing angiosperm genera, has three or four cotyledons in most of the embryos. The possible significance of dicotyledonous embryos in an assessment of the relationships of the Gneticae to the angiosperms is presently dubious.

The reduction and loss of the archegonium in the female gametophyte has evidently occurred within the confines of the Gneticae. *Ephedra* has normal archegonia. In *Welwitschia* a specialized cell which is probably homologous with the archegonial initial functions directly as an egg. In *Gnetum*, as we have noted, fertilization takes place while the gametophyte is still in free-nuclear condition. Since we have already pointed out that any connection between the angiosperms and the Gneticae must involve pre-Gnetican ancestors, the reduction and loss of the archegonium in the Gneticae must have occurred subsequent to the split. The similarity in this respect, then, is due to parallelism, and merely reflects the general tendency toward reduction of the gametophyte among the vascular plants as a whole.

We have left, then, as indicative of a relationship between the Gneticae and the angiosperms, the double integument of the ovule, the net-veined leaves of *Gnetum*, and the habital similarity of *Gnetum* to angiosperms. That is not enough, especially in view of the evidence that the similarities in other respects must be secondary rather than original. Indeed the evidence from vascular anatomy strongly controverts any suggestion of a relationship.

## CYCADICAE

Having eliminated the Pinicae and Gneticae as prospective ancestors to the angiosperms, we turn to a consideration of the Cycadicae. The Cycadicae are here considered to consist of three classes and four orders. The Cycadales are the only order of the class Cycadatae, the Bennettitales are the only order of the class Bennettitatae, and the Caytoniales and Lyginopteridales collectively make up the class Lyginopteridatae. The Lyginopteridales might with some justification be broken up into several orders, but that is immaterial to our purpose here.

Among these four orders, only the Cycadales are represented by living species. The rest are known only from fossils of Cretaceous or earlier age. Fossil Cycadales are also known from as far back as the Triassic period, long enough ago so that age alone poses no barrier to the possibility of deriving the angiosperms from them.

The Cycadales, Bennettitales, and Caytoniales may be excluded as likely ancestors to the angiosperms, because each has some specializations which are not to be expected in the proto-angiosperm.

The Cycadales typically have sympodial but otherwise apparently un-branched stems. Sympodial inflorescences are common enough in the angiosperms, but the vegetative stems are usually (and primitively) mono-podial. Some cycads do have a sparsely branched, monopodial stem, how-ever, so that growth habit does not pose an insuperable barrier to the derivation of angiosperms from cycads. The structure and arrangement of the reproductive parts of cycads pose more serious difficulties to such a concept. The cycads are dioecious, with well developed male and female cones on different individuals. Any cycad-like ancestor of angiosperms would need to have bisporangiate strobili and branching, monopodial stems, characteristics which would make the plant not a cycad but more nearly a pteridosperm. The cycads themselves are generally conceded to have evolved from pteridosperms.

The Bennettitales have bisporangiate strobili with the microsporophylls below the megasporophylls, as a good pre-angiosperm should, and some of the species have freely branched, monopodial stems. The megasporophylls, however, are wholly different from the angiosperm carpel. Each mega-sporophyll is narrow and undivided, with a single terminal ovule. These megasporophylls may well be phyletically reduced from the compound, foliaceous megasporophylls of the seed ferns, but even under that interpre-tation the protoangiosperm must have had more leaflike sporophylls than any which are known in the Bennettitales. On the positive side, it is of some interest, though perhaps of no great importance, that both the Cycadales and Bennettitales have dicotyledonous embryos, but as we have noted the primitive angiosperm may have been polycotyledonous. It would not be too surprising, however, if polycotyledonous forms were to be found among the Bennettitales and the fossil cycads. Reduction in the number of cotyledons seems to be a a general trend among seed plants.

The Caytoniales are in their own way too specialized to be likely ances-tors of the angiosperms. In the Caytoniales the ovules are semi-enclosed in small pouches which are pinnately arranged along the axis of a megasporo-phyll. These pouches have been compared to the ovaries of angiosperms but the structures are not strictly homologous. The primitive angiosperm ovary consists of a single complete carpel, folded along the midrib; one megasporophyll (carpel) forms one ovary. The ancestor must have had open megasporophylls with exposed ovules, as do the seed ferns. Thus, on close scrutiny, the ovary-like pouches of the Caytoniales, which caused them to be compared to angiosperms, provide ample reason why the Cay-toniales cannot be ancestral to the angiosperms.

Only the pteridosperms (Lyginopteridales) remain as potential ancestors of the angiosperms. It is a long way, morphologically, from any known seed fern to an angiosperm, but each of the differences could logically be bridged in the course of evolution. A long chain of unknown ancestors must be postulated, but there is nothing inherently unlikely about any of the requisite evolutionary changes.

The pteridosperms had large, usually pinnately compound leaves, and some of the species were distinctly net-veined. The stems were either simple or branched, the branching being of the ordinary, monopodial sort. Many of the species had a definite cambium, although the proportion of secondary tissues was usually not high. The megasporophylls and microsporophylls of the pteridosperms were often borne on the same plant. They were not aggregated into definite strobili, but the fact that the Cycadales and Bennettitales do have such strobili shows that the evolutionary potentiality for strobili existed in the pteridosperms, which are generally considered to be ancestral to the other orders of Cycadicae. Both the microsporophylls and the megasporophylls were compound, but here again the capacity for reduction is shown by the nature of the sporophylls in both the Bennettitales and the Cycadales. Even the modern cycads show a series of stages in reduction of the megasporophyll from a compound leaf to a tack-shaped cone-scale.

The origin of sepals from leaves, and of petals from both sepals and stamens, are amply documented within the angiosperms, so that these characteristic angiospermous structures pose no barrier to associating the angiosperms with any group of gymnosperms we wish to consider. The characteristic double fertilization in angiosperms, with the attendant development of a triploid endosperm, evidently relates to the extreme reduction of the female gametophyte; it provides no indication at all of the relationships of the group. Reduction of the gametophyte is a general trend throughout the vascular plants as a whole; this trend has merely been carried farther in the angiosperms than in other groups.

The seeds of pteridosperms were much more primitive than those of any known angiosperm. They evidently had a large female gametophyte, like most other gymnosperms, and not one pteridospermous seed has yet been found to contain an embryo. Possibly the embryo developed well after the apparent ripening of the seed, as in the modern *Ginkgo*. The ovule had only a single integument, like the ovules in all other gymnosperms except the Gneticae.

The origin of the second integument in angiosperm ovules is not yet fully understood. The single integument of gymnospermous seeds in general is believed to have originated by fusion of a set of leaf segments beneath the megasporangium, at a time in evolutionary history when the leaf itself was scarcely more than a somewhat modified set of branching stems, i.e., when the evolutionary differentiation of the typical megaphyll from the branching stem system of the rhyniophytes had not yet been completed. The pteridosperms also commonly have one or several ovules collectively subtended by a cupule that is believed to be of syntelomic origin like the integument. It is not difficult to suppose that a cupule subtending only a single ovule could become further modified to form a second integument. That is only a supposition, but no other reasonable supposition comes to mind.

The ovules of pteridosperms were mostly borne on the margins of the megasporophylls, but in some species they were on one or the other surface. The primitive position of the ovules on the angiosperm carpel has in the past usually been considered to be marginal, but more recently it has been suggested that they were on the upper surface instead. This question is considered in the following chapter. Apparently either position is compatible with an origin of angiosperms from seed ferns, although the marginal position provides a larger set of possible ancestors. The recent interpretation of some material of the Permian seed fern *Glossopteris* as having reproductive structures arising from the upper surface of the leaf is interesting, but its significance can scarcely be evaluated until the structural details are clarified.

We have already noted that the primitive angiosperm probably did not have vessels, and that the tendency to develop vessels exists in diverse groups of vascular plants. The absence of vessels from the pteridosperms is therefore not at all incompatible with their being ancestral to the angiosperms.

Thus, all of the characters in which the angiosperms differ from seed ferns could have evolved from a pteridospermous ancestry.

Seed fern fossils are most abundant in Carboniferous deposits, but some of them occur in rocks as late as the Jurassic. Unequivocal angiosperm fossils are not known from deposits of earlier than Cretaceous age, but we have noted that the group may well have originated earlier. The time factor therefore presents no problem to the hypothetical derivation of the angiosperms from seed ferns.

It is a long step from saying that the angiosperms could have evolved from the seed ferns to saying that they actually did. As long as the fossil record provides no suggestion of a connection, the concept that the angiosperms evolved from the seed ferns must be regarded with some reserve. On the other hand, all other groups of gymnosperms, both living and fossil, seem highly unlikely or even impossible as potential ancestors of the angiosperms. Most students of phylogeny therefore provisionally accept the seed ferns as the probable ancestors of the angiosperms.

## SELECTED REFERENCES

Andrews, H. N., Early seed plants. Science 142: 925–931. 1963.

Brown, R. W., Palmlike plants from the Dolores Formation (Triassic) Southwestern Colorado. U.S. Geol. Surv. Prof. Pap. 274H: 205–209. 1956.

Camp, W. H., and M. M. Hubbard. On the origins of the ovule and the cupule in the lyginopterid pteridosperms. Am. Jour. Bot. 50: 235–243. 1963.

Maheshwari, P., Evolution of the ovule. 7th Sir Albert Charles Seward Memorial Lecture. Birbal Sahni Institute of Paleobotany. Lucknow, India. 1960.

Scott, R. A., E. S. Barghoorn, E. B. Leopold, How old are the angiosperms? Am. Jour. Sci. 258A: 284–299. 1960.

Takhtajan, A. L., Origins of the angiospermous plants. Translated by Olga Gankin. Am. Inst. Biol. Sci. Washington. 1958. (Original in Russian published by Soviet Sciences Press, 1954.) Designed primarily for student use in Russia, this work cites mostly Russian language sources, just as the book here presented emphasizes works in English.

## Chapter 3

*Flowers of* Phacelia purshii, *a woodland species of eastern U. S., with an insect visitor. Adaptation to diverse sorts of pollinators has played a major role in the evolutionary diversification of flowering plants. Photo by Charles C. Johnson.*

# The Evolution
## of Characters

## The Determination of Primitive Characters and Evolutionary Trends

The idea that organisms show a series of progressive modifications from a basic type long antedates any general belief in organic evolution as such. Naturalists of the 17th and 18th centuries were preoccupied with the establishment of a *Scala Naturae* in which every group would have its proper place according to its relative complexity or advancement. As long ago as 1813 de Candolle[1] referred to primitive and advanced characters ("Pour connaître le veritable nombre absolu des organs d'une plant, il faut, par la théorie des soudures, ou celle des avortemens, le ramener au nombre qui paraît le type primitif de sa classe. . . ."), and he proposed a taxonomic "serie lineaire et par conséquent artificielle."

Present day biologists, steeped in evolutionary tradition, find it hard to see how their predecessors could entertain such thoughts without also grasping the concept of organic evolution. ("We think our fathers fools, so wise we grow; our wiser sons no doubt will think us so.") There were, of course, some stirrings of evolutionary thought long before Darwin, but if de Candolle and others like him had any inklings of evolution, they carefully avoided saying so. They wrote, and seemingly thought, in terms of logically successive modifications from a basic type, sometimes as if they were trying to retrace the steps in the mind of the Creator.

All efforts at a natural system assume that there is a way, if we can discover it, to arrange things into a scheme which is more logical than other possible schemes. We have noted that the establishment of natural groups requires the perception of multiple correlations. The methods of estab-

[1] A. P. de Candolle, *Théorie Elementaire de la Botanique* (1813), p. 135.

**49**

lishing a "natural" sequence among groups of the same rank in such a scheme are purely subjective and intuitive. Differing schemes, once proposed, are subjectively judged by how well everything fits into place, and by the number and importance of the internal contradictions in the scheme.

As we have noted in an earlier chapter, such concepts of natural relationships are readily transformed into evolutionary concepts. The necessity for every taxon to have a logically possible antecedent is thereby re-emphasized, and a new light is cast on the significance of fossils. However, the consideration of fossils does not change the fact that the propriety of a scheme is judged by how well it provides for all the available evidence.

The well known system of Adolph Engler, although pre-evolutionary in origin, was considered by its author to reflect evolutionary relationships among the families and orders of angiosperms. That view, although defensible at the time, is no longer tenable. The system and its concepts do not now meet the test of providing for all the evidence. In the 12th (1964) edition of the Engler Syllabus, Engler's successors have in effect admitted that a major reorganization will be necessary, and indeed they have undertaken such a reorganization of the monocotyledons.

The greatest weakness of the Englerian system is that it does not distinguish adequately between primitive simplicity and simplicity through reduction. Inasmuch as most students of the subject now agree that floral reduction has been a pervasive (though not exclusive) trend within the angiosperms, the flaw is fatal.

By 1926[2] Engler had realized that the flowers of the Amentiferae are simplified rather than primitively simple, and he argued that their extreme reduction indicated the great antiquity of the group. Such an argument misses the whole point of a phylogenetic system. An essential requirement of any phylogenetic system is that one start with the groups which are least modified from the ancestral prototype, rather than with those which have undergone the most change. All groups are of equal age, if one takes in all the ancestors as well as the members of the group. It is only if one bases concepts of age on the members that would actually be referred to a particular group that groups differ in age and a phylogenetic system becomes possible.

The search for a general arrangement and set of principles that will permit everything to fall into place has led taxonomists to revive, modify, and expand the concepts first clearly expounded by de Candolle in 1813. The treatment by Bentham and Hooker in the Genera Plantarum was a lineal descendant of that of de Candolle. Although it was published over a period of years from 1862 to 1883, it was pre-Darwinian in concept, and

[2] Adolph Engler, *Die natürlichen Pflanzenfamilien* (2nd ed., 1926). 14a: 136–137.

FIGURE 3.1

Some botanists who have made distinguished contributions to the system of angiosperms. Top, left, Augustin Pyramus de Candolle (1778–1841); top, right, Adolf Engler (1844–1930); bottom, left, George Bentham (1800–1884); bottom, right, Armen Takhtajan (1910–).

the authors never claimed anything else. It was, however, an important historical link in the progression from de Candolle's natural system to the avowedly phylogenetic system of C. A. Bessey.

In 1915 Bessey published his epochal paper, "The phylogenetic taxonomy of flowering plants" (Ann. Mo. Bot. Gard. 2: 109–164), in which he set forth the principles on which a phylogenetic system should be founded, a list of putatively primitive characters, and an outline of a system incorporating these ideas. Although the system itself is now generally conceded to be faulty in execution, his principles (stated as dicta) are widely accepted. (The most notable exception is that alternate leaves are now usually considered to be more primitive than opposite, in contrast to Bessey's view.) We are all — or nearly all — Besseyans.

It is conceivable that some completely new set of concepts, for example, that of Melville,[3] will at some future time displace the modified Besseyan concepts now in vogue. The ultimate failure of the highly useful and highly regarded Englerian system to be compatible with the accumulating information should inspire some caution about the durability of present ideas. Yet the prognosis is for evolution rather than revolution. Even the replacement of Englerian concepts by Besseyan ones was not such a great change as it might seem. Large blocks, and groups of blocks, of the Engler system remain in all present systems, merely rearranged with respect to each other.

One of the principles which Bessey enunciated and which modern taxonomists stress is that within any one phylad the evolution of different organs may proceed at different rates. At any one time any particular group will probably present a mosaic of relatively advanced and relatively primitive characters. One result of this fact is that no one family of angiosperms has all of the primitive characters which are known among modern angiosperms; each of the "primitive" families is more advanced in some respects than in others.

On the other hand, highly primitive and highly advanced characters are seldom mixed in helter-skelter fashion. Within very broad limits, primitive characters do tend to be associated, and advanced characters likewise. A plant which is at the bottom of the ladder in one respect is not likely to be on the top rung in others. The Ranalian complex was postulated as primitive on grounds of floral morphology long before the primitive nature of the xylem in so many of its members was even suspected. A nice statistical correlation among a number of putatively primitive characters in angiosperms has been demonstrated by Sporne.[4] However, his statistical

[3] R. Melville, A new theory of the angiosperm flower. I. The gynoecium. Kew Bull. 16: 1–50. 1962.

[4] K. R. Sporne, Statistics and the evolution of dicotyledons. Evolution 8: 55–65. 1954.

analysis would suggest that the primitive flower was unisexual, whereas most botanists now believe it must have been bisexual.

To recapitulate, all ideas of what is primitive and what is advanced depend ultimately on how well these ideas fit together into a coherent scheme in which every character has a logically possible evolutionary history, and every group has a logically possible ancestor. Fossils have a potentially very important role in the establishment of such a scheme. In the angiosperms, however, the fossil record has up to now been of relatively little help, partly because so many of the critical taxonomic characters are seldom preserved. Phylogenetic ideas about the angiosperms therefore depend very largely on comparisons of living species.

## Characters of the Primitive Angiosperms

If the evolution of the angiosperms from their presumed pteridospermous ancestors followed the now familiar pattern of parallel developments in a series of closely related lines, there may never have been an original angiosperm which was ancestral to all other angiosperms. Each of the several features which collectively mark the angiosperms as a natural group may well have evolved separately in different lines, and separately in each line from the other critical characters, with only a loose overall correlation among the different advances.

Indeed this is exactly what seems to have happened with regard to the xylem, phloem, and the gross structure of the flower. Diverse surviving members of the primitive subclass Magnoliidae retain one or more primitive (presumably ancestral) features, but are more advanced in other respects. The Winteraceae and some other small families do not have vessels, but do have sieve tubes and companion cells. *Austrobaileya* has vessels but has a very primitive, gymnosperm-like phloem, with elongate, overlapping sieve elements that do not form typical sieve tubes, and published studies to date do not disclose the presence of companion cells. *Degeneria* has laminar stamens with unsealed carpels that have the ovules borne on the adaxial surface, but it has vessels, sieve tubes, and companion cells.

Double fertilization, the extreme reduction of the female gametophyte, and germination of the pollen grain at some distance from the ovule are characteristic angiosperm features which are regularly present in primitive as well as advanced living members of the group. These features are not known in living gymnosperms, and they may be regarded as definitive angiosperm characters. There is no good reason to believe, however, that they evolved in any different way from the anatomical and gross floral characters. Germination of the pollen grain remote from the ovule is obviously concomitant with the closure of the carpel, and the later stages of this process can still be seen among living angiosperms. The pollen grains may also have germinated at a point removed from the ovule in the Caytoniales, which had the ovules semi-enclosed in pouches. As we have seen, the Caytoniales are not to be regarded as ancestral to the angiosperms, and the possible similarity in this respect would have to be due to evolutionary parallelism. Things which are unexpected are not necessarily always impossible, but it is not to be expected that the fossil record can ever provide evidence on the origin of double fertilization. Stages in the reduction of the female gametophyte might conceivably be preserved, however, in some as yet undiscovered fossils connecting the angiosperms to the presumably ancestral pteridosperms.

In spite of the fact that such a plant probably never existed, it is useful to think of a hypothetical primitive angiosperm which had all the primitive features now shown in diverse living angiosperms, but which did not have any characters not known in living angiosperms. The evolutionary trends may then be considered character by character, and we may eventually arrive at a better understanding of the relationships among the families and orders of the group.

This hypothetical primitive angiosperm was an evergreen tree (or perhaps a large shrub) of moist, tropical places with an equable climate. It had alternate, simple, entire, stipulate, pinnately net-veined leaves. The stomates were ranunculaceous; i.e., the guard cells adjoined ordinary, unspecialized epidermal cells. The nodes were probably unilacunar with two leaf traces, or perhaps trilacunar with three traces. It had an active cambium, with elongate, vertically overlapping initials, but the wood had no vessels and no pronounced annual rings. The tracheids were very long and slender, with long, tapering ends and numerous scalariform-bordered pits — altogether fernlike. The wood parenchyma was diffuse, and both uniseriate and multiseriate rays were present. The phloem had sieve elements, but no true sieve tubes and perhaps no companion cells; it had no sclerenchyma. The sieve elements were elongate and vertically overlapping, with lateral sieves areas but no terminal sieve plate.

The flowers were borne singly at the ends of leafy branches, and were relatively large, with numerous, spirally arranged, somewhat leaflike tepals. One might say, with equal correctness, that the perianth was not differentiated into calyx and corolla, or that the perianth consisted wholly of sepals, with no petals. The stamens were numerous, spirally arranged, relatively large, and laminar, with no differentiation into filament and anther. The microsporangia were embedded in the broad, flat, sessile blade of the stamen. The tepals, stamens, and carpels may all have had three leaf traces and three primary veins each, from trilacunar nodes, or some of them may have had the more primitive double trace from a unilacunar node, like the leaves.

The pollen was uniaperturate, with a single long sulcus. It was binucleate at the time of transfer; the division of the generative cell to form two sperms occurred after the germination of the pollen grain. Pollination was effected by beetles which chewed and ate parts of the flower. There were no nectaries and no nectar.

The flower had numerous separate carpels, each probably stipitate at the base, the stipe being equivalent to the petiole of the leaf. Each carpel was folded along the midrib so that the margins were brought together and the morphologically upper (adaxial) surface was concealed. The margins of the carpel were merely loosely appressed to each other rather than anatomically joined, and the ovules were probably borne on the inner surface of the unsealed carpel. The margins of the carpel bore a tangle of glandular hairs so that the carpel, although anatomically unsealed, was effectively closed. These hairs formed, in effect, an elongate stigma on which the pollen grains germinated. The fruit opened at maturity, releasing the seeds.

The ovules were anatropous, each with two integuments and a massive nucellus. There were several or rather many potentially sporogenous cells in the young ovule, although only one of these went on to produce an embryo sac. The embryo sac was of the familiar, monosporic, eight-nucleate type. Double fertilization occurred, leading to the formation of a copious, triploid endosperm and a small embryo that was still immature at the time of ripening and discharge of the seed. The endosperm was probably of the nuclear type, i.e., it had a free-nuclear stage in early ontogeny. The embryo had two or more likely several cotyledons, and germination was epigeal, i.e., the cotyledons were brought above the ground during germination.

## Habitat and Growth Habit

It seems clearly established that the moist tropics are the original home of the angiosperms. By far the largest number of species occur in the tropics. A large proportion of these species are highly frost-sensitive. The frost line, south of which (in the northern hemisphere) freezing temperatures never occur at ordinary altitudes, is a very real barrier to the northward spread of tropical plants. To most species, the barrier is insurmountable. At sea level the frost line commonly lies a little north of the Tropic of Cancer, and it crosses Florida near the southern tip of the state. A similar frost line lies near the Tropic of Capricorn in the southern hemisphere. Between these two frost lines the angiosperms are represented by great hordes of species, and every botanical expedition into the less frequented areas still brings back its quota of "new" species. As one crosses the frost line and moves toward the pole, the number of species of angiosperms per unit area progressively decreases, and probably not more than a thousand species occur north of the Arctic Circle. A similar thinning out occurs in the southern hemisphere.

Resistance to cold is clearly an adaptation in the angiosperms, not an ancestral characteristic. This adaptation takes two principal forms, both of which are commonly shown by the same species. The first and most important is a physiological resistance to the harmful effect of freezing temperatures. The second is some means of reducing transpiration when the soil is frozen. Another adaptation found in many species is the development of strengthening tissue which reduces the likelihood of mechanical injury when the leaf is physiologically wilted.

Direct resistance to the effect of low temperature is a continuously variable character among the angiosperms as a whole, ranging from complete frost-sensitivity to an ability to survive temperatures many degrees below zero (on either the Fahrenheit or the Centigrade scale) in arctic and antarctic species. Several factors contribute, in varying combinations, to this type of resistance. Among them are: (1) increase in osmotic potential and thus direct resistance to freezing; (2) increased resistance to coagulation of the protein when desiccated; (3) increased permeability of the plasma membrane to water, permitting ice to form in intercellular spaces instead of within the cell; and (4) decreased structural viscosity of the protoplasm, making it less susceptible to mechanical injury by ice crystals.

Freezing of the soil during the winter makes the soil water much less available to plant roots and thus causes a physiological drought. Winter-killing of plants is often due to desiccation resulting from this drought, rather than to the direct effect of low temperature. Anything that reduces transpiration therefore promotes resistance to frost-injury.

The deciduous habit, the herbaceous habit, and the sclerophyllous habit are all adaptations to a climate in which vegetative activity cannot be carried on uninterruptedly throughout the year. Leaves give up their water to the air much faster than does the stem, which has a smaller surface-volume ratio and is likely to be more fully waterproofed. If green leaves are to be retained throughout the winter (or the dry season), these leaves must be able to go into a dormant state and withstand a high degree of desiccation. Such leaves commonly have a high proportion of strengthening tissue and are known as sclerophylls. Without this strengthening tissue they would doubtless be much more subject to fragmentation during the dry season.

In moist, tropical regions with an equable climate, growth is often continuous. New leaves are always forming, and old ones are always falling. The terminal bud never goes into a prolonged state of dormancy, and it has no specialized bud scales. In moist, temperate regions, or in tropical

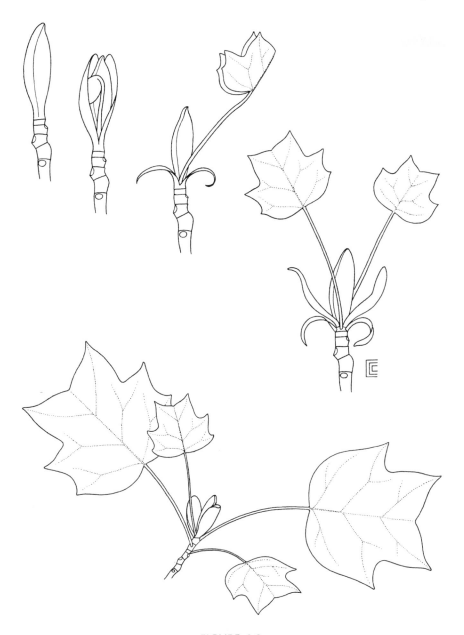

FIGURE 3.2

*Liriodendron tulipifera*, the tulip tree. Successive stages in the development of a winter bud at the beginning of the growing season.

regions with alternating wet and dry seasons, the deciduous habit allows trees to make the best use of the effective growing season. If the dry season is too severe, or if the wet season is not wet enough, the deciduous habit does not furnish an adequate buffer against desiccation, and perennial herbs are favored instead. In desert country, where the dry season may be both severe and prolonged, annual herbs are favored. These can survive the dry season, or several years of drought, in the form least vulnerable to death by desiccation — as seeds. The sclerophyllous habit in angiosperms is favored chiefly in regions with a Mediterranean climate, one characterized by moist, mild winters and hot, dry summers. Warm deserts also frequently have many small-leaved sclerophylls.

Trees in temperate regions typically produce all their leaves in a burst of growth in the spring. The leaves are preformed in the winter bud, and they quickly expand to full size at the beginning of the growing season. Following this flush of growth, the apical meristem goes dormant and forms a winter bud that is covered and protected by characteristic modified leaves called bud scales. The leaves that have been formed in the spring remain on the tree all summer and are shed at the approach of winter.

The common tulip tree (*Liriodendron*, family Magnoliaceae) occurs in temperate regions and shows winter dormancy, but it betrays its tropical ancestry by its pattern of growth. The well developed stipules cover and protect the growing bud, and there are no other bud scales. New leaves are produced throughout the growing season, and the first-formed leaves of the year are already dropping in June. Growth merely stops in the fall, and is resumed in the spring. The pair of stipules that covers the bud all winter becomes somewhat sclerified, but they are otherwise unmodified. The first two or three leaves that open in the spring do not attain full size, but they are otherwise normal.

The Salicaceae show a closer approach to the typical behavior of deciduous trees. They have ordinary winter buds with one or more bud scales and with preformed embryonic leaves, but after these leaves have expanded in the spring the bud continues to grow, producing new leaves during much or all of the growing season. All the leaves are usually retained until the general leaf-fall in the autumn.

It seems obvious that *Liriodendron* and the Salicaceae represent successive way-stations between the typical tropical growth pattern and the typical temperate one. This is not to suggest that the two groups are closely related; indeed they are not. Here as elsewhere, similar changes have occurred repeatedly in diverse groups.

An interesting factor which tends to confirm both the woody habit and the tropical habitat of the ancestral angiosperms comes in the probable relationships of herbaceous and temperate-zone groups to woody and tropical ones. Over and over, it appears that herbs are derived from trees, that temperate groups are derived from tropical groups, and, in combination, that herbaceous groups of temperate regions are derived from woody tropical groups. Thus the largely temperate and herbaceous Umbelliferae come from the largely tropical and woody Araliaceae, and there is indeed no very sharp line between the two families. Within the Leguminosae, the most primitive of the three subfamilies, the Caesalpinoideae, is largely tropical and woody, and the most advanced subfamily, the Faboideae, is mostly temperate and herbaceous. Likewise the Cruciferae, mostly herbaceous and of temperate regions, relate to the more primitive family Capparidaceae, a largely tropical and subtropical group with many woody members. Botanists whose experience is concentrated in the North Temperate zone may think of the Boraginaceae as a temperate, herbaceous group with a gynobasic style, but the more primitive members of the family, such as Cordia, are tropical woody plants with a terminal style.

It should be emphasized that the evolution of herbs from trees and the adaptation of tropical groups to colder climates have occurred not just once, but many times. In general, it may be said that the herbaceous groups of angiosperms relate not directly to each other, but separately to the woody "core" of the division. (Hutchinson disagrees, and believes there was an early and fundamental dichotomy between herbaceous and woody angiosperms. In the opinion of many botanists, this concept is the fatal flaw in his system.[5])

These separate herbaceous groups are of all ranks, from species to class. The genus Cornus, with about 45 species, is mostly woody, but it has two essentially herbaceous species, C. canadensis and C. suecica, which clearly have a woody ancestry within the broadly defined genus. Here the evolution of herbaceous species has occurred within the temperate zone, rather than concomitantly with a migration from the tropics. The large family Rubiaceae is chiefly tropical and woody, but two small tribes, the Rubieae and Spermacoceae, are chiefly herbaceous and have many temperate as well as tropical species. The Primulaceae are chiefly herbaceous and occur mainly in temperate and warm-temperate regions, but the other two families (Myrsinaceae and Theophrastaceae) generally referred to the same order are both tropical and woody. The order Campanulales is largely herbaceous and occurs in both tropical and temperate regions. The ancestry of this well marked group is to be sought in the complex of orders often called the Tubiflorae, which has woody, herbaceous, tropical, and temperate members, the woody forms being mostly tropical. The whole class Liliateae (monocotyledons) is largely herbaceous, although there are some woody members of a special type, such as the palms. The origin of

---

[5] J. Hutchinson, *The Families of Flowering Plants*, Vols. I and II, 2nd ed. (Oxford: The Clarendon Press, 1959).

FIGURE 3.3

*Cornus florida*, the flowering dogwood (left), an arborescent species, and *Cornus canadensis*, the dwarf cornel (right), an herbaceous species. It is generally believed that *Cornus* is primitively woody, and that herbaceous species such as *Cornus canadensis* are derived from woody ancestors within the genus. Some botanists would divide *Cornus* into several closely related genera, but that does not change the pattern of relationship. Photo of *C. canadensis* by Charles C. Johnson.

the monocots is to be sought in the primitive subclass Magnoliidae of the dicotyledons. The more primitive members of the Magnoliidae are woody.

Although we have disparaged the significance of the fossil record in interpreting the evolutionary history of angiosperms, it does provide some evidence that herbs are derived from woody ancestors, rather than vice versa. Fossil angiosperm pollen from pre-Miocene deposits, when it can be identified at all, belongs almost entirely to woody groups. From the beginning of the Miocene period onward, herbaceous fossil pollen is common, and indeed the presence of herbaceous pollen is coming to be considered as a marker of Miocene and post-Miocene deposits. This correlates with the fact that the evolutionary explosion of the Hymenoptera, Dipthera, and Lepidoptera, which are so important as pollinators of herbs, did not come until after the close of the Cretaceous period, probably at the beginning of the Miocene. The woody Magnoliales, which we think of as primitive within the angiosperms, are pollinated largely by beetles, which were abundant and varied long before the Cretaceous period.

Because herbaceous angiosperms are evidently derived from woody angiosperms, it is easy to fall into the error that all woody plants are primitively woody, and therefore more primitive than any herbaceous relatives they may have. We have become accustomed to the idea that evolutionary trends are essentially irreversible, and that a character once lost is never regained, at least not in the same form. These comfortable thoughts have a factual basis, but they are not strictly and universally correct.

Evolutionary trends have all degrees of stability, from those which are so vague and subject to reversal that they can scarcely be recognized as trends, to those which fasten an inescapable grip on the destiny of the group. Reduction of the gametophyte appears to be a universal trend in the vascular plants, although a new twist is given to this trend in those angiosperms which have more than eight nuclei in an embryo sac. Many stable and persistent trends are known in the animal kingdom; these are often associated with progressive specialization and adaptation to an ecological niche, and the trend is obviously driven by natural selection. As we have seen, the correlation of major taxa of angiosperms with ecologic niches is often obscure, and irreversible trends of specialization are correspondingly less frequent and conspicuous. It is interesting to note that one of the most tantalizing and thoroughly documented evolutionary reversals in the animal kingdom relates to a character whose adaptive significance is obscure — the progressive coiling and uncoiling of the shells of certain ammonites during the Paleozoic era.

Returning to the question of woody versus herbaceous angiosperms, it is clear enough that the herbaceous habit has its advantages for certain habitats, and the change from woody to herbaceous could well be governed by natural selection. However, we have noted that the evolutionary barriers between different niches for angiosperms are relatively low and easily crossed. So long as they retain the genetic potentiality for secondary growth, herbs have the evolutionary opportunity to reverse their field and become woody plants. The possibility of such a reversal depends on how much of the genetic mechanism has been lost. Some herbs have no cambium at all, some have cambium within the vascular bundles but not between them, and others have a complete cambial cylinder which connects the vascular bundles as well as passing through them.

Many herbaceous Compositae fall into this last group, with a complete cambial cylinder, and furthermore the cambium is very active and produces a considerable amount of secondary tissue. Arthur Eames once remarked to me that *Solidago* is a perfectly good woody plant, so far as the cambium and the secondary xylem are concerned. It is an herb only because the stem dies down to the ground every year. There is a lovely story about an American wood anatomist of several decades ago who claimed to be able to recognize any arborescent genus in the world by its wood. Another botanist, with deliberate dissimulation, cut down a vigorous specimen of the common sunflower (*Helianthus annuus*) late in the season and sent a piece of xylem for identification. He was delighted to learn that his plant was an African leguminous tree.

It should not be surprising if some herbs which have retained most of the necessary genetic information should revert to the woody habit, and indeed this appears to have happened several times in the Compositae, to mention only one family. The more obvious examples, however, have only gone so far back as to become shrubs (e.g., *Chrysothamnus, Olearia, Baccharis*, spp. of *Artemisia*). If any large dicot trees have herbaceous ancestors within the angiosperms, it remains to be demonstrated. Among the monocots, the palms very probably have an herbaceous ancestry, but their anatomy is wholly different from that of typical trees.

It has been suggested that the herbaceous habit originates from the woody one essentially by neoteny, i.e., the plant flowers while still vegetatively juvenile, and the later stages in vegetative ontogeny are curtailed or postponed out of existence. It is certainly true that it is the secondary xylem that makes woody stems woody, and that in the very youngest stages, before the cambium has begun to function, a woody stem often has essentially the same anatomy as an herbaceous one, with epidermis, cortex, and a ring of vascular bundles separated by medullary rays which connect the cortex to the pith. Conceptually, then, a tree evolves into an herb by blooming and dying (or dying back to the ground) while still vegetatively immature.

Neoteny is not the whole story, however. The evolutionary progression is commonly from middle-sized or large trees to small trees to arborescent shrubs to smaller shrubs to perennial herbs. Thus there is a progressive decrease in the size of the mature plant. To some extent this change in size reflects a decrease in the age at which the growth curve flattens out, and thus it might still be interpreted as a sort of neoteny, but such an extension of the meaning of the term tends to rob it of its conceptual utility.

Neoteny can also be invoked to explain the change from herbaceous perennial to annual duration. Most perennials do not flower until they are several years old. Some flower the first year, however, and year after year thereafter. Such plants are potentially annual, for if they were killed at the end of the first year the life cycle would continue unbroken and they would be functioning as annuals. The cultivated tomato is such a plant.

Perennial in its native habitat, it is cultivated in temperate regions as an annual, succumbing to frost at the approach of winter. The common perennial plantain (*Plantago major*) also blooms the first year and thus can function as an annual.

## Leaf Structure and Arrangement

The choice of the simple, entire leaf as primitive within the angiosperms is less certain than the woody habit and the moist, tropical habitat. Most pteridosperms had compound leaves, but some of them, such as *Glossopteris*, had simple, entire, pinnately veined leaves. Without the fossils to prove the point, a connection to the simple-leaved pteridosperms can be postulated about as easily as a connection to the compound-leaved ones.

Some simple-leaved dicots clearly have compound-leaved ancestors, but these are not among the more primitive groups. Thus, in the legumes, the compound leaf has repeatedly been reduced to a unifoliate leaf which is jointed to the petiole, and the joint itself may then be suppressed, so that the leaf is truly simple. All stages, for example, exist in the single genus *Crotalaria*. The Malpighiaceae, which have simple leaves, betray their compound-leaved ancestry by the fact that many of them have a joint in the middle of the petiole.

On the other hand, some compound-leaved species obviously have simple-leaved ancestors. It can hardly be otherwise with the compound-leaved types in *Bidens*, *Coreopsis*, and *Cosmos*, of the Compositae, *Pedicularis*, of the Scrophulariaceae, and *Sambucus*, of the Caprifoliaceae. These too are advanced rather than primitive taxa, so that no inference as to the primitive condition in angiosperms as a whole can be drawn.

The main reason for taking the simple, entire leaf as primitive is that this is by far the most common type in the woody members of the primitive subclass Magnoliidae. A plant may be primitive in some respects and more advanced in others, but on a statistical basis primitive characters do tend to hang together, and a plant which is primitive in several respects has a good statistical chance of being primitive also in the next respect to be considered. It therefore seems reasonable to suppose, in the absence of any other convincing evidence, that the common leaf type in the woody members of the Magnoliidae is the primitive type for the angiosperms as a whole.

Net venation is obviously primitive within the angiosperms, as contrasted to parallel venation. (These terms are here used in their accustomed sense; parallel-veined leaves in the monocots normally have a fine reticulum as well, but they are not referred to as net-veined.) The probable origin of the monocot leaf as a phyllode, essentially a flattened, bladeless petiole, is discussed in a later chapter. Similar phyllodial leaves also occur in some genera of dicots, such as *Eryngium*, in the Umbelliferae.

If the reasoning leading to the choice of the simple, entire leaf as primitive in the angiosperms is sound, then the primitive venation was probably pinnate. Pinnate venation is by far the commonest type in simple, entire leaves, both within and without the Magnoliidae. Pinnate venation is also fairly standard in the pteridosperms and indeed in the whole cycadophyte group, and many of these plants also have a vascular reticulum connecting the main veins.

The choice of pinnate venation as more primitive than palmate is also more appealing on a priori conceptual grounds. Pinnate venation is readily homologized with the typical pattern of shoot growth in post-Rhyniophyte vascular plants, with a central axis from which lateral branches arise at intervals. Palmate venation may then be seen as resulting from enlargement of the first pair of lateral veins, or enlargement of the first two or three such pairs and suppression of the intervals between them.

The dichotomous, free, Ginkgo-like venation of the leaves of *Kingdonia*, an herb referred to the Ranunculales, is more easily explained as a modification which simulates an ancient pattern than as a primitive survival.

The nature and evolutionary origin of stipules are obscure. They have sometimes been interpreted as the vestigial basal lobes or leaflets of an ancestrally lobed or pinnately compound leaf, but that is pure speculation. So far as modern species are concerned, they are organs *sui generis*, and the fossil record tells us nothing.

Some families which are generally regarded as exstipulate have species with stipule-like structures which appear to be petiolar flanges or displaced parts of the blade (e.g., *Aster undulatus*, in which all transitional forms can be seen in different leaves, and *Lonicera hispidula*). Structures of this appearance would doubtless be interpreted as stipules if they occurred in families that are normally stipulate. Whether the clearly laminar nature of these pseudostipules has any bearing on the nature of true stipules is an open question.

In *Galium*, of the Rubiaceae, the stipules have been modified into sessile leaves; the whorled leaves of *Galium* represent the normal opposite leaves of the family plus the enlarged interpetiolar stipules. In *Galium bifolium* the situation is clear enough; the stipules are still smaller than the true leaves, so that the whorl of four "leaves" at the node appears to consist of two larger leaves alternating with two smaller ones. In species with four equal leaves at a node the stipules have lost their identity as stipules and become leaves like other leaves, except for the probable absence of axillary buds. Species with six or eight leaves at a node have doubtless been derived from the four-leaved species by saltatory changes in the symmetry of the shoot; different numbers of leaves sometimes occur on different plants of the same species, or even at different nodes on the same stem. The situation in *Galium* should serve as a warning against too formal and rigid an approach to evolutionary morphology.

In a number of tropical trees the stipules of the new leaves cover and protect the terminal bud, just as they do in *Liriodendron*. This may well be the primitive function of stipules. In some other groups the stipules are variously modified into prickles which may conceivably furnish some protection against grazing animals. We have already noted that in *Galium* they are transformed into ordinary leaves. In most groups, however, the stipules have no evident function, except sometimes for a small amount of photosynthesis. They appear to be vestigial rather than functionally important structures.

It seems clear that stipules are on an evolutionary downgrade. They are more common in primitive families than in advanced ones. Many (not all) of the families of the Magnoliidae have stipules, whereas only a few of the sympetalous ones (notably the Rubiaceae and Loganiaceae) do. It has been estimated that 40 percent of the woody species of dicots have stipules, whereas only 20 percent of the herbs do. Even among species that have stipules, they are apt to fall off while the leaf is still young, as in many species of *Salix*, or they may be reduced to a mere ridge on the stem, as in many Rubiaceae.

The concept that angiosperm leaves are primitively alternate rather than opposite or whorled rests essentially on the fact that alternate leaves appear to be basic for most vascular plants, with the notable exception of the Equisetophyta. Derivation of monopodial branching from the ancestral dichotomous branching of the rhyniophytes, by unequal growth and overtopping, leads naturally to an alternate arrangement of the appendages of the stem.

Among the angiosperms as a whole, alternate leaves are much more common than opposite ones, and the opposite-leaved forms are scattered at various places on the evolutionary tree rather than being concentrated in the Magnoliidae or in any other one group. The main framework of the family tree of angiosperms consists of alternate-leaved rather than opposite-leaved groups. Whorled leaves are still less common and need not be considered as potentially primitive within the angiosperms.

The origin of opposite leaves from alternate leaves is not an immutable trend. Like some other trends, it is subject to reversal for no obvious reason. Within the large family Compositae, it is perfectly clear that opposite leaves are primitive and alternate leaves are advanced. Many species, such as the common sunflower (*Helianthus annuus*), represent a transitional stage, with the lower leaves opposite and the upper ones alternate.

Although most of Bessey's famous dicta on evolutionary trends in the angiosperms are still regarded as sound, he erred (or so we now believe) in considering opposite leaves to be primitive. His conclusion reflects an overemphasis of the principle of recapitulation applied to the fact that the cotyledons of dicots are opposite. The recapitulation principle ("Ontogeny is a recapitulation of phylogeny") often provides useful hints as to the course of evolution, but it is so beset with exceptions and special cases that it should never be regarded as definitive. The opposite position of the cotyledons in the dicots may reflect nothing more than suppression of the internodes in the very condensed embryo.

The adaptive significance of the changes in leaf structure which we have discussed is obscure. They do not evidently correlate with the environment; both simple and compound leaves are found in widely varying environments, as are pinnate and palmate venation, and alternate, opposite and whorled arrangements. The cyclic change from simple to compound to unifoliolate to simple leaves is particularly difficult to explain in terms of survival value.

The evolutionary history of the stomatal apparatus in angiosperms is not entirely clear. I here suggest that the primitive type is the anomocytic, or ranunculaceous type, in which the guard cells are bounded by ordinary epidermal cells, without specialized subsidiary cells. Evolutionary specialization led to the differentiation of two, three, four, or more subsidiary cells in various arrangements. Stomates with two equal subsidiary cells flanking the guard cells are said to be paracytic, or rubiaceous. Stomates with two more or less similar subsidiary cells that have a common wall at right angles to the long axis of the guard cells are said to be diacytic, or caryophyllaceous. Stomates with three subsidiary cells, one evidently smaller than the other two, are said to be anisocytic, or cruciferous. Other possibilities also exist and have been given special names, and there are intermediate types which are difficult to refer to a particular category. In some instances the paracytic type may be derived from an anisocytic type by suppression of the smallest of the three subsidiary cells, but other paracytic stomates probably do not have an anisocytic ancestry. It also appears that the subsidiary cells may be lost at any time, so that the stomate reverts to the ranunculaceous type. Ranunculaceous stomates in such advanced families as the Compositae and Caprifoliaceae may well be secondary rather than primitive. In the Liliatae there appears to be a reduction series from three or four to two to no subsidiary cells. Certainly in this group the absence of subsidiary cells is advanced rather than primitive.

The adaptive significance of these changes in the stomatal apparatus is still speculative. One would assume a priori that increasing anatomical differentiation has some functional importance, but direct evidence is notably lacking. The significance of the loss of supporting cells in some of the Liliatae is equally obscure.

## Vascular Structure

NODAL ANATOMY

The nodal anatomy of angiosperms and its taxonomic significance have been much discussed, but some uncertainties remain. Branch stems characteristically have two traces from a single gap in the stele, not only in angiosperms but also in ferns and gymnosperms. In these other groups the leaves generally also have two traces departing from a single stelar gap. Two-trace, unilacunar nodes are uncommon in the angiosperms, but they do exist in several families, notably in some members of the Magnoliales (Amborellaceae, Austrobaileyaceae, Calycanthaceae, Illiciaceae, Lactoridaceae, Lauraceae, Monimiaceae, Schisandraceae, Trimeniaceae). It seems reasonable to suppose that here the two-trace, unilacunar node is a primitive survival. It is more difficult, however, to see a primitive survival in similar two-trace, unilacunar nodes found in some members of more advanced families such as the Verbenaceae, Labiatae, and Solanaceae.

If the two-trace, unilacunar node is regarded as primitive, then it seems reasonable to visualize a phylogenetic progression to the one-trace, unilacunar node, and thence to the other types, such as the three-trace unilacunar, three-trace trilacunar, and multilacunar. Unfortunately, this interpretation presents some problems. Some one-trace, unilacunar nodes are clearly at the end of a reduction series from three-trace trilacunar

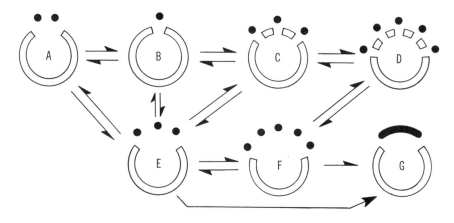

FIGURE 3.4

Possible phylogeny of nodal anatomy. A, two-trace, unilacunar; B, one-trace, unilacunar; C, trilacunar; D, multilacunar; E, three-trace, unilacunar; F, multi-trace, unilacunar; G, compound single-trace, unilacunar. Adapted from James E. Canright, The comparative morphology and relationships of the Magnoliaceae: IV. Wood and nodal anatomy. Jour. Arnold Arboretum 36: 119–140. 1955.

through three-trace unilacunar to one-trace unilacunar. It therefore becomes necessary to visualize the derivation of the single-trace, unilacunar node via two different routes. Indeed most of the unilacunar groups outside the Magnoliales would seem to have a trilacunar ancestry.

Furthermore, the three-trace, trilacunar node is commonly associated with stipulate leaves, which we have seen on other grounds to be probably primitive. Two of the families with two-trace, unilacunar nodes also have stipules (the Chloranthaceae and Lactoridaceae), but the several Magnolialian families with single-trace, unilacunar nodes uniformly lack stipules. If the presence of stipules is primitive, and if the evolutionary progression in nodal anatomy is as postulated above, then we apparently must assume an ancestry for the stipulate, trilacunar families of the Magnoliidae that included a combination of characters (stipules and one-trace, unilacunar nodes) that does not now exist. That is not impossible, especially in view of the fact that most of the existing families of this group are isolated end-lines, but it leaves one a little uneasy. How this problem is to be resolved remains to be determined.

Very recently, David Benzing has reopened the question of the phylogeny of nodal anatomy. (Developmental patterns in stem primary xylem of woody Ranales. I and II. Am. Jour. Bot. 54: 805–813; 813–820. 1967.) He concludes that the two-trace, unilacunar node is probably not primitive in the angiosperms and suggests instead that "the primitive node in the Angiospermae, and possibly in all seed plants, is one-trace, unilacunar or trilacunar." Acceptance of this point of view might call for some rearrangement of the sequence of families in the order Magnoliales in the system here presented, but it would have no other important effect.

The adaptive significance of most of the changes in nodal anatomy of angiosperms is obscure. The multilacunar node, as seen for example in the Araliaceae and Umbelliferae, may merely reflect the large size of the leaf. Reduction of the leaf blade in some of the Umbelliferae has not affected the petiole, which remains broad and sheathing, with many traces from the base.

Xylem

Concepts of primitive and advanced xylem structure are based on the xylem itself and do not depend on preconceived notions of the relationships among the families of angiosperms. Here, as in floral evolution, there seems to be only one general way of organizing the bewildering diversity into a coherent evolutionary pattern. It is reassuring to find a general, though loose, correlation between primitive xylem and primitive floral structure.

The primitive angiosperm tracheid is slender and elongate, with long, tapering, overlapping ends and numerous scalariform-bordered pits — very much like a typical fern tracheid. Phyletically, each of these transversely elongate pits becomes divided into a horizontal row of shorter, rectangular pits. These rectangular pits then become more rounded, and eventually they become spirally or irregularly arranged.

Tracheids function both as strengthening and conducting elements. Specialization towards strength, at the expense of conduction, leads to the xylem fiber. Specialization towards conduction, at the expense of strength, leads to the xylem vessel.

Fibers are usually more slender than tracheids, with a thicker wall and a smaller (often nearly evanescent) lumen. The pits are usually fewer and smaller than in tracheids, and are often imperfectly developed. They are always of the circular rather than the scalariform or rectangular type. Fibers in which these modifications have been carried to the greatest extreme are called libriform fibers; cells intermediate between typical fibers and typical tracheids are called fiber-tracheids. Wood with highly specialized fibers is likely to be highly specialized in other ways as well.

The primitive angiosperm vessel-segment is much like the primitive tracheid, differing only in the absence of the closing membrane from some of the numerous scalariform pits of the long, slanting end-walls. Phyletic specialization leads to shorter, broader vessel elements with somewhat thinner walls and with less sloping, more nearly transverse end-walls with fewer and larger perforations. Ultimately there is a single large perforation extending fully across the wholly transverse end of the vessel segment, so that the vessel might be compared to a series of barrels stacked end on end, with the end-walls knocked out. Concomitantly with these changes, the pits on the lateral walls undergo the same set of evolutionary modifications as they do in tracheids, so that eventually the lateral pits are round and scattered. Vessel segments with a single large pore in each end are said to be simply perforate; those with two or more perforations at each end are said to be multiperforate, or (if the perforations form a ladder-like configuration) scalariform. There are also some other possible variations in vessel structure which would appropriately be considered in a more detailed treatment.

Scalariform vessels are known in more than a hundred families of angiosperms, some of them, such as the Ericaceae and Caprifoliaceae, rather advanced on other grounds. In most of these families, simply perforate vessels are intermingled with scalariform ones. Families with only scalariform vessels are much less numerous, and most of them (e.g., Austrobaileyaceae, Canellaceae, Eupomatiaceae, Illiciaceae) belong to the Magnoliidae. With few exceptions, all vessels in herbs of all families of dicots are simply perforate. This is another line of evidence which strengthens the belief that within the angiosperms herbs are derived from woody ancestors.

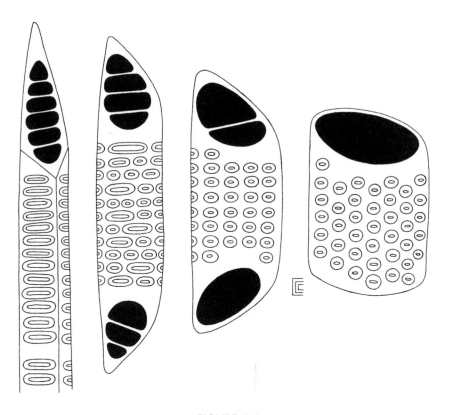

FIGURE 3.5

Evolution of vessel types. Left, scalariform; right, simply perforate, with circular bordered pits.

Only a few angiosperms (Amborellaceae, Tetracentraceae, Trochodendraceae, Winteraceae, and *Sarcandra*, in the Chloranthaceae) are primitively vesselless, and all of them are woody plants belonging to the Ranalian complex. (The Tetracentraceae and Trochodendraceae are here referred to the subclass Hamamelidae, but they are primitive within their subclass and evidently allied to the Magnoliidae. They may still be regarded as belonging to a loosely defined Ranalian complex.) A few herbaceous aquatic angiosperms both within and without the Ranalian complex lack vessels (e.g., Ceratophyllaceae, Nymphaeaceae, Lemnaceae), but this absence probably reflects a secondary loss, related to the aquatic habitat.

It may be significant that all the herbaceous angiosperms which lack vessels also lack a functional cambium. In the dicotyledons vessels are at first (phyletically) restricted to the secondary xylem, after which they appear earlier and earlier in the ontogeny of the primary xylem. In a plant having vessels only in the secondary xylem, suppression of cambial activity would also suppress the formation of vessels. The relationship of the aquatic habit to the absence of vessels will be further discussed in another chapter dealing with the ancestry of the monocotyledons.

Primitive xylem has well scattered vertical files of wood parenchyma cells, and has both uniseriate and multiseriate rays. Evolutionary advance leads to a clustering of the vertical files and their restriction to certain parts of the xylem, e.g., to the last-formed wood of the season, or to sheaths surrounding the vessels. Concomitantly with these changes, one or another of the two types of rays may be suppressed. The secondary wood of many herbs lack rays entirely.

The several different evolutionary advances in xylem structure tend to proceed more or less in unison, but this is only a tendency rather than a definite rule. Even more than with some other characters, evolutionary modification of the xylem has independently followed the same course in many different lines, so that the sharing of one or several advanced features provides but little indication of relationship. Wood anatomy is useful to phyletic studies mainly in negating relationships that might otherwise be suggested. A group with primitive xylem is not likely to be derived from a group with highly specialized xylem.

PHLOEM

Phloem undergoes a series of evolutionary changes comparable to those of the xylem, and, as with xylem, families which are considered to be advanced on other grounds tend to have specialized phloem. Specialized phloem and specialized xylem also tend to go together.

The primitive sieve element in angiosperms is long and slender, with groups of minute pores forming scattered sieve areas along the longitudinal walls and the very oblique end-walls. The sieve elements are progressively modified to be shorter, with more definite end-walls which come to be perpendicular to the axis. The sieve areas are progressively more restricted to the end-walls and more sharply defined, with larger pores, and the sieve areas on the lateral walls become obscure or vestigial. The several sieve areas on the end-wall eventually coalesce (phyletically) into a single sieve area. Highly advanced phloem has sieve tubes composed of short sieve tube elements stacked end to end, with each end-wall consisting of a single transverse sieve plate with large openings. The similarity of the changes in the evolutionary origin of sieve tubes and vessels is obvious.

Companion cells are reported to be absent from the phloem of *Austrobaileya*, and they may prove to be absent from some other members of the Magnoliales when all the genera have been carefully examined. Otherwise companion cells are universally distributed among the angiosperms. They show no obvious evolutionary trends, except that the sieve elements of more advanced phloem are likely to have two or more companion cells, as viewed in cross section, instead of only one.

Primitive angiosperm phloem has a considerable amount of parenchyma intermingled with the sieve tubes and companion cells, but has few or no hard cells. More advanced phloem has little or no parenchyma, but often has bands of fibers. "The story of evolutionary modification seems to be one of increase in complexity from a simple type, consisting wholly or largely of 'soft cells' to one that has also sclerenchyma cells of one or more types variously arranged, and to a secondarily simple type, consisting largely or wholly of sieve tubes and companion cells." [6]

Some angiosperms have phloem internal to the primary xylem as well as external to it, so that the vascular bundles are bicollateral. The presence of internal phloem is a useful taxonomic character, but, like other characters, it must be evaluated with some caution. It is fairly consistently present in certain families and even orders (e.g., the Myrtales), and absent from other groups. The families with internal phloem are all fairly well or highly advanced, as contrasted with the Magnoliidae, but they do not form a single taxonomic group. Internal phloem has evidently originated several times.

The adaptive significance, if any, of internal phloem is obscure. It confers no obvious advantage, and the species which have it differ so much in other respects and occur in such a wide variety of habitats that it is hard to think of it as merely an incidental side-effect of the genes controlling some more important character. The evolution of sieve tubes and vessels, on the other hand, is clearly correlated with increasing efficiency in conduction.

## Cambium and Vascular Bundles

The length of cambial cells is correlated with the length of the xylem and phloem cells formed from them. The progressive evolutionary decrease in the length of vessel segments and sieve tube elements is accompanied by (and ontogenetically due to) a decrease in the length of the fusiform cambial initials.

Aside from the change in length of the fusiform initials, the evolutionary history of cambium in angiosperms is mainly one of reduction in activity and area. The change from woody to herbaceous habit is governed by a number of other factors in addition to cambial activity, but there is a further reduction series among the herbs. Some herbs have a complete cylinder of active cambium and form a complete cylinder of secondary xylem and phloem, but in many others the secondary vascular tissues are formed only within the bundles, and the cambium between the

[6] Arthur J. Eames, *Morphology of Angiosperms* (New York: McGraw-Hill, 1961).

bundles forms only parenchyma cells or is completely inactive. Most monocots and a considerable number of herbaceous dicots have no functional vascular cambium at all.

A few monocots, notably some of the Agavaceae, have a specialized, atypical cambium which produces complete vascular bundles and other tissues. This cambium appears to be a newly originated tissue, derived by a continuation of activity in a part of the primary thickening meristem. It is analogous to typical cambium, rather than homologous with it. This is an example of the fact that a structure once completely lost is not likely to be regained in exactly the same form, even under strong selective pressure.

The primary vascular structure of primitive herbs is essentially like that of woody plants. In cross section, the stem shows a single ring of vascular bundles, separated by relatively narrow medullary rays. Evolutionary modification leads to an increase in the proportion of soft tissues in the primary body, accompanied by a more definite separation of the vascular bundles (i.e., an increase in the width of the medullary rays). This change in proportions may well relate to the changed functional and structural requirements of an herbaceous stem as compared to those of a tree.

Stems with scattered vascular bundles, as seen in monocots and a few dicots (e.g., Nymphaeaceae), are clearly advanced, in comparison to stems with the bundles in a single ring. The adaptive significance of the change to scattered bundles is obscure.

### Inflorescences

The diversity of inflorescences is seemingly endless. Certain patterns or tendencies can be observed, but no scheme of classification of inflorescences is even approximately satisfactory. Another facet of the same problem is that the evolutionary history of inflorescences is not at all clear. If it were clear, a satisfactory classification would be more easily devised; or if a satisfactory classification had been devised, the evolutionary history might be more easily discerned. The most familiar and at the present time the most useful classification of inflorescences recognizes two major types, the cymose (determinate) and the racemose (indeterminate). The analysis of probable evolutionary history here presented is made in terms of this classification.

Although an inflorescence as a whole may have either determinate or indeterminate (or mixed) sequence of flowering, the flower is always terminal to its own shoot; growth of the shoot terminates in the formation of a flower. Given a possible ancestry of the angiosperms in megaphyllous pteridosperms, it seems reasonable to suppose that primitively each flower was relatively large and terminal to its own leafy branch. The flowers were not aggregated into distinct inflorescences. The same organization could be expressed by saying that the plant had several or many uniflorous inflorescences well removed from each other, each terminal to its own shoot. Large, solitary, terminal flowers are commonly seen in the Magnoliaceae and some other families of the Magnoliales.

Termination of apical growth, attendant on the development of a flower, may well have encouraged lateral branching not far beneath the flower. Apical dominance is well known in vegetative shoots: Remove or inactivate the terminal bud, and one or more suppressed lateral buds begin to grow. In any case, three-flowered and two-flowered cymules (dichasia and monochasia respectively) are very common components of inflorescences.

If one can assume dichasia and/or monochasia as basic organizational units, it becomes possible to develop a series of other types by logical modification from these. Whether such a logical series conforms in all respects to the course of evolution is uncertain, but at least it provides a reasonable conceptual framework for the organization of our knowledge. The Ranalian hypothesis of floral evolution was no more than such a set of logically progressive modifications from a conceptually basic type when it was first advanced by de Candolle in pre-Darwinian times.

If a primitive dichasium is considered to be relatively loose, with the two side-branches originating at different levels on the shoot axis, then a loose compound cyme can readily be evolved by repeated branching of the successive orders of branches. Such a compound cyme, with a more or less pyramidal or ovoid shape, can be converted into a classical (strictly racemose) panicle by a progressive shift in the growth pattern, so that the lower branches develop before the upper ones. The classical panicle is in fact not a very common type of inflorescence. Most so-called panicles actually show a mixture of determinate and indeterminate growth, often with the terminal flower of each branch developing before the lateral flowers. Compound inflorescences with mixed determinate and indeterminate growth have appropriately been called mixed panicles.

One family which very often does have a typical panicle is the grass family, but it should be noted that the grass panicle is actually a secondary inflorescence; it is the spikelets which are paniculately arranged. The sequence of flowering within the individual spikelets is also racemose, so that the grass inflorescence is (at least typically) without taint of determinate sequence.

Proliferation of a more compact dichasium, with opposite or subopposite lateral branches, leads to compound dichasial cymes such as are seen in many members of the Caryophyllaceae and Rubiaceae, among other families. This type of inflorescence is seen especially in plants that have opposite leaves, but opposite-leaved plants may also have other types of inflorescences.

Condensation of the internodes of a compound cyme, without reduction of the ultimate pedicels, leads readily to an inflorescence having the form of an umbel but with a more or less distinctly determinate sequence of flowering. Such inflorescences are well exemplified by *Allium* and many other members of the Liliaceae, especially those genera which have often been segregated as the Amaryllidaceae. Careful anatomical study confirms the basically cymose nature of these determinate umbels. There is no doubt that here the relationship of the two types of inflorescence is truly evolutionary rather than merely conceptual.

Reduction of the pedicels as well as the branches of an originally cymose inflorescence leads to heads of the sort seen in the Dipsacaceae and certain Rubiaceae (e.g., *Cephalanthus*), among other groups. These heads typically retain a more or less distinctly determinate sequence of flowering, but some of them (e.g., *Cephalaria*) are mixed, with the outermost (lowermost) row of flowers blooming immediately after the terminal flower. Here too there is no doubt that we have a truly evolutionary progression. The possible relationship between this type of inflorescence and that of the Compositae, in which centripetally flowering heads are arranged in a cymose secondary inflorescence, is discussed under the order Asterales.

In the ancestry of the Compositae, it appears that condensation of the inflorescence into a head preceded the change from cymose to racemose sequence of flowering. In some other groups the reverse seems to be true. The heads in *Trifolium*, for example, seem to be merely condensed racemes.

A special type of modified cyme which superficially resembles a raceme or spike is the sympodial cyme, in which the apparent main axis consists of a succession of short axillary branches. Each flower is terminal to its true axis, but the side branch originating just beneath the flower produces its own terminal flower with another side branch just beneath it, so that theoretically the inflorescence can elongate indefinitely by producing a branch just below each flower.

Sympodial cymes are of two sorts, helicoid and scorpioid. In a helicoid cyme the branches are all on the same side of the developing axis, so that the inflorescence is curved or circinately rolled. Often the inflorescence is conspicuously circinate when young and progressively straightens and unrolls as flowering proceeds. In a scorpioid cyme the branches are alternately on opposite sides of the developing sympodial axis, so that the axis has a zig-zag appearance. Helicoid cymes are commoner than scorpioid cymes and are especially frequent in the Boraginaceae and Hydrophyllaceae.

In the less modified forms of sympodial inflorescences, each of the successive lateral branches which make up the apparent axis is subtended by a bract, so that the axillary nature of these branches is not difficult to see. Superficially, it appears that the main bract is on the opposite side of the stem from the flower, instead of directly beneath the flower.

In more highly modified sympodial inflorescences these bracts are suppressed, and only the coiled or zig-zag form of the apparent axis remains to distinguish the inflorescence from a true raceme or spike. Bractless zig-zag racemes, as seen for example in *Smilacina stellata*, are really modified scorpioid cymes.

Indeterminate inflorescences may be classified as racemes, panicles, corymbs, umbels, spikes, or heads. All of these types have in common the fact that the lowest or outermost flowers bloom first, and flowering progresses toward the top or center. A typical raceme has an elongate central axis from which the individual pedicels arise. Each pedicel is axillary to a bract. Flowering progresses evenly from the bottom upwards, and theoretically the axis may continue to elongate and produce new flowers indefinitely. It is as if an ordinary vegetative axis had been modified in just two respects: The leaves are reduced to bracts, and every axillary bud develops into a short lateral branch with a single terminal flower.

Such a simple and nicely symmetrical inflorescence as a raceme ought logically to be a primitive type, but apparently it is not. True racemes are not notably abundant in the more primitive families. They are very common instead in such more advanced families' as the Cruciferae, Leguminosae, and Scrophulariaceae. In a number of instances racemes pretty clearly have a cymose ancestry. In species of *Campanula* and *Saxifraga*, for example, racemes appear to be derived from mixed panicles by suppression of the secondary branches, and we have already noted that zig-zag racemes are likely to be sympodial cymes.

On a conceptual, if not always an evolutionary basis, the other types of indeterminate inflorescences may all be considered as modified racemes. Thus a panicle is a branched raceme, a corymb is a raceme with elongate lower pedicels and relatively short axis, an umbel is a raceme with the internodes suppressed, a spike is a raceme with the pedicels suppressed, and a head is a raceme with both the internodes and the pedicels very short or suppressed.

Each of the forms of racemose inflorescence has its cymose counterpart. Within the racemose type, it appears that racemes may be derived from panicles, and vice versa, that spikes, corymbs, and umbels may be derived from racemes, and that heads may be derived from racemes or spikes or umbels, but some of these types can also be derived from their cymose counterparts. The Rose family is particularly noteworthy in having a wide variety of both racemose and cymose inflorescences, with many doubtful or transitional types.

The adaptive significance of the evolutionary changes in the inflorescence is obscure. If it makes any difference to the plant whether its inflorescence is racemose or cymose, or whether it is monopodial or sympodial, the fact is not obvious. Certainly there is no evident correlation between these different types and ecological niches. Something might be made of the difference between relatively large, well separated flowers which individually attract pollinators, and smaller flowers which are grouped into showy inflorescences, but this distinction cuts across all the others and brings together groups which are dissimilar in other respects and not closely related. The culmination of the emphasis on the inflorescence as the attractive unit is seen in such disparate groups as the Compositae and the genus *Euphorbia*. The advantage, if any, of such pseudanthia over ordinary buttercup flowers is not clear.

## Flower Structure

GENERAL CONSIDERATIONS

The primitive angiosperm flower may be visualized as having numerous, spirally arranged, rather large and firm tepals, numerous, spirally arranged, laminar stamens, and numerous, spirally arranged, unsealed carpels. All of these characteristics can be found individually among modern members of the Magnoliales, but not combined in the same species or even the same family.

Some members of the Magnoliaceae, such as *Magnolia stellata* and *Aromadendron* spp., have the tepals, stamens, and carpels all spirally arranged, with large, more or less petaloid tepals that are not differentiated into sepals and petals. *Degeneria* has laminar stamens and unsealed carpels. If these primitive floral features of *Magnolia stellata* and *Degeneria* could be combined into one flower, this flower would closely approach, in its external morphology, the theoretical archtype for the angiosperms. We should note at once, however, that the Winteraceae have more primitive (vesselless) xylem than either the Magnoliaceae or the Degeneriaceae.

Many of the tendencies in floral evolution can be grouped under the general heading of aggregation and reduction. The receptacle is shortened, so that parts of all kinds are brought closer together, in a series of cycles rather than an elongate spiral. The number of parts of a given kind is reduced from many to few, the parts of a given kind become connate instead of free, and parts of different kinds may become adnate to each other. Parts of any given kind may even be lost completely. We have already noted that a repetition of this same set of aggregative and reductive processes leads to the formation of pseudanthia such as those of the Compositae and Euphorbiaceae, but here we are concerned only with evolutionary changes within the individual flower.

Another set of tendencies is toward elaboration and differentiation of parts. These two seemingly opposed tendencies may both be expressed in the same structure. Sympetaly reflects the evolutionary union of ancestrally distinct petals, but many sympetalous flowers are also irregular, with some of the corolla lobes different from the others.

There is in fact only the very loosest correlation among the various advances in floral structure. The different advances do not in general depend on each other, and it may be expected that any particular kind of flower chosen at random will show some relatively primitive and other relatively advanced features. Sympetalous flowers may also be polysepalous (Sapotaceae), and polypetalous flowers may be synsepalous (Silene). The calyx may be synsepalous but regular in flowers with polypetalous but irregular corolla (many legumes). Dioecism occurs sporadically from the Magnoliales to the Asterales, and also in the monocots. Sympetalous plants may have separate carpels (Kalanchoe), and polypetalous plants may have united carpels (many families).

The lack of close correlation among the various advances in floral structure extends to advances in vegetative structure as well. There is, however, a very loose and general correlation among the various advances. The Ranalian complex well illustrates both the fact that the correlation exists and the fact that it is only very loose. The group was established as the most primitive among the angiosperms largely on the basis of traditional characters of floral morphology, but it is also in this group that the most primitive xylem and phloem in the angiosperms has been discovered, and many members have primitive (uniaperturate) pollen as well. On the other hand, every individual member of the Ranalian complex shows a combination of primitive and more advanced features, and some members, such as the Canellaceae, even have a compound pistil with parietal placentation.

The "morphological integration" which often permits students of vertebrates to reconstruct an entire animal with considerable accuracy from one or a few bones simply does not exist among the angiosperms. Another facet of this same situation is that higher plants, being nonmotile and lacking a nervous system, do not have their structure so precisely prescribed by the environmental requirements as do the animals. This brings us back to the fact, mentioned in the first chapter, that the higher taxa of angiosperms are in general not well correlated with ecologic niches, and are correspondingly difficult to recognize at a glance.

## Unisexuality

We have noted that the primitive flower is perfect, having both stamens and pistils. Unisexual flowers are derived from perfect ones. An intermediate type has more or less well developed parts of both sexes, but with only one sex functional. The Sapindaceae very often have seemingly perfect but functionally unisexual flowers. Many maples, in the related family Aceraceae, have staminate flowers with an evident but abortive ovary, and functionally pistillate flowers with more or less well developed stamens. A special case of unisexuality is shown in a number of genera of Compositae in which the disk flowers are functionally staminate. In the Compositae the style elongates at anthesis, pushing the pollen from the anther-tube, and it retains this function even when the ovary has been phyletically lost.

Monoecious and dioecious groups are derived from perfect-flowered ancestors through a variety of intermediate states which often involve the presence of perfect and unisexual flowers on the same individual. A number of families, such as the Compositae and Cyperaceae, run the gamut from perfect flowers to monoecism and dioecism, with diverse transitional stages. Dioecious species of *Carex* clearly have a monoecious ancestry within the genus. Dioecism in the Simaroubaceae, on the other hand, has evidently been attained via polygamo-dioecism, in which some individuals of a species bear staminate and perfect flowers, and others bear pistillate and perfect flowers.

The adaptive significance of the change from perfect to unisexual flowers and eventually unisexual individuals is difficult to assess accurately. Obviously it is a means of restricting and eventually preventing self-pollination. The advantages of cross-pollination must be balanced, however, against the possible reduction in seed set through failure of pollination.

Unisexuality is merely one of several means by which angiosperms meet the evolutionary need to maintain a high level of outcrossing, and it may be assumed to share the same advantages as, for example, self-sterility. Thus one could see some utility in monoecism or even polygamo-dioecism, which might provide an adequate mix of outcrossing and high seed-set. It is not so easy, however, to see how survival value could spark the further change to dioecism. In addition to the decreased likelihood of pollination, the species is burdened with the necessity of producing all its seeds on only half of its individuals. Whatever may be the reasons why dioecism is favored in higher animals, it is difficult to see how they could operate in angiosperms, in which no active cooperation between different individuals is required or even possible. The fact that something is hard to see or understand does not prove that it is only mythical, but a bit of skepticism is in order. The long-doubted Homeric legend of Troy turned out to have a solid basis in fact, but the once widespread belief in unicorns did not.

It is interesting to note that some dioecious genera, such as *Antennaria* (Compositae), have taken to apomixis, with chiefly or wholly female populations. One is tempted to speculate that apomixis in *Antennaria* is an evolutionary response to the disadvantages of dioecism, but the widespread

occurrence of apomixis in groups with perfect flowers makes such speculation hazardous. At least, one must admit that *Antennaria* has surrendered
any possible advantage to be obtained from its dioecism.

Neither monoecious nor dioecious angiosperms, nor the two groups
collectively have any habitat requirements or other features in common to
suggest that they are particularly suited to any ecologic niche in contrast to
species with perfect flowers.

## THE PERIANTH

It is clear enough that the tepals of the more primitive angiosperm
families are modified leaves, and like ordinary leaves they often have three
vasular traces. In this group it is often difficult to distinguish between the
tepals or sepals and the reduced leaves (bracts) subtending the flower.
The bracts pass into the tepals as one proceeds distally on the pedicel.

It appears that sepals are in nearly all cases homologous with the tepals
of a primitively undifferentiated perianth. In some cases the ancestral undifferentiated perianth has become differentiated into two cycles, the calyx
and the corolla. In other cases the ancestral undifferentiated perianth has
been reduced to a single cycle, the calyx, whereas the corolla, if present,
has a different ancestry. (The possibility that the apparent sepals of the
Portulaceae and Basellaceae are actually bracts is noted in a later chapter.)

Petals are of two different origins. Some are modified tepals, representing the inner members of an ancestrally undifferentiated perianth. Others
are essentially staminodes. Convincing examples of each of these two
modes of origin of petals can readily be found. The single genus *Magnolia*
provides all transitional stages from an undifferentiated perianth with an
indefinite number of tepals, to a well differentiated perianth with a definite
number of sepals and petals, each in a single cycle. Other examples may
be taken from other members of the Magnoliidae.

Petals in the Ranunculaceae and many other families, on the other hand,
appear to be staminodial. In the Ranunculaceae the petals have the additional pecularity of serving as nectaries. The extra petals of "double"
flowers of many different kinds are obviously staminodial, and it may well
be that in such families all the petals are eventually of staminodial origin.
*Mentzelia* provides an especially instructive example. All species of *Mentzelia* have numerous stamens. Some species have five petals and no staminodes. Others have five petals and five more or less petaloid staminodes
which may have a vestigial anther at the summit. *Mentzelia decapetala*
has ten petals, all essentially alike, although close examination shows that
the inner five are a bit narrower than the outer five. *Nymphaea* has often
been cited as a genus showing complete transition between stamens, staminodes, and petals in a single flower.

An anatomical character which it was once hoped might distinguish sepals from petals, or at least distinguish staminodial petals from others, has proven faulty. It was noted that sepals commonly have three traces, like ordinary leaves, whereas petals commonly have one trace, like stamens, and this was thought to provide a critical distinction between sepals and petals. Unfortunately, the hypothesis was based on false premises. Stamens in some families, such as the Degeneriaceae, have three traces. In *Umbellularia californica* (Lauraceae) the perianth is imperfectly differentiated into sepals and petals.[7] The sepals have three traces, but the petals have one fully developed median trace and a pair of imperfectly and irregularly developed lateral traces. The next step would appear to be complete loss of the lateral traces. Thus the presence of three traces in a petal does not preclude the possibility that it is of staminodial origin, nor does the presence of only a single trace prove that it must be staminodial. Certainly the presence of three traces in a perianth member does not show conclusively that it cannot be a petal.

It therefore becomes necessary to revert to a pragmatic definition of sepals and petals, based primarily on position, but taking other factors into account as seems necessary to the case. There is a general tendency in angiosperms for the perianth to be divided into two functionally and morphologically differentiated series. The outer series is called the calyx, and the inner series the corolla.

In flowers that have both a calyx and a corolla, the sepals are customarily green or inconspicuously colored and more or less foliaceous in texture, serving primarily to protect the developing flower bud; the petals, on the other hand, are brightly colored and serve to attract pollinators. This distinction does not always hold true, however. Sometimes, as in many of the lilies, the sepals are showy and essentially like the petals. In some groups the sepals are petaloid in texture and function but different in structure from the petals, as in *Aquilegia*. The sepals may even be larger and more showy than the petals, as in *Delphinium* and *Aconitum*. These last three genera all belong to the Ranunculaceae, in which the petals function as nectaries.

In some groups the sepals have other specialized functions, as in *Valeriana*. *Valeriana* has epigynous flowers with the sepals at first inconspicuous and circinately inrolled. At maturity the sepals are plumose and fully extended, serving to facilitate the distribution of the fruit by wind.

When the perianth consists of only a single cycle it is customary to call that cycle a calyx and its members sepals (although it is not incorrect to call them tepals). This is true even when the members of this single cycle are very petaloid, as in *Clematis* and *Anemone*. Only when there is clear evidence that the true calyx has been lost, as in some members of the

---

[7] B. Kasapligil, Morphological and ontogenetic studies of *Umbellularia californica* Nutt. and *Laurus nobilis* L. Univ. Calif. Pub. Bot. 25: 115–240. 1951.

Umbelliferae and Compositae, are the members of a single perianth-cycle called petals (or collectively the corolla). This terminology is in part arbitrary, but it is not out of harmony with the probable evolutionary history. In all cases in which the perianth consists of a single cycle called sepals, it is to be presumed that these are descended from the tepals of an undifferentiated perianth.

When the flowers have no corolla, the calyx is often corolloid. The family Nyctaginaceae is strictly apetalous, but most of the species have a corolloid calyx. In some species of Mirabilis, including the common four-o'clock (Mirabilis jalapa), this corolloid calyx is subtended by a calyx-like involucre of five basally united bracts. One would be tempted to interpret such flowers as having an ordinary calyx and corolla, were it not that some other species of Mirabilis have two or more flowers within the calyx-like involucre. In both structure and function, the flowers of Mirabilis jalapa are comparable to ordinary flowers of other families which have a differentiated calyx and corolla, but they have achieved this condition by a different route. Because of the existence of a series of transitional forms, it is useful here to define the parts on the basis of homologies rather than on appearance and function.

In floral parts of all kinds the general evolutionary progression is from a large and indefinite number of parts to a small and definite number. Stages in the change from spirally arranged perianth members of indefinite number to cyclically arranged ones of definite (and usually smaller) number can be seen in various members of the Magnoliidae. Magnolia presents a good series from one type to the other among its various species. Here, as elsewhere, the same evolutionary tendency is independently expressed in many lines, so that the common possession of an advanced character is no guarantee of relationship.

Reduction in the number of tepals from numerous and indefinite to few and definite in number commonly results in the formation of a single cycle (compressed spiral) of sepals and a single cycle of petals. The two cycles are not continuous; instead they are offset so that the petals and sepals spread on alternate radii. The petals are said to be alternate with the sepals.

Two sets of sepals, or two sets of petals, offset from each other in the manner of a set of sepals and a set of petals, are much less frequently seen. Liriodendron (Magnoliaceae) and many of the Annonaceae have three cycles of three tepals each; the outer set is usually interpreted as sepals, and the two inner sets as petals. Some of the Sapotaceae, such as Manilkara and Mimusops, have two cycles of sepals, and some species of Mentzelia have two cycles of petals.

Except for members of the Magnoliidae, the possession of more than one cycle of sepals, or more than one cycle of petals, is always secondary rather than primitive. In the Sapotaceae the bicyclic calyx apparently results from an increase in the number of sepals, correlated with a change in their arrangement. The number of petals is also increased in these genera, but the corolla is sympetalous and the increased number of petals is therefore maintained in a single cycle. In some other members of the Sapotaceae, such as *Calocarpum* (a segregate from *Pouteria*), the number of sepals only is increased, but the basically spiral arrangement is maintained, and there is no distinction into two sets. We have already noted that the petals of the second cycle in *Mentzelia* are essentially staminodes; the more primitive species of *Mentzelia* have only five petals in a single cycle.

It should be noted that cycles of perianth parts are, in origin, compressed spirals. Often this can be seen on close inspection, as in the very common quincuncial arrangement of sepals, which represents a very tight spiral on a 2/5 phyllotaxy. The two outermost sepals are wholly exposed, the two innermost ones are wholly included, and one sepal has one margin exposed and the other included.

The tendency for the perianth to consist of an outer, protective part (the calyx) and an inner, attractive part (the corolla) is obviously functional. The frequent secondary reduction or loss of the petals, often accompanied by abandonment of insect pollination, is more difficult to explain. Perhaps it results in some cases from pleiotropic effect of otherwise useful genes, and in others from happenstance counterselective fixation in small populations. We have noted in Chapter 1 that a change in a single gene can break a biosynthetic chain and prevent the development of a complex, multigenically controlled structure.

The change from numerous, spirally arranged perianth parts to fewer, cyclic ones reflects the general trend toward condensation which permeates floral evolution. Conceivably this particular manifestation of the trend might reflect simple economy, but it is a bit difficult to see enough advantage in it to provide much selective force. Perhaps an abundant supply of the right mutations, with or without a small selective advantage, would provide the requisite impetus.

As Stebbins has recently emphasized, the number of floral parts of a kind may also undergo an evolutionary increase. Most such increases affect parts that have not yet been reduced to definite cycles, so that the number is commonly given in botanical descriptions merely as "numerous." Such phyletic increases in parts that are already numerous easily escape attention. Increases in the numbers of parts which have already been reduced to one or two cycles are much less common, but they undoubtedly occur. Some probable examples are mentioned elsewhere in this work, and others are documented by Stebbins.[8]

[8] G. L. Stebbins, Adaptive radiation in trends of evolution in higher plants. Vol. I, pp. 101–142, in Th. Dobzhansky, M. K. Hecht, and W. C. Steere, eds., *Evolutionary Biology* (New York: Appleton-Century-Crofts, 1967).

## Fibonacci Numbers

The most common number of sepals or petals in the dicots is five, in the monocots three. Sets of four also occur, less commonly, in both groups. Other numbers, except for the large and indefinite numbers in some of the more primitive families, are relatively rare. It is probably significant that the numbers 3 and 5 both belong to the Fibonacci series.

The Fibonacci series is a mathematical series in which, starting with 1, each succeeding number is the sum of the two previous numbers. The series therefore proceeds 1, 1, 2, 3, 5, 8, 13, 21, 34, 55, 89. . . .

It is well known that Fibonacci numbers are involved in phyllotaxy, typical phyllotaxies being 1/3, 2/5, 3/8, 5/13, etc. These arrangements are not so different as might at first appear. If they are put in terms of the arc of a circle which they represent, it can easily be seen that the higher phyllotaxies progressively approach a figure of about 137.5 degrees. Two-fifths of a circle is 144 degrees; three-eighths of a circle is 135 degrees; five-thirteenths of a circle is 138.46— degrees; eight-twenty-firsts of a circle is 137.14+ degrees, etc.

Attempts to explain why phyllotaxy follows such a pattern usually center about the geometry of close packing. I am not fully satisfied with any of them, but I have nothing better to propose. This is a problem in morphogenesis rather than in taxonomy or phylogeny per se, and it will not be further pursued here.

Inasmuch as the various sorts of floral appendages are in the last analysis all homologous with leaves, it seems reasonable to consider their arrangement as a facet of phyllotaxy. It is easy enough to see that the quincuncial arrangement of sepals and petals in many flowers reflects the presence of two complete spirals of parts of each kind in a 2/5 phyllotaxy. When the two spirals of sepals are compressed into what amounts to a single cycle, the sepals are nicely spaced at distances of 72 degrees around the circle. If one more sepal were added in the same sequence, it would fall directly in front of the first sepal, and subsequent numbers in the same progression would likewise fall in front of other sepals. Such additional sepals might be expected to be superfluous. Given a 2/5 phyllotaxy, 5 is the only number of sepals which can be symmetrically arranged to give uniform coverage without superfluity. The same sort of argument applies equally to the number of petals. The number 3, so common in monocots, reflects a 1/3 phyllotaxy. Symmetrical arrangements of 8 and 13 are also theoretically possible, and indeed they are occasionally seen in species of *Ranunculus* and some other genera.

Except for some primitive kinds of flowers with indefinite numbers of parts, the sepals and petals do not form a single continuous spiral. Instead the petals are in their own compressed spiral separate from that of the calyx, and the two series are offset so that each petal is midway between two sepals. Occasionally there may be the same sort of separate cycles for parts of the same kind. Thus flowers with four petals, as in the large families Cruciferae and Onagraceae, probably have two cycles of two petals each, arranged in a 1/2 phyllotaxy, the second cycle being offset at 90 degrees from the first. In a few kinds of plants, such as some species of the sapotaceous genus *Pouteria*, some flowers have five quincuncial sepals, whereas other flowers on the same plant have four sepals arranged in two pairs like those of the Cruciferae. The change in number and position of the sepals is basically a reflection of change in phyllotaxy.

Fibonacci numbers also appear frequently in inflorescences, as in the number of rays in an umbel, and in the number of phyllaries, ray flowers, or total flowers in the heads of the Compositae. Here, as in numbers of perianth parts, these numbers apparently reflect the phyllotaxy. Numbers above 5 are often unstable, however, and actual counts in a series of individuals may show a considerable range of variation, with the Fibonacci number merely being the high point on a more or less normal curve of distribution.

Sometimes there is even a bimodal curve, with the two high points being adjacent Fibonacci numbers, such as 8 and 13, or 13 and 21. Here we have the seeming anomaly of a bimodal curve which lacks taxonomic significance. *Senecio canus* shows such a bimodal curve in the number of principal involucral bracts, the two high points on the curve being 13 and 21. I was at first tempted to try to distinguish two species on this basis, but on further study it became evident that the more vigorous individuals, or the larger heads in a given inflorescence, tended to center about 21, whereas the less vigorous individuals, or the smaller heads in an inflorescence, tended to center on 13.

Bimodal curves of Fibonacci numbers sometimes do reflect specific differences, however. *Bidens frondosa* has five to ten (typically eight) outer involucral bracts, whereas the closely related *B. vulgata* has ten to 16 (typically 13). *Tragopogon dubius* characteristically has 13 involucral bracts on the primary head in well developed plants, the number sometimes varying down to eight in later heads or on depauperate plants. The closely related *T. pratensis* generally has eight involucral bracts, regardless of vigor.

## Hypogyny, Perigyny, and Epigyny

One expression of the tendency for floral parts of different kinds to become attached to each other is the frequent fusion of the basal parts of the calyx, corolla, and androecium. The compound structure thus formed is called a hypanthium, or floral tube. The hypanthium often resembles the calyx in texture, so that the petals and stamens appear to be inserted on the calyx tube. In some plants, such as many legumes, the proper calyx tube extends well beyond the hypanthium, with no obvious change in texture. In others, such as *Prunus* and many members of the Rosaceae and Saxifragaceae, the hypanthium is coextensive with the apparent calyx tube; the stamens, petals, and apparent calyx lobes all diverge from the summit of the hypanthium. The different floral organs making up the hypanthium may also be adnate to different levels, so that the stamens appear to be attached to the calyx tube lower down than the petals.

It is customary and convenient to define the hypanthium in terms of external descriptive morphology rather than on evolutionary homologies. Vascular anatomy suggests that the lower part of the hypanthium in *Rosa* is actually the hollowed out edge of the receptacle, and that only the upper part consists of the fused bases of the sepals, petals, and stamens. In external appearance, however, the hypanthium of *Rosa* is a single discrete structure, well differentiated from the other parts of the flower, and bearing the sepals, petals, and stamens at its summit.

Flowers which have a hypanthium and a superior ovary are said to be perigynous. Perigynous flowers become epigynous (phyletically) by adnation of the hypanthium to the ovary. Sometimes, as in many of the Saxifragaceae, only the lower part of the ovary is adnate to the hypanthium, and the ovary is only partly inferior, the flower partly epigynous. Or, as in many Onagraceae, the hypanthium may extend well beyond the wholly inferior ovary.

Epigyny may also originate in other ways. In the Rubiaceae and Campanulaceae it appears that the bases of the calyx, corolla, and androecium are all fused to the ovary wall, without any history of a separate hypanthium. The Rubiaceae, with inferior ovary, are generally admitted to be closely related to but more advanced than the Loganiaceae, with superior ovary. Two genera with hypogynous flowers, *Pagamea* and *Gaertnera*, form a sort of connecting link between the two families. On formal morphological characters they should be referred to the Loganiaceae, but they find their nearest relatives in the Rubiaceae instead. If epigyny in the Rubiaceae has a perigynous ancestry, there is no indication of it in the species alive today. Although most of the Campanulaceae have the ovary wholly inferior, some few members are essentially hypogynous, and others have the ovary partly inferior in varying degrees. Here again there is no history of a separate hypanthium.

Less commonly, the ovary may become inferior by submergence in the receptacle. This appears to be the case in the Santalaceae. A complicated variation of receptacular epigyny has been proposed to explain the floral structure of the Cactaceae.[9]

[9] N. Boke, The cactus gynoecium: A new interpretation. Am. Jour. Bot. 51: 598–610. 1964.

SEQUENCE OF MATURATION OF FLORAL ORGANS

Inasmuch as a flower is essentially a condensed short shoot with specialized leaves, one might a priori expect the floral appendages to mature in serial sequence from the outside (or bottom) to the inside (or top). A little investigation shows that the situation is often much more complex. The first question is how maturity is to be judged. The stamens are mature when they shed their pollen, but the carpels are not truly mature until the fruit is ripe. In *Ranunculus* and some other genera, the calyx is commonly deciduous at about the time of anthesis. In many others it persists but does not enlarge materially after anthesis. In still others, such as *Hyoscyamus*, it is both persistent and accrescent, investing the mature fruit; the carpels and sepals thus mature simultaneously, long after the petals and stamens have fallen.

If one judges maturity by the time of anthesis, there are still problems. The calyx obviously must open before the corolla can expand, but the stamens and carpels do not necessarily follow in sequence. The androecium and gynoecium may mature at essentially the same time, or either one may mature before the other. Both protandry and protogyny are well known in many widely differing families. The flowers may even be self-pollinated in bud, long before the perianth has opened. These numerous deviations from a simple serial sequence doubtless represent "advances" from the ancestral type, but they occur in so many different groups as to provide little or no significant information for the general system at the level of families and orders.

When the floral parts of a given kind form only a single cycle, it is usually difficult or impossible to see any sequence of maturation among them. When they are more numerous and form a spiral or several cycles, they often mature (or at least are initiated) in serial sequence. Here again one might expect the sequence to be from the outside inwards, and here again the facts do not wholly conform to the expectation.

Most members of the Magnoliidae have stamens that are initiated in centripetal sequence, but the Limnocharitaceae (Alismatidae) and the multistamened members of both the Caryophyllidae and the Dilleniidae all have more or less distinctly centrifugal stamens, so far as current information shows. It may reasonably be doubted that this single character, which has no obvious functional significance, will be entirely constant throughout the groups in which it has been found, but clear-cut exceptions remain to be demonstrated. One order which has been considered to embrace both types (the Rhoeadales) is now generally divided on other grounds into two orders. One of these, the Papaverales, turns out to have centripetal stamens, whereas the other, the Capparales, has centrifugal stamens. (The possibility that the centrifugal sequence is correlated with a secondary increase in the number of stamens is discussed on p. 91.)

Although the developmental sequence appears to have some taxonomic usefulness in groups with spirally arranged stamens, it is perhaps more variable in groups with cyclic stamens. In both *Oxalis* and *Geranium* it is reported by Eames that the sequence of development of the two whorls may be in either direction.

## STAMENS

We are so accustomed to seeing "typical" stamens, with sharply differentiated anther and slender filament, that it comes as something of a shock to see the laminar stamen of *Degeneria*. Here there is neither anther nor filament, but a broad, somewhat petal-like, sessile lamina, with two pairs of elongate pollen sacs embedded in the abaxial side, well removed from the tip. This stamen furthermore has three vascular traces from the base. All transitional stages between laminar, three-trace stamens and typical, single-trace stamens with filament and anther can be seen in various members of the Magnoliidae — indeed even within the family Magnoliaceae. The prominently exserted connective seen in so many members of the Annonaceae is merely a vestige of the ancestral lamina.

There is now little doubt that the laminar stamen with embedded pollen sacs, as seen in *Degeneria* and some other Magnolialean genera, is a primitive type. It is not certain, however, whether the primitive position of the pollen sacs is on the adaxial or the abaxial side. Both positions can be

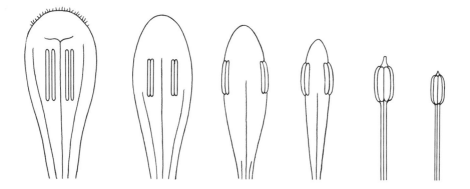

FIGURE 3.6

Idealized series from the laminar stamen of *Degeneria* (left) to the typical angiosperm stamen with filament and anther. Transitional forms such as those shown here exist among several families of the Magnoliales.

seen in laminar stamens: abaxial in *Degeneria* and *Himantandra*, adaxial in *Austrobaileya* and *Magnolia*. Conceivably both types may be primitive within the angiosperms.

Once the stamen has been reduced to the familiar anther-filament form, the orientation of the pollen sacs is easily modified. Introrse and extrorse orientation can sometimes be seen in different genera of the same family, as in the Lauraceae. The position may even change ontogenetically; certain palms are reported to have the stamens introrse in bud but extrorse at anthesis. Introrse dehiscence is much more common in "typical" stamens than extrorse, and probably most extrorse stamens are secondarily rather than primitively so.

Reduction of the androecium from many stamens to few has occurred a number of times, and possibly in more than one way. The ordinary way is for the indefinite spiral to be organized into several cycles, after which whole cycles may be dropped out at once. Thus the progression might go from numerous and spiraled, to four cycles, to three cycles, to two cycles, to a single cycle. These several cycles are offset from each other in the same way that the cycle of petals is offset from the cycle of sepals. Thus the outermost set of stamens is normally alternate with the petals (and opposite the sepals), the next set is opposite the petals, and so on.

Either the inner or the outer cycle of stamens may be the first to be lost. In several families (e.g., Caryophyllaceae, Geraniaceae) which commonly have two sets of stamens, the outer set is opposite the petals, indicating that a third set, outside the other two, has been lost. When the stamens are reduced to a single cycle, these are usually alternate with the petals, indicating that they belong to either the first or the third ancestral set (counting from the outside). Sometimes, on the other hand, the single cycle of stamens is opposite the petals, as in the Primulaceae, indicating that only the second cycle (or the fourth) has been retained.

The genus *Mitella*, in the Saxifragaceae, provides an especially instructive example of the reduction of two sets of stamens to one. Some species, such as *M. nuda*, have ten stamens, in two sets, one set opposite the sepals, the other set opposite the petals. Other species, such as *M. pentandra*, have a single set of five stamens, these opposite the petals. Still other species, such as *M. breweri*, have a single set of five stamens, opposite the sepals. The two different groups of five-stamened species of *Mitella* reflect the loss of different sets of stamens. The segregation of the five-stamened species as a distinct genus, which has been proposed by some taxonomists, is therefore indefensible. The two groups of five-stamened species relate not directly to each other, but separately to the ten-stamened ancestral type.

Another way that numerous, spirally arranged stamens can be reduced to a single cycle is for the stamens to be grouped into fascicles, and the number of stamens in each fascicle then progressively reduced by fusion or abortion. Each such fascicle is usually alternate with a petal, opposite a sepal.

An early stage in this sort of reduction is shown by *Paeonia*, in which the tendency toward fasciculation can be demonstrated only by careful ontogenetic anatomical studies.[10] In the Dilleniaceae, the single genus *Hibbertia* shows all conditions from numerous, free stamens, to five fascicles of stamens, to five separate stamens, and thence to only a single stamen. In the Guttiferae, *Hypericum* has a similar series from numerous, separate stamens to five fascicles of three stamens each. It is noteworthy that at least in *Paeonia* and those Dilleniaceae that have numerous, separate stamens, fusion of the vascular traces antedates connation of the filaments, so that a number of separate but adjacent stamens all have their vascular trace arising from a common stamen-trunk.

Some botanists have thought that the evolutionary sequence in the androecium of the Paeoniaceae, Dilleniaceae, and Guttiferae is the reverse of that here described, that the progression has been from few stamens in a single cycle, to fascicled stamens, to numerous separate stamens. I would give short shrift to this idea, were it not that these several groups also have centrifugal stamens. Inasmuch as the evolutionary origin and significance of centrifugality are still unknown, any possible correlation with any other character or trend deserves serious consideration. If centrifugality were always associated with a secondary increase in number, we would at least have a starting point for an explanation. The origin of centrifugality in the Caryophyllidae has received less attention, but it may be noted that the Phytolaccaceae, the most primitive family in the group, have fewer stamens than the more advanced families Aizoaceae and Cactaceae.

On the other hand, Eames[11] has interpreted the anatomical evidence in the Paeoniaceae, Dilleniaceae, and Guttiferae to indicate a reduction series, and Wilson[12] has expounded this point of view at some length for *Hibbertia*, in the Dilleniaceae. Furthermore, a similar reduction series associated with a grouping into fascicles exists in the Myrtaceae, which have centripetal stamens. If the series in the Myrtaceae has ever been interpreted as a secondary increase, the fact has escaped my attention.

Recently it has been noted[13] that two genera of the Limnocharitaceae have a centrifugal androecium with the same sort of compound stamen

[10] Paul Hiepko, Das zentrifugale Androeceum der Paeoniaceae. Ber. Deuts. Bot. Ges. 77: 427–435. 1964.

[11] Arthur Eames, *Morphology of Angiosperms* (New York: McGraw-Hill, 1961), pp. 96–99.

[12] Carl Wilson, The floral anatomy of the Dilleniaceae. I. *Hibbertia* Andr. Phytomorph. 15: 248–274. 1965.

[13] R. B. Kaul, Development and vasculature of the androecium in the Butomaceae. Am. Jour. Bot. 52: 624. 1965. (Abstract)

trunks that have been observed in the Dilleniales and Theales, although the stamens themselves show no trace of fasciculation. *Butomus* and a third genus (*Tenagocharis*) of the Limnocharitaceae, with nine stamens, have the antepetalous stamens arising later and slightly higher on the receptacle than the paired antesepalous stamens, but Kaul interprets this condition as derived by reduction from a centrifugal androecium with numerous stamens.

My present thought is that stamens of indefinite number may increase or decrease, but that once the process of organization into cycles and/or fascicles has begun, significant reversals are rare. I do not understand the origin of centrifugality, but I do not believe that it results from a secondary increase in the number of stamens.

Once the stamens have been reduced to a single cycle, the number of stamens in the cycle may be further reduced, so that there are fewer stamens than petals. This has happened, for example, in the Scrophulariaceae and Labiatae, both of which typically have five corolla-lobes and only two or four functional stamens. In both of these families it is perfectly clear that one or more stamens have been modified or lost, and the missing stamens are often represented by vestigial parts. In some of the Scrophulariaceae such as *Penstemon*, the fifth stamen has a well developed, more or less modified filament but no anther.

Increase in the number of stamens, once they have been confined to a single cycle, is much less common. In some of the Sapotaceae the number of stamens has probably increased concomitantly with an increase in the number of sepals and petals, but the number is still not large. In the Adoxaceae each stamen has apparently been divided longitudinally; the stamens are borne in pairs at the sinuses of the corolla, and each stamen has only a single pollen-sac.

## POLLEN

Differences in the shape of the pollen grains and more especially in the structure and ornamentation of the wall provide many useful taxonomic characters. The most useful of these characters, from a broad-scale viewpoint, is the number of germ furrows. These furrows are elongate strips in which the wall is relatively thin and elastic. Usually each furrow contains a particularly thin (or rupturable) spot, the germinal pore, through which the pollen tube may emerge. Sometimes these germinal pores are dissociated from the furrows, or the pollen tube may emerge from a germinal furrow that shows no specialized germ pore.

Most angiosperm pollen grains fall into only two general types, uniaperturate and triaperturate, from which some less common types have been derived. Uniaperturate pollen grains have a single germinal furrow or pore, usually across the end opposite the contact zone of the members of the tetrad. Triaperturate grains primitively have three germinal furrows radiating like the lines of a trilete mark, but at the opposite end of the grain from where a trilete mark would be if there were one. (A trilete mark is the three-parted ridge or scar which makes the contact lines of a tetrahedral tetrad of spores.) Typically the three germ furrows of a triaperturate grain are shortened so that they stand at the angles of the grain and do not closely approach each other.

Uniaperturate pollen grains are characteristic of the monocots, some members of the Magnoliidae, and all Cycadicae, including the pteridosperms. Triaperturate grains occur in the bulk of the dicots and are unknown elsewhere. Each of these two main types appears to have given rise to multiaperturate and nonaperturate types, which are found in both monocots and dicots. The multiaperturate type results from cross-partitioning of one germinal furrow into several. The biaperturate type found in certain monocots also reflects cross-partitioning. The nonaperturate type results from progressive shortening and eventual elimination of the furrow.

In more precise descriptions of pollen grains it is customary to distinguish between strictly tricolpate grains, in which the germ furrows lack germ pores, tricolporate grains, in which each furrow has specialized germ pore, and triporate grains, which have germ pores but not furrows. More loosely, the term tricolpate is sometimes used to cover all triaperturate types. A similar distinction is made among uniaperturate pollen grains, with the term monosulcate sometimes being substituted for monocolpate sens. strict. The terminology is still developing and changing, and the terms monotreme and tritreme are sometimes used instead of uniaperturate and triaperturate. The consensus is that the evolutionary progression is from colpate to colporate to porate, but there are many grades, refinements, and special conditions.

It is generally agreed that the uniaperturate type of pollen grain is primitive, and that the triaperturate type is derived from it. Real intermediates between the two types are few and difficult to interpret, however, and there has been some controversy about how the one type is derived from the other. Perhaps the most likely hypothesis is that of Wilson,[14] which was inspired by his observations in the Canellaceae. In all of the members of the Canellaceae which have been studied, most of the pollen is of an undistinguished, uniaperturate type. Some species, however, have a small percentage of other sorts of grains intermingled with the more common ones. Some of these unusual grains have the single furrow bent at the middle, so as to be broadly V-shaped. Others differ from the V type in having a short side-spur at the point of the V, so that the furrow is Y-

[14] T. K. Wilson, Comparative morphology of the Canellaceae. III. Pollen. Bot. Gaz. 125: 192–197. 1964.

shaped (trichotomosulcate). It is but a short step from such a Y-shaped furrow to three separate furrows, but it is this step which remains in question. Several families of monocots, including the palms, have trichotomous furrows in the pollen of some species, but here this has not led to the typical triaperturate grains so commonly seen in the dicots.

The adaptive significance of the changes in number and distribution of the germ furrows and germ pores is obscure. With a little stretching of the imagination one might suppose that grains with germ pores are more efficient than grains with only germ furrows, but the competitive advantage, if any, must be vanishingly small. If such an interpretation were accepted, a considerable ingenuity — or ingenuousness — would be required to explain the repeated origin of nonaperturate types against the selective gradient. The possibility that these changes are side-effects of genes selected for other functions merits consideration, but we are in Stygian darkness about the possible nature of these other functions.

In most angiosperms the pollen grain has two nuclei at the time it is shed from the anther. One of these, the generative nucleus, later divides to produce two sperm nuclei. In a considerable number of taxa the generative nucleus divides before the pollen is shed, so that the grains are trinucleate. This early division of the generative nucleus apparently represents one more step in the progressive compression of the gametophyte that characterizes vascular plants in general. In considering binucleate and trinucleate pollen types, the student should remember that both types produce mature male gametophytes with three nuclei; the difference lies in when the generative nucleus divides.

Some whole orders, such as the Caryophyllales, have trinucleate pollen. Other orders and even individual families include both binucleate and trinucleate types. No taxonomic grouping can properly be based on this character alone, any more than on any other single character. It does appear, however, that the change is irreversible. There are no binucleate groups which clearly have a trinucleate ancestry.

The number of nuclei in the pollen grain is strongly (though not perfectly) correlated with a syndrome of physiological features. In general, binucleate pollen has a high storage longevity, and can easily be germinated in vitro. Genetic incompatibility operates through substances in the ovary or style which act only after the pollen grain has germinated. Trinucleate pollen, on the other hand, usually has a low storage longevity, and it is not easily germinated in vitro. Genetic incompatibility usually operates at the stigma, preventing the pollen from germinating. Brewbaker[15] has suggested that

[15] J. L. Brewbaker, The distribution and phylogenetic significance of binucleate and trinucleate pollen grains in the angiosperms. Am. Jour. Bot. 54: 1069-1083. 1967.

The syndrome of differences between II and III pollen can best be explained by assuming that the second mitotic division deprives the pollen grain of reserves essential for germination, for prolonged storage, and for the growth (albeit limited) into styles prior to inhibition by incompatibility alleles. These reserves are perhaps made available to pollen only by genetically compatible stigmas, perhaps only following the action of pollen-released enzymes. . . .

The physiological consequences (or associates) of the trinucleate condition in pollen are so far-reaching that they must be a part of an intricate mechanism associated with adaptation and survival. For the most part, however, the details of the operation of the mechanism still escape us. One might speculate that inhibition at the stigma would reserve the style for compatible pollen tubes, but it is not clear that this would really make any difference to the plant.

Trinucleate pollen obviously should have a strong survival value in plants which produce flowers beneath the surface of the water, where binucleate pollen might well germinate without reaching a stigma. Effective underwater pollination by trinucleate pollen occurs in some or all genera of at least 19 families of diverse affinities, but there are also several genera which get along with binucleate pollen under similar conditions. In some groups the evolution of trinucleate pollen evidently preceded the evolution of submersed flowers; the pollen was preadapted to submersion, rather than selected by it.

Before the physiological consequences of the trinucleate condition had been discovered, the change from binucleate to trinucleate appeared to be unrelated to adaptation and survival. Here we have an example of the hidden significance of a seemingly trivial morphological character. The proportion of other seemingly trivial characters which will turn out to have a real importance remains to be discovered.

## POLLINATION

Flowers of modern angiosperms are pollinated by such diverse agents as wind, water, insects of various orders, birds, and even bats. They may also be self-pollinated. When the Amentiferae were generally regarded as primitive angiosperms, wind-pollination was also regarded as primitive, but with the ascendancy of the Besseyan school of thought, wind-pollination has come to be considered a secondary condition, usually associated with floral reduction.

It is now generally believed that the primitive angiosperms were pollinated by beetles that chewed and ate bits of the perianth and androecium. Several of the existing Magnoliid families (Calycanthaceae, Eupomatiaceae, Himantandraceae, Magnoliaceae, Nymphaeaceae) are still chiefly or wholly pollinated in this way. These mostly have little or no nectar and no obvious nectaries, but some of them have more or less specialized, non-secretory "food bodies" consisting of surface pads or apical knobs of succulent tissue on the tepals, stamens, and staminodia. These food bodies may superficially resemble nectaries.

The nectaries of diverse angiosperm families have little in common beyond the production of nectar. In some Magnoliid genera (*Magnolia*, *Nymphaea*), nectar is secreted, but not in localized areas; it apparently diffuses out from the surfaces of the tepals, stamens, and even the carpels. In many of the Ranunculaceae some of the stamens are modified into nectariferous staminodia. These are apparently homologous with the petals of *Ranunculus*, which have a basal nectary. In some "apetalous" families, such as the Proteaceae and Salicaceae (*Salix*), the nectaries probably represent a reduced and modified corolla; in many other families the nectaries are actually reduced stamens. Parts of the carpel may also function as nectaries, especially among the monocots.

The inference to be drawn is that the use of nectar to attract pollinators evolved several times independently after the angiosperms were already highly diversified. This brings us back to the thought that the primitive pollinators of the angiosperms were chewing insects such as beetles, rather than nectar-sippers such as bees, flies, and butterflies. Pollination by birds and bats is also doubtless more recent than beetle-pollination, because here again the principal attractant is nectar. Birds may have been physically available as pollinators early in angiosperm history, but bats apparently did not even exist until the angiosperms were already well diversified.

The Coleoptera (beetles) are well known as fossils as far back as the Permian period. Since the angiosperms are presumably post-Permian in origin, the Coleoptera were physically available as possible pollinators from the beginning. The Diptera and Hymenoptera date from the Jurassic period, but only in forms such as midges and crane flies (Diptera) and saw flies (Hymenoptera), which are not and presumably never were important pollinators. The bees and the higher Diptera, which are now important pollinators, first appear in the fossil record in early Tertiary time, but they may actually have arisen during the Cretaceous period; the paucity of Cretaceous fossil insects of all sorts makes any conclusion hazardous. However, most or all of the early Tertiary bees belong to extinct genera, and one may legitimately speculate that the evolution of modern bees was intimately related to the evolution of bee-pollinated flowers during the Tertiary period. The Lepidoptera were sufficiently diverse in early Tertiary time to suggest that they may have arisen considerably before, but here again the important pollinators are apparently not ancient types.

With the aid of a little imagination, one can set up a series of morphological types of flowers which represent progressive adaptation to more discriminating pollinators. Starting with relatively amorphous flowers with numerous tepals, as in *Magnolia*, one progresses to flowers with a small and definite number of petals that are all about alike and all in more or less the same plane, as in many polypetalous flowers, to flowers which are regular but markedly three-dimensional, such as daffodils and phloxes, to flowers which are not only three-dimensional but also irregular, such as orchids and snapdragons.

In a very loose way, such a series does actually reflect progressive evolutionary advance, but there are many exceptions and special cases. The flower heads of many of the Compositae compare in form to primitive Magnoliid flowers, but the Compositae also produce abundant nectar, and they are visited by a wide array of pollinators. Some of the Compositae, such as species of *Bidens* and *Coreopsis*, have five or eight petal-like rays and simulate ordinary polypetalous flowers. Even more than with many other characters, the different classes of pollinators reflect extensive parallelism.

Wind-pollination has evolved from insect-pollination many times, usually associated with reduction of the perianth, as in the Amentiferae. Although a large part of the traditional Amentiferae probably hangs together, the group as a whole comes from at least four different sources, each of which independently adopted wind-pollination in association with floral reduction. Even within the single family Compositae, taxonomically far removed from any of the Amentiferae, wind-pollination has evolved more than once. The most immediately obvious examples are the subtribe Ambrosiinae of the tribe Heliantheae, and the genus *Artemisia*, in the tribe Anthemideae. In these groups also, wind-pollination is associated with floral reduction.

Flowers with the perianth much reduced are not always wind-pollinated. *Populus* and *Salix*, in the same family, both have highly reduced flowers, but *Populus* is wind-pollinated, whereas *Salix* has well developed nectaries and is (at least in part) insect-pollinated. It remains to be seen whether *Salix* has a continuously entomophilous ancestry, or whether it has reverted to entomophily after having passed through an anemophilous evolutionary stage. It may be significant that in at least two respects (number of stamens and number of bud scales) in addition to its usually more reduced growth form, *Salix* appears to be more advanced than *Populus*. However, the combination of primitive and advanced features in one plant is so common in angiosperms in general that no conclusion can safely be drawn here without further evidence. It may also be noted that although the Fagaceae have reduced flowers and are usually wind-pollinated, insects are attracted to the conspicuous inflorescences of *Castanea* and some other genera. Here it may well be that wind-pollination has followed floral reduction instead of preceding it.

Wind-pollination and insect-pollination are not necessarily mutually exclusive. The change from one to the other can take place gradually, without any sudden jumps, especially if the adaptation to insect-pollination is generalized and does not involve complex or unusual structure associated with a particular kind of pollinator. The well known genus *Solidago*, like a great many other Compositae, is generally insect-pollinated but has no special adaptations for particular pollinators. Some of the species, such as *S. speciosa*, are reported to release a considerable amount of pollen into the air as well. *S. speciosa* is not less well adapted to the attraction of insects than other goldenrods; indeed it is one of the most showy species of the genus.

There is some correlation between the surface ornamentation of the pollen grain and the method of pollination. Wind-distributed pollen tends to be small, thin-walled, smooth, and nonadhesive, in contrast to insect-distributed pollen, which tends to be larger, thicker-walled, variously sculptured, and sticky. The correlation is obviously functional and need not reflect common ancestry or mutation pressure.

The vast majority of angiosperms are ordinarily cross-pollinated, and there are many mechanisms which restrict or prevent selfing. The most effective of these is dioecism, which we have already discussed. Among species with perfect flowers, self-sterility, heterostyly, exserted styles, exserted stamens, and protandry-protogyny are common methods of restricting selfing. Each of these devices has evolved repeatedly in different groups, and the term self-sterility masks a variety of different physiological responses. Neither taxonomic nor ecologic affinity unites the diverse species which share any one mechanism for restricting selfing.

It is perfectly obvious that evolutionary progress depends on exchange of genes as well as on mutations per se. On the other hand, selfing is the only *certain* method of pollination, and any mechanism which inhibits selfing carries the inherent danger of reducing seed-set. Inasmuch as plants cannot seek each other out for pollinating purposes, any restriction of self-pollination puts them in thrall to their pollinators. Anyone who has ever tried to grow just a few hills of corn, well separated from any large cornfield, will know that the problem of ensuring pollination is real.

Although frequent gene-exchange is clearly an evolutionary desideratum, it is not necessary that selfing be completely eliminated. A mixture of crossing and selfing can keep the evolutionary pot bubbling nearly or quite as well. A single cross-pollination can provide the materials to produce new and different gene-combinations through several generations of selfing. Some plants, such as species of *Panicum* and *Viola*, meet the problem of ensuring both exchange of genes and ample seed set by having some open-pollinated flowers and some cleistogamous, self-pollinated ones. One may wonder why this solution has not been more widely adopted. Another interesting combination of selfing and outcrossing is seen in some small-flowered western American species of *Gilia*, which are visited by pollinating insects only in dry years, when the supply of larger flowers with more abundant nectar is insufficient for their needs.

The variety and ubiquity of the mechanisms inhibiting self-pollination amply indicate the evolutionary usefulness of maintaining a considerable degree of outcrossing. On the other hand, self-pollinated species or genera are widely scattered in diverse taxonomic groups, and some of them are highly successful. It appears that small differences in the conditions of selection may shift the competitive advantage back and forth between selfing and outcrossing.

Self-pollinators are not restricted to any one habitat or set of habitats. It is evident that species well adapted to a particular habitat may survive indefinitely without outcrossing, but the genetic mix necessary for the production of new species is much more likely to occur in outcrossing groups. Obligate self-pollination is the same sort of evolutionary dead end as obligate apomixis.

## CARPELS AND PLACENTATION

The morphological nature of the carpel has been much discussed, and new hypotheses continue to be proposed. The most generally accepted view for many years has been that it is essentially a megasporophyll, and this view continues to command widespread support. The concept of the carpel as a megasporophyll fits very well into the general Ranalian theory of angiosperm evolution, and indeed any other interpretation would seem to require an extensive reconsideration of the facts and concepts of floral morphology and phylogeny which now seem to be fitting into a harmonious, internally consistent pattern.

For many years it was believed that the primitive, pre-angiospermous open carpel probably had a row of ovules along each margin, and that closure of the carpel brought the two rows of ovules together into a single row along the ventral suture. This view may still prove to be correct, but it now seems more likely that the primitive, open carpel had ovules scattered over its upper surface, and that restriction of the ovules to the apposed margins of the carpel represents an early modification from the ancestral condition. This latter view dates from a paper by I. W. Bailey and B. G. L. Swamy[16] in 1951 and reflects a reconsideration of the evidence, sparked by the discovery of the primitive genus *Degeneria* in the Fiji Islands a few years before.

It seems perfectly clear that one way in which the typical closed angio-

[16] The conduplicate carpel of dicotyledons and its initial trends of specialization. Am. Jour. Bot. 38: 373-379.

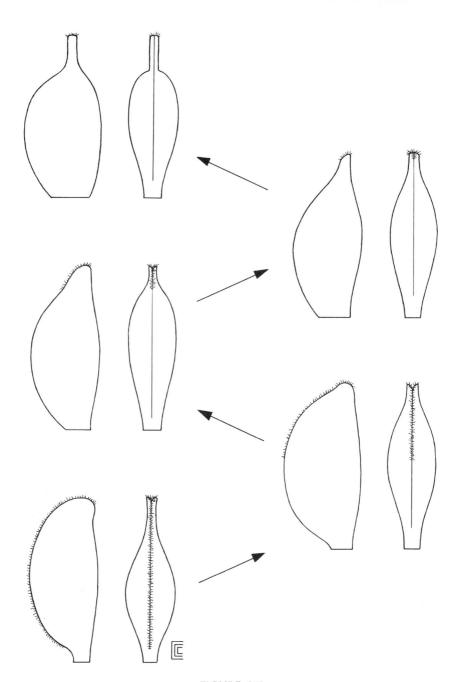

FIGURE 3.7

Idealized series from the conduplicate, unsealed carpel of *Degeneria* to a typical simple pistil with ovary, style, and terminal stigma. Transitional forms such as those shown here exist in the Winteraceae and some other families.

spermous carpel originated was by conduplicate folding of an open carpel and subsequent closure of the open margin. Among modern members of the Magnoliales, some species of *Drimys* (Winteraceae) have thin, unsealed carpels that are merely folded along the midrib, the ovules being borne on the two inner surfaces of the folded carpel. The carpels of some species of *Bubbia* (Winteraceae) and of *Degeneria* are very much like the *Drimys* carpels just mentioned, except for being abaxially somewhat deformed. In these genera a mat of tangled hairs running the length of the carpel serves as an elongate stigma on which the pollen grains germinate. Stages in the development of the typical single pistil, with closed ovary, style, and terminal stigma are still preserved among various living members of the Magnoliales.

Whether closure of the carpels may have taken place in any other way is still debatable. It was at one time thought that in many angiosperms parietal placentation reflected connation of several open carpels by their lateral margins. Most of the groups to which this concept had been applied now seem better interpreted as derived from forms with axile placentation. After these advanced types have been disposed of, there remain a few taxa in the Magnoliales which have syncarpous, parietal placentation but which do not have near relatives with axile placentae. The Canellaceae, for example, do not seem related to any group with axile placentation. Even more strikingly, two genera (*Isoloma, Monodora*) of the Annonaceae, an otherwise apocarpous family, have a compound pistil with parietal placentation. Among the primitive monocots, the Hydrocharitaceae, with laminar-parietal placentation in a compound ovary, belong to an otherwise apocarpous complex. In such taxa it seems not improbable that conduplicate but unsealed carpels have become connate by their margins, after which the initially deeply sulcate ovary form became more smoothly rounded.

Primitively the angiosperm gynoecium must have consisted of several or many carpels, spirally arranged on a more or less elongate receptacle. Each of these carpels doubtless had more or less numerous ovules, and it may reasonably be assumed that the carpels spread open at maturity, freeing the seeds. Gynoecia of this type exist in many of the more primitive angiosperms today, both monocots and dicots. Sometimes these spirally arranged carpels became more or less fused into a syncarp, as in some species of *Magnolia*, but this line of evolution (as opposed to the union of several carpels to form a compound pistil) did not lead to any further advance.

Reduction of the number of carpels to a single whorl opened new possibilities for further evolution. One theoretical possibility which seems never to have been exploited is the fusion of the carpels into several separate compound pistils within the flower. The carpels in known angiosperms may be separate, or loosely united, or firmly united, but if they are united at all they form a single pistil.

Given insect pollination and the tendency for several or many grains to be transported collectively in more or less of a lump, a compound pistil is more efficient than several separate pistils. This greater efficiency may not be the whole cause of the evolutionary fusion of the carpels, however. This fusion is merely part of a general tendency toward reduction of flowers which permeates the whole angiosperm phylad, and some aspects of this tendency, such as the repeated evolution of wind-pollinated flowers with no perianth, certainly cannot be ascribed to selection for greater efficiency in the entomophilous habit. Nor can entomophily be assigned any clear role in the change from spirally arranged carpels to cyclic carpels, the necessary precursor to the evolution of the compound pistil.

It is interesting that phyletically the ovaries fuse before the styles and stigmas, so that it is not at all unusual to have a single ovary with separate styles and stigmas, or a single ovary and style with separate stigmas. Indeed it is unusual to find the stigmas so completely fused as to leave no external trace of the carpellary units. Although the phyletic fusion of the ovaries and possibly the styles might be explained in terms of pollinating efficiency, the subsequent partial or even complete fusion of the stigmas in various groups can hardly be explained in the same way; indeed it would seem instead to be counterselective. Here, as with some other evolutionary trends in the angiosperms, it would be convenient to invoke the concept of evolutionary momentum, if the concept itself had not fallen into disrepute.

We have noted that although the primitive carpel may have had the ovules scattered over the upper surface, the restriction of the ovules to marginal rows was an early evolutionary step. Thus the presence of several carpels in a whorl, each with a single marginal placenta, is still a fairly primitive condition. Fusion of the members of such a whorl gives a compound pistil with axile placentation. This step has occurred repeatedly in different groups. The partitions in such a compound pistil are, at least phyletically, double membranes, each composed of the connate walls of two adjacent carpels.

Many of the Apocynaceae and Asclepiadaceae have two separate ovaries with a common style or stigma. Here the separation of the bodies of the carpels appears to be secondary rather than primitive. We are not witnessing stages in carpellary fusion from the top down. Whatever may have been the cause of this secondary separation, it apparently could not overcome the selective advantage of having a common pollinating surface for both carpels, and the stigma has remained intact. At least we may reasonably so speculate about the forces involved.

The most common types of placentation in compound ovaries are axile,

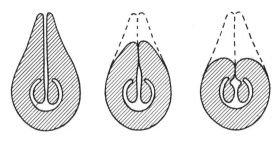

FIGURE 3.8

Idealized series of carpellary cross-sections showing derivation of marginal placentation in a closed carpel (right) from laminar placentation in a conduplicate, unsealed carpel (left). Adapted from I. W. Bailey and B. G. L. Swamy, The conduplicate carpel and its initial trends of specialization. Am. Jour. Botany 38: 373–379. 1951.

parietal,[17] free-central, basal, and apical. The axile type is ancestral to all the others. As we have noted, in some few groups parietal placentation may also result from marginal fusion of folded carpels, without the intervention of a truly axile condition.

Parietal placentation typically arises from a failure of the partitions to join in the center of the ovary. These incomplete partitions are often called deeply intruded placentae. The partitions may then be further reduced and eventually eliminated, so that the placentae are actually on the wall of the strictly unilocular ovary. On the other hand, the placental edges of the partial partitions may separate and curve away from each other. If these partitions then meet and fuse again in the center of the ovary, axile placentation is restored. This sort of return to axile placentation may actually have occurred in some of the Gentianaceae and Hydrophyllaceae.

Free-central placentation arises from axile by abortion of the partitions, accompanied by failure of the placental column to reach the top of the ovary. The Caryophyllaceae are usually said to have free-central placentation, but the early ontogenetic stages are typically axile, and vestigial partitions are often visible at the base of the mature capsule. In some other families, such as the Primulaceae, the origin of free-central placenta-

---

[17] The term parietal placentation has often been used broadly to apply to both simple and compound pistils. It is here restricted to compound pistils. Simple pistils with the ovules in a line formed by the apposed edges of the carpel, as in the Legumonosae, are here said to have marginal placentation.

tion from axile is not so immediately obvious, but the basic structure is the same.

Basal placentation is easily derived from free-central by reduction of the placental column, usually accompanied by a decrease in the number of ovules to only one or a few. Basal placentation may also arise from axile or parietal by progressive restriction of the placenta.

Apical placentation arises from axile or parietal by progressive restriction of the placenta from the base upwards. A slight further shift of position of the remaining part of the placenta makes it fully apical.

These several types of placentation do not exhaust the possibilities. The Cruciferae and some of the Papaveraceae (e.g., *Macleaya*) have the ovules attached to the margins of the partition in a bilocular ovary. The evolutionary morphology of such fruits has occasioned much controversy and is still not settled to the satisfaction of all investigators. It may be significant that the placentation in other genera of the Papaveraceae ranges from axile through parietal with intruded placentae to ordinary parietal.

The adaptive significance of changes in placentation is obscure. In a few taxa the elastic dehiscence of the fruit requires a particular type of placentation, but in general it makes no obvious difference to the plant whether the placentation is axile, parietal, or free-central. Likewise in one-seeded, indehiscent fruits, it is hard to see any biological importance in whether the ovule is attached at the top, the bottom, or along the side of the ovary. If these features confer any selective advantage, either in general or in correlation with a particular habitat or way of life, the fact remains to be demonstrated.

Fusion of carpels to form a compound pistil may take place in flowers with two, three, four, five, or more carpels. After the initial fusion has taken place, the number of carpels may be progressively reduced down to two or even one. Decrease in carpel number may be saltatory, by a change in symmetry, or one or more carpels may be gradually reduced and lost. Several families and genera have a pseudomonomerous gynoecium which may or may not show vestiges of lost carpels.

Gradual reduction in carpel number appears to be taking place in *Navarretia*, of the Polemoniaceae. Most species of *Navarretia* have three stigmas and a tricarpellate ovary, like other members of the family. In *N. divaricata* two of the three stigmas are often partly connate. The capsule of *N. divaricata* is loculicidally dehiscent by three valves, but it is commonly only two-locular by complete failure of one of the partitions, the odd valve also being narrower than the others. *N. minima* and *N. intertexta* have bivalved, bilocular capsules and usually only two stigmas.

Gradual reduction of some of the carpels can also be seen in some of the Caprifoliaceae and the related family Valerianaceae. *Linnaea*, in the Caprifoliaceae, has a trilocular ovary, but two of the locules contain several abortive ovules each, whereas the third one has a single normal ovule. The fruit is one-seeded and unequally three-locular. In the Valerianaceae, *Valerianella* has a tricarpellate pistil with a three-lobed stigma; the fruit is trilocular, but two of the locules are sterile and poorly developed. *Valeriana* also has a trilobed stigma and a basically tricarpellate ovary, but two of the carpels are vestigial. In *Plectritis* the stigma is two-lobed or rarely three-lobed, and the ovary is wholly unilocular, with no evident vestige of the two sterile carpels.

Saltatory reduction in carpel number can also be seen in the Caprifoliaceae, among other families. Among the American species of *Lonicera*, some, such as *L. ciliosa* and *L. involucrata*, are tricarpellate, whereas others, such as *L. caerulea* and *L. utahensis*, are bicarpellate. *Triosteum*, in the same family, has three-, four-, and five-carpellate ovaries, sometimes even in the same species, without intermediate stages.

Saltatory changes in the number of carpels are usually downward, but are sometimes upward instead. *Nicotiana multivalvis*, an Indian cultigen of northwestern United States, has six to eight carpels. It is generally believed to be derived from *N. bigelovii*, a wild species which has two carpels, the typical number for the genus and family. In the vicinity of Logan, Utah, there is a form of *Leptodactylon watsoni* (Polemoniaceae) that has six calyx lobes, six corolla lobes, six stamens, and four carpels, instead of five, five, five, and three, as in other individuals of the species and as in the family in general. One might suspect that the tropical American tree *Chrysophyllum caimito*, with six to eight carpels, has a five-carpellate ancestry.

The adaptive significance of such changes in carpel number is obscure. The downward changes, which are more frequent than the upward ones, fit into the general pattern of reduction which is expressed in so many different ways in floral evolution. We have seen that some manifestations of this general pattern might well reflect selection and survival value, but others apparently do not. There is not a close enough correlation between the evidently or probably adaptive and the seemingly nonadaptive manifestations of the pattern to warrant the thought that both types are governed by the identical set of genes. If the explanation is to be found in pleiotropic effect of genes valuable for other functions, the nature of the other functions is yet to be discovered.

## Ovules

In all living gymnosperms the ovule typically has the micropyle at the end opposite the stalk. All theories of ovular evolution take this as the primitive orientation. In contrast to these straight, orthotropous ovules of most gymnosperms are the anatropous ovules of most angiosperms, in which the ovule is bent back on itself to bring the micropyle alongside the

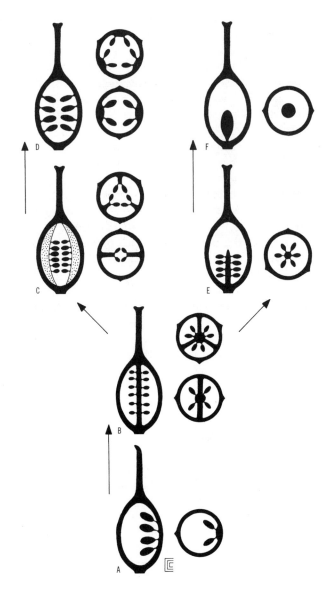

FIGURE 3.9

Evolutionary relationships among some types of placentation. A, marginal placentation in a simple pistil. B, C, D, E, and F all show compound pistils; B, C, and D each show two and three carpels in different cross-sections. B, axile placentation. C, parietal placentation with intruded placentas. D, parietal placentation. E, free-central placentation. F, basal placentation. Placentation like F can also be derived directly from A.

funiculus (stalk). It seems perfectly clear that the anatropous condition is eventually derived from the orthotropous.

Anatropy must have originated very early in the evolution of the angiosperms, because among the living members of the group this appears to be the primitive state. (No evidence is available from fossils, because details of this sort are almost never preserved.) Anatropous ovules are standard among the more primitive families of both monocots and dicots, as well as being widely distributed among the more advanced families.

Within the angiosperms, orthotropous, campylotropous, and amphitropous ovules are specialized types derived from an anatropous ancestry. Orthotropous ovules are found chiefly in families in which the ovule is basal and solitary in the ovary — advanced types such as the Juglandaceae, Piperaceae, Polygonaceae, Restionaceae, and Urticaceae, which otherwise have little in common. In at least some instances the orthotropous condition is attained by progressive phyletic migration of the funiculus. The Moraceae have anatropous ovules suspended from the top of the ovary. In the related but more advanced family Urticaceae, *Urtica* and *Boehmeria* have orthotropous, basal ovules, and other genera show various transitional stages between the two types.

Campylotropous and amphitropous ovules, with the micropyle pointing more or less at right angles to the funiculus, are derived from the anatropous type by differential growth, and an ovule which is campylotropous or amphitropous at the time of fertilization may be anatropous at an earlier ontogenetic stage. Families with these specialized types of ovules are not primitive and are not necessarily closely related among themselves. Campylotropous ovules are found, for example, in such diverse families as the Capparaceae, Caryophyllaceae, Geraniaceae, Apocynaceae, and Verbenaceae.

The primitive angiosperm ovule evidently had two integuments. Fusion of the two integuments into one has occurred in many different families, and intermediate stages of fusion can often be seen. The Ranunculaceae, Rosaceae, Salicaceae, Ericaceae, and Palmae, among other families, include both unitegmic and bitegmic types. Even when the two integuments are perfectly distinct at the micropylar end of the ovule, the distinction tends to fade out at the chalazal end. In anatropous ovules the outer integument also tends to be adnate to the funiculus, so that in vertical section the ovule appears to have only one integument on the side next to the funiculus. Fusion of the two integuments into one may be accompanied by a reduction in the thickness, so that the single integument is no thicker than either of the two ancestral ones; or, as in most of the Asteridae, the single integument may be massive, as thick as or thicker than both of the ancestral integuments together.

The unitegmic condition may also be achieved by abortion of one or the other integument. In *Populus*, for example, the inner integument is sometimes present but poorly developed, and sometimes completely wanting. (J. Mauritzon, Die Bedeutung der Embryologischen Forschung für das natürliche system der Pflanzen. Lunds Univ. Arsskr. Avd. 2. 35 [15]: 1–70. 1939 [p 22].)

In the more primitive families of angiosperms the nucellus is usually several or many cells thick and well differentiated from the integuments, recalling the massive nucellus of many gymnosperms. Evolutionary advance leads to tenuinucellate ovules, with a nucellus only one cell thick. In many of the Astéridae the integument appears to abut directly on the embryo sac, and the nucellus can be distinguished only as a single layer of a few cells forming a cap over the micropylar end of the embryo sac. Reduction of the nucellus, like other evolutionary advances in the angiosperms, has occurred independently in many different groups.

The specialized cell or group of cells within the young ovule which gives rise to or matures into the spore mother cell is called the archesporium. In gymnosperms and some angiosperms the archesporium is characteristically multicellular, even though only one cell actually functions as a megaspore mother cell. In most angiosperms the archesporium is unicellular; there are no wasted potential megaspore mother cells. Clearly this difference is in pattern with the evolutionary loss of other excess baggage in the reproductive apparatus of the angiosperms.

Within the angiosperms, crassinucellate ovules commonly have a multicellular archesporium, and tenuinucellate ovules usually have a unicellular archesporium, but there are exceptions. The Compositae, for example, have tenuinucellate ovules, but some few of them are reported to have a multicellular archesporium. The loose correlation between nucellar and archesporial type fits the pattern of a loose correlation among primitive characters in the angiosperms in general. It seems plain enough that within the angiosperms the unicellular archesporium is derived from the multicellular. This is not to deny the possibility of reversion; the multicellular archesporium of some Compositae may well be secondary.[18]

Among living angiosperms, the "typical," *Polygonum*-type embryo sac with eight nuclei derived from a single megaspore appears to be primitive. It seems perfectly obvious that this type is derived by progressive reduction from a multicellular gametophyte such as is found in gymnosperms. The next step in such a reduction series leads to a four-nucleate embryo sac such as that of *Oenothera*. Inasmuch as three of these nuclei (one egg

---

[18] In a book published too recently to be fully utilized in the preparation of this work, Gwenda L. Davis proposes to redefine the terms crassinucellate and tenuinucellate on the basis of the ontogeny rather than the mature structure of the ovule. In most cases there is no change in the actual application of the terms, but a few ovules which are tenuinucellate under one definition are crassinucellate under the other, and vice versa. It remains to be seen whether her definitions will be widely adopted. (Gwenda L. Davis, *Systematic Embryology of the Angiosperms*. New York: John Wiley & Sons, 1966.)

and two polar nuclei) are involved in double fertilization, the prospects of any further reduction of the female gametophyte, short of bypassing it altogether, do not seem good. In fact no more reduced type is known.

Other variations in type result from the participation of more than one megaspore in the development of the embryo sac. We have here an example of the fact that nature is not bound by our ideas of classification of knowledge. In evolutionary theory, the embryo sac is the female gametophyte of angiosperms. As such, it is a distinct, separate individual which retains its identity even though it is completely enclosed by the sporophyte. It should have *n* chromosomes and should be derived from a single megaspore. In actual fact, however, the embryo sac of angiosperms is not a distinct individual; it is merely a stage in the reproduction of the (sporophyte) plant. As such, it is free from some of the restrictions which would be convenient to impose on it from the standpoint of evolutionary theory. The participation of two or four megaspores in the formation of the embryo sac poses no threat to its reproductive function; neither do variations in the chromosome number of its nuclei, so long as one of these nuclei has *n* chromosomes and can serve as the egg. Thus, from a functional standpoint, the existence of bisporic and tetrasporic embryo sacs should not be surprising, and indeed several variations of these two types do exist. One of the tetrasporic types, called the *Peperomia* type, has 16 nuclei. Considered on the basis of the number of nuclei alone, this type should be more primitive than the typical, eight-nucleate, monosporic embryo sac. However, all tetrasporic embryo sacs must ultimately be derived from monosporic ones; furthermore, it take one less set of divisions to produce a 16-nucleate tetrasporic embryo sac than an eight-nucleate monosporic one.

Most of the evolutionary trends in ovular structure fit into the general pattern of floral reduction which pervades the angiosperms. The resulting economy of structure is on the plus side for general efficiency and survival value, but it is hard to believe that the difference is great enough to provide much driving evolutionary force. Without invoking any discredited mechanisms, it might be possible to think of these reductions as another example of the well known evolutionary tendency for unused structures, or unnecessary complications of structure, to degenerate. Aside from simple economy, this degeneration is believed to result from the accumulation of "loss" mutations which may or may not be useful for any of many other functions, these other functions being generally unspecified and indeed unknown.

## Fertilization, Seeds, and Seedlings

Double fertilization and the development of a copious endosperm from a triple fusion nucleus are clearly primitive characters within the angiosperms. The formation of an endosperm nucleus by fusion of a sperm cell with two cells of the embryo sac is such a standard accompaniment to typical fertilization in angiosperms that the term "double fertilization" has been coined to cover the two processes collectively. Even when the mature seed has no trace of endosperm, an endosperm nucleus is generally formed coincident with fertilization. Aside from apomicts, the most notable exceptions to this rule are provided by some of the orchids, in which the seeds are extremely small, with tiny, undifferentiated embryos and no food storage tissue as such. Orchids are highly advanced in other respects and the absence of even an endosperm nucleus in some of them is clearly secondary rather than primitive.

In primitive angiosperms the endosperm is a food storage tissue whose nutrients are absorbed by the embryo during germination. In more advanced types the embryo absorbs the food from the endosperm before the seed is ripe, so that in the mature seed the food is stored in the cotyledons or elsewhere in the embryo, and the endosperm is missing. This absorption may take place rather late in ontogeny, as in many of the Sapotaceae, in which the endosperm is obvious in young seeds but wanting or scanty in mature ones. (Some other genera of the Sapotaceae have a copious endosperm at maturity.) In still more advanced types, the endosperm may degenerate early in the ontogeny of the fertilized ovule, and most of the stored food in the mature seed has never been in the endosperm. These changes, like other evolutionary advances in the angiosperms, have occurred repeatedly in different groups.

The nucellus or integuments may also serve as food storage tissue in the mature seed, forming what is called a perisperm. This is clearly an advanced condition. Many perispermous seeds also have an endosperm, and during germination the food is absorbed from the perisperm by the endosperm and thence is passed on to the embryo.

The evolutionary origin of double fertilization is obscure. It appears to have no forerunner among the gymnosperms. It may be suggested that inasmuch as two sperm nuclei are regularly delivered to the embryo sac, it should not be surprising that the one not involved in ordinary fertilization should also find something to fuse with. Why it should fuse with two other nuclei instead of only one, and why this triple fusion nucleus should initiate food storage tissue, are still wholly speculative.

One may speculate that the difference in ploidy-level between the zygote and the endosperm nucleus facilitates the physiological distinction between the two in subsequent ontogeny. Against (or perhaps tangential to) this thought is the fact that triple fusion which does not involve a sperm nucleus does not lead to the formation of an endosperm. In some tetrasporic embryo sacs (Lilium type) three of the four megaspores fuse soon after being produced. This triple fusion nucleus gives rise eventually to half of the nuclei of the embryo sac. One of these triploid nuclei and one haploid nucleus of the embryo sac then fuse with a sperm nucleus to form a pentaploid endosperm nucleus. The origin of the *Lilium* type embryo sac may itself have been facilitated by habituation of the embryo sac to triple fusion in the formation of the endosperm nucleus.

There is no obvious reason why the precursors of the angiosperms should have continued to deliver two sperms to the female gametophyte before a function had been found for the second one. It may be noted, however, that modern gymnosperms produce two or more sperms per pollen tube, even though only one can function successfully. Reduction of the female gametophyte in protoangiosperms to a size at which it could no longer function effectively as a food storage tissue may have made it possible for the second sperm to find a function instead of being wasted. Why the female gametophyte became so reduced that the origin of a new food storage tissue was necessary remains to be explained. Presumably the explanation would have to invoke either evolutionary momentum (the persistence of a trend beyond the point of usefulness) or some benefit from other effects of the genes governing a decrease in the size of the female gametophyte. As we have noted, the concept of evolutionary momentum is not in favor among most modern students of evolutionary mechanisms, who feel that any evolutionary trend must ultimately be related to survival value.

Endosperm is divided into three general types on the basis of its ontogeny. If the formation of cell walls is postponed until after a number of nuclear divisions have occurred, so that there is a free-nuclear stage in early ontogeny, the endosperm is said to be nuclear. If cell walls are formed more or less coincident with most or nearly all of the nuclear divisions, the endosperm is said to be cellular. A peculiar intermediate type, common in the Alismatidae (Helobiae), is called helobial. In this type a cell wall is laid down between the first two nuclei of the developing endosperm, after which development proceeds along the cellular pattern in one half of the endosperm, and along the nuclear pattern in the other. The distinctions among these three types are not as sharp as the descriptions would suggest; there are many transitional conditions.

There is a continuing difference of opinion about whether nuclear or cellular endosperm is the more primitive. Within the family Gentianaceae, it seems clear enough that nuclear endosperm is primitive and cellular endosperm is advanced. The family as a whole has nuclear endosperm, but two genera, *Voyria* and *Voyriella*, have cellular endosperm. These two genera are mycotrophs without chlorophyll, and they also have specialized pollen. It is highly unlikely that these advanced genera would have retained a primitive character (endosperm ontogeny) that has been lost in the remainder of the family. What is true for the Gentianaceae may or may not be true for the angiosperms in general. It may well be that the ontogeny of the endosperm is subject to reversal, and that in some other family cellular endosperm is primitive and nuclear advanced. There is, however, a loose correlation between nuclear endosperm and other primitive characters, as has been pointed out by Sporne (Phytomorphology 4: 275–278. 1954). Furthermore, it may be significant that in the living gymnosperms both the female gametophyte and the embryo have a free-nuclear stage in early ontogeny.

The adaptive significance of the different types of endosperm ontogeny is wholly unknown.

None of the many fossil seeds of pteridosperms has yet been found to contain an embryo. It is therefore inferred that at the time the seeds were shed the embryo must have been very small or even still unformed, as in the modern *Ginkgo*. Fertilization in *Ginkgo* is long delayed after pollination, sometimes until after the ripe seed has fallen to the ground.

Since pteridospermous embryos are unknown, we do not know how many cotyledons they may have had. Modern cycads have two cotyledons, as do the Gneticae. Most of the conifers, on the other hand, have several cotyledons. Although most angiosperms have only one or two cotyledons, sporadic tricotyledonous embryos occur in many genera. One genus, *Degeneria*, is typically tricotyledonous; about 7/8 of the embryos have three cotyledons, and nearly all the rest have four, with only rare individuals having only two. Inasmuch as *Degeneria* is also primitive in many other respects, it has been suggested that its embryos are also primitive and that the ancestral stock of the angiosperms had several cotyledons instead of only two. This suggestion may well be correct, but at present it is hardly more than a speculation.

Regardless of whether embryos with three or four cotyledons are considered to be primitive or advanced, as compared with dicotyledonous ones, the change from one number to the other apparently takes place at a single jump. It results from a simple change in the symmetry of organization, rather than from a gradual reduction of one of the cotyledons.

It is now agreed by all that monocotyledonous embryos are derived from dicotyledonous ones. The change may take place in any of several ways. The following examples are all drawn from the class Magnoliatae ( = dicotyledons), with the intent of reserving discussion of the origin of monocotyledony in the Liliatae for a subsequent chapter.

The simplest way for monocotyledony to originate from dicotyledony is by abortion of one of the cotyledons. Like the change from three cotyledons to two, this change can take place at a single step. *Centranthus*, in the advanced family Valerianaceae, has one, two, or three cotyledons in the embryo. Several genera of the Compositae (e.g., *Ambrosia*, *Calendula*, *Dimorphotheca*) as well as scattered genera in other families (e.g., *Impa-*

*tiens, Raphanus*) produce occasional monocotyledonous embryos, with or without a vestige of a second cotyledon. Here the change in number evidently requires only a single step. It should be noted that the presence of a vestige of the second cotyledon proves nothing about the number of steps required to achieve the reduction.

The second cotyledon may also be gradually reduced by a series of short steps. All stages of this reduction can be seen among the dicotyledons. In *Trapa, Eranthis,* and *Mamillaria,* representing three different subclasses, one cotyledon is smaller than the other. Such widely differing genera as *Claytonia, Corydalis, Pinguicula,* and *Ranunculus* embrace species with two unequal cotyledons and species which are essentially monocotyledonous.

Two cotyledons may become one by fusion along the lateral margins. This is well illustrated in the Nymphaeales. *Nymphaea* has two separate cotyledons. *Nuphar* has two unequal cotyledons, which are fused by one margin in some species and around the base in others. *Nelumbo* has two similar cotyledons fused by both margins to form a cup-shaped, bilobed unit surrounding the epicotyl.

Finally, monocotyledony can arise by gradual differentiation of two cotyledons for different functions. This sort of change is illustrated by species of *Peperomia. Peperomia pellucida* has two equal cotyledons, both of which are withdrawn from the seed during germination and become the first functional leaves of the seedling. *P. peruviana* likewise has two initially equal cotyledons, but only one of them is withdrawn from the seed and becomes foliar; the other remains within the seed coat and functions only as an absorbing organ. *P. parvifolia* has two initially unequal cotyledons. The larger one remains within the seed coat and functions as an absorbing organ; the smaller one becomes the first leaf of the seedling. The situation in *P. parvifolia* is hardly to be distinguished from that of many monocots, in which the structure resembling the second cotyledon of *Peperomia parvifolia* is interpreted as the first true leaf.

It is probably significant that newly ripe seeds of most of the more primitive families of angiosperms have a small, poorly differentiated embryo embedded in a copious endosperm. The embryo continues to grow and differentiate during the after-ripening period, but even when the seed is ready to germinate the embryo is small in comparison to the endosperm. Evolutionary progress from the primitive type of seed leads first to endospermous seeds with well differentiated embryo, and then, as we have noted, to nonendospermous seeds with the food stored in the embryo. Perisperm in addition to or in place of endosperm is also advanced rather than primitive. These changes, like other evolutionary advances in the angiosperms, have occurred independently in many different lines.

The primitive cotyledon serves both as a food-absorbing and a food-making organ. It absorbs food from the endosperm during germination, and then it is withdrawn from the seed coat and becomes photosynthetic. Evolutionary specialization of the cotyledons emphasizes food absorption (and sometimes also storage) at the expense of photosynthesis. In the garden bean the cotyledons are chiefly absorbing and storing organs, but they are also brought above the ground during germination and they become weakly photosynthetic for a short time before withering and falling off. In the garden pea the photosynthetic function has been dispensed with and the cotyledons remain below ground within the seed coat. Among the monocotyledons the cotyledon is often complex, with one part remaining in the seed and another part becoming aerial.

Although the seed coat of angiosperms is usually firm and dry, it sometimes has a fleshy covering. This fleshy covering may arise by modification of the integuments (or outer integument), or by elaboration of the funiculus. The first of these types of fleshy covering is called a sarcotesta, the second an aril. The term aril has often been loosely used to include the sarcotesta, and many taxa which are said to be arillate have a sarcotesta instead of a true aril. The same seed may have both a sarcotesta and a true aril, and careful ontogenetic study is often necessary to distinguish the one from the other.

It is uncertain whether the presence of either a sarcotesta or an aril is more primitive than its absence. Corner has maintained, in a series of papers,[19] that the truly arillate seed is primitive in the angiosperms, but his view has not been widely adopted. Other botanists have often viewed the aril and sarcotesta, which are obviously adaptations relating to seed dispersal, as having originated separately in diverse groups. Corner's view is part of a more comprehensive theory, the taxonomic consequences of which have not yet been fully evaluated. It is, however, at least compatible with most of the views here expressed on the evolutionary history of other organs.

## Fruits and Seed Dispersal

One of the most important functional characters related to fruit structure is whether the fruit opens to release the individually dispersed seeds, or whether the fruit remains closed and is dispersed whole with the included seeds. The former condition is certainly the more primitive. Follicles, legumes, and capsules generally have individually dispersed seeds. Most other fruits, including achenes, grains, samaras, and the various sorts of fleshy fruits, are generally dispersed intact. In a few common families

[19] E. J. J. Corner, The durian theory extended. II. The arillate fruit and the compound leaf. Phytomorph. 4: 152–165. 1954.

with a compound pistil the carpels (Umbelliferae) or half-carpels (Boraginaceae, Labiatae, Verbenaceae) separate at maturity and serve as one-seeded disseminules.

It is perfectly clear that the primitive fruit is a follicle, consisting of a single carpel which opens along the ventral suture at maturity, releasing the individually dispersed seeds. The seeds of many follicles have no specialized means of dispersal. Others are plumed (as in *Asclepias*) and are distributed by wind, or have a fleshy outer layer and are eaten by birds or other animals, passing through the digestive tract unharmed. We have noted that it is uncertain whether a fleshy covering on the seed is primitive or advanced. Plumed or winged seeds are clearly secondary as compared to seeds with no specialized means of dispersal.

A simple modification of the follicle is the typical legume, which opens down both the dorsal and the ventral sutures. Many follicles open for a short distance along the distal part of the dorsal suture, as well as along the whole length of the ventral suture. It requires only a progressive elongation of the dorsal opening to convert a follicle into a legume. Many legumes dehisce explosively, throwing the seeds several feet. This explosive dehiscence may well be useful, but its usefulness could hardly take effect before the transition from follicle to legume had been essentially completed. The ancestors of the legumes could not have foreseen that progressive modification of a follicle could eventually produce a fruit that could dehisce explosively. The factors governing the evolution of the follicle into a legume remain obscure.

Unicarpellate dry fruits which dehisce along both sutures (i.e., typical legumes) are very common in the family Leguminosae. Similar fruits occur in a few families, such as the Connaraceae, but are not generally called legumes. As a matter of practical convenience, a legume is defined as the fruit of a member of the Leguminosae. Although these are always unicarpellate, they are not always dehiscent. Indehiscent legumes have the same evolutionary opportunities as other indehiscent dry fruits, and some of these opportunities have been exploited. Some legumes (e.g., *Astragalus crassicarpus*) have even become fleshy. A special type of indehiscent legume is the loment, which breaks transversely into one-seeded segments. Loments are often provided with hooked hairs which adapt them to distribution by animals, or they may be thin and samaroid.

We have noted that there is a general evolutionary tendency for separate carpels to become joined into a compound pistil. This change most commonly occurs in plants with follicular fruits, and the result is to turn the fruit into a capsule. The most primitive type of capsular dehiscence is doubtless septicidal. The carpels merely separate when ripe and open in follicular fashion along the ventral suture. *Veratrum* and scattered other genera have a septicidal capsule, but this is not a common type. From the standpoint of functional engineering, loculicidal dehiscence is much more efficient than septicidal, particularly in pistils with firmly united carpels. Regardless of the causes of the continuing trend toward carpellary fusion, loculicidal dehiscence appears to be the evolutionary answer to the problem it presents in how to liberate the seeds. Some loculicidal capsules are explosively dehiscent, as in *Impatiens*, but others open more gently and merely expose the seeds.

Poricidal dehiscence is obviously derived from loculicidal by restricting the length of the opening. This tends to limit the number of seeds which are liberated at any one time, extending the period of release to several days or even weeks. The survival value of such a change is not well understood, but in any case it has occurred independently in a number of families, such as the Papaveraceae and Campanulaceae.

Circumscissile dehiscence is superimposed on other types of dehiscence, rather than resulting from stepwise modification of them. From a purely morphological standpoint there is no reason why a capsule should not at once be circumscissile and loculicidal, as it can be both septicidal and loculicidal, but in fact such combinations are rare. The advantage, if any, of circumscissile dehiscence over other types is obscure.

Indehiscent fruits have evolved from dehiscent fruits many times, both in unicarpellate and multicarpellate types. The typical achene, as seen in many members of the Ranunculaceae, is derived from the follicle by failure of dehiscence and reduction of the number of seeds to one. Achenes can also have two or more carpels, as in the Polygonaceae and Compositae — the former with superior ovary, the latter with inferior ovary.

Dehiscent fruits with only one seed, and indehiscent fruits with more than one seed are less frequently seen. The infrequency of one-seeded dehiscent fruits is easy enough to explain in teleological terms: It is pointless for the fruit to open and release a single seed; the whole fruit might just as well be the disseminule. Translating this into more acceptably mechanistic terms, one might say that the dehiscence mechanism, having no great usefulness, is easily lost because mutations affecting it are not selected against. This change is in pattern with the general evolutionary principle that useless structures tend to deteriorate through the accumulation of loss mutations. Because genes so commonly have pleiotropic effects, a mutation which is a loss mutation with regard to one character may be useful for some other effect.

The infrequency of several-seeded indehiscent dry fruits is more directly selective. In most instances it is more efficient for the seeds to be dispersed separately than collectively. Anyone who has ever had to thin beets will recognize the problems inherent in having several seeds germinate at the same place. Beets produce aggregate fruits, each derived from several flowers, but the dispersal problem is the same as that of indehiscent simple fruits containing several seeds.

The achene is the most generalized type of dry, indehiscent, one-seeded fruit. There are several possibilities for further evolution within this broad category. The achene may develop hooks or barbs which suit it to distribution by animals, e.g., the hooked stylar beak of some species of *Ranunculus*, and the retrorsely barbed pappus-awns of most species of *Bidens*. It may become flattened and winged, forming a samara. It may become enlarged and thick-walled, forming a nut. All of these changes are readily explained in terms of adaptation and survival value, and all have taken place several or many times independently in different lines.

Another type of modified achene is the caryopsis (grain), in which the seed coat is adnate to the pericarp. The vast majority of grasses have this type of fruit. Some grass fruits, such as those of *Sporobolus*, have the seed loose within the pericarp and are by morphological criteria achenes, but purely as a matter of terminology it is customary to define the caryopsis in taxonomic rather than morphologic terms. A caryopsis is the fruit of a member of the Gramineae. Fruits of a few species or genera in other families have the seed coat adnate to the pericarp but are still called achenes. It is not easy to see survival value in adnation of the seed coat to the pericarp, although the change does no harm; in general it should be selectively neutral. If selection is to be invoked as the explanation, it must be in terms of pleiotropic effect.

It should be noted that the fruit of *Sporobolus* does not appear to be primitive within the grass family. Instead it is a reversion to an ancestral type, in this one character of loose seed within the fruit. In many other respects *Sporobolus* is advanced within the family, and it is most unlikely that one primitive character persisted while so many others were changed. Many of the evolutionary trends and developments in the angiosperms appear to be selectively neutral, or merely move the plant from one available ecologic niche to another, and the frequency of reversals should not be surprising.

Fleshy fruits have evolved from dry fruits many times, both in unicarpellate and multicarpellate types. Evolution of fleshiness may proceed concurrently with suppression of dehiscence, or a dry, indehiscent fruit may evolve into a fleshy one. The drupe of *Prunus* and the coherent individual drupelets of *Rubus* are doubtless derived from dry, indehiscent fruits that had only a single seed (i.e., from achenes). On the other hand, the berries of the Solanaceae and many other families appear to be derived from capsules, with the two changes proceeding more or less simultaneously. The berry of *Actaea*, in the Ranunculaceae, is a modified follicle. Neither the dry, indehiscent, many-seeded fruit nor the dehiscent, fleshy fruit has a felicitous combination of characters, and these types are rare.

Fleshiness of fruits is at least ordinarily an adaptation to dissemination by animals (including birds). In most cases the seeds are swallowed along with the pulpy pericarp and pass through the digestive tract unharmed.

Obviously this method of seed dispersal entails the development of a protective covering for the seed that is resistant to digestion. This may be the seed coat, or the inner part of the pericarp, or even the whole pericarp. The fleshy covering of the drupe-like fruits of the Elaeagnaceae is formed by the hypanthium, and the true pericarp is wholly stony. Drupes ordinarily have only a single seed, less often two or several. Only one seed can usefully germinate at a given time and place, and natural selection does not favor many-seeded drupes. Seeds with a sarcotesa or aril, mentioned in a previous section, are ecologically similar to one-seeded, fleshy fruits.

### Interpretation of Evolutionary Trends

The nature of the forces governing evolutionary trends in the angiosperms is not entirely clear. The orthodox position among students of evolutionary theory per se is that all long-term trends must be explained in terms of survival value. Either the trend is directly useful, or the genes governing it are favored because of other effects. There have been few public challenges to this position since the flowering of the neo-Darwinian school of thought some three decades ago. The neo-Darwinian mechanism (mutation, selection, and fixation in small populations) is now so well worked out, and it so evidently provides the explanation for so much of what is going on, that one can easily believe it provides the full and complete answer.

Easily, that is, until one gets down to cases. Working plant taxonomists during the past several decades have usually paid lip-service to the concept that all evolutionary trends are selective, but their discussions of the trends in the groups they study have commonly lacked any attempt at a selectionist interpretation. Privately, and not for attribution, many of these same people will express doubt as to the universality of survival value as the only guiding force in evolution.

We have noted in our survey of evolutionary trends in the angiosperms that some of them are clearly selective, others more doubtfully so, and others appear to be selectively neutral or even mildly counter-selective. Hidden survival value and pleiotropic effect will have to be worked very hard to provide a neo-Darwinian cover for many of the changes affecting the inflorescence and the structure of the gynoecium, although some few of these changes might well be selective. The difficulty is compounded by the fact that a given trend often operates in a widely differing set of environments, so that it is most difficult to find a common environmental feature which could conceivably be governing. The problem becomes particularly acute when one deals with a trend that seems merely to lead 'round and 'round the mulberry bush. For example, all students of the Compositae believe that aggregation and reduction are widespread (though not irreversible) trends in the inflorescence of the family. In several different genera, in different tribes, one sees composite composites, with one-flowered, individually involucrate heads aggregated into a secondary head with its own common involucre. The flower head of *Dipsacus* is probably also a secondary head, consisting of an aggregation of uniflorous heads.

In order to remember, understand, and interpret what we see, our minds constantly seek patterns, not just in the diversity among angiosperms, but in everything that we consider. In this very paragraph I am trying to elucidate a pattern in human behavior. Sometimes the mind will impose a pattern where none exists, or it will arbitrarily choose one of two or more equally valid possibilities to the exclusion of others. Probably most readers have seen a drawing of a box deliberately made to be mentally reversible, so that one can at will visualize the interior or exterior view of the box from the same drawing. There are many tile floors in which one can see different patterns of blocks, according to one's wishes. In these simple cases we recognize the arbitrariness of the choice. In more complex cases the existence of alternative possibilities may escape our attention.

In the section dealing with inflorescences we noted that no existing classification of inflorescences is even approximately satisfactory. We then considered the probable evolutionary history of inflorescences in terms of the traditional classification. If some other classification of inflorescences had been used instead, the analysis of evolutionary trends would necessarily have been quite different. Certainly, then, there is an arbitrary element in our concept of evolutionary patterns in inflorescences.

Should we therefore give up all attempts at logical analysis? Of course not. The mind cannot operate without them. But we should be alert to the possibility of alternative interpretations, and we should not be disappointed if we can impose only a less than perfect order on the endless diversity of nature. A bit of skepticism about the ultimate correctness of any rational scheme will do no harm.

How do these caveats apply to a consideration of evolutionary trends in angiosperms? We have noted that many of the evolutionary trends bear little apparent relation to survival value, that the same thing often happens in a number of different groups, and that there are some reversals. It might be claimed that among the multitude of evolutionary changes which occur, our minds fasten upon those which can be arranged in stepwise progression, and that the resulting evolutionary trends are mere creations of the mind. This approach implies that each step in such an artificial sequence is useful either for its own sake or through pleiotropic effect, but that the sequential progression of steps, being purely mental and arbitrary, needs no explanation. The occurrence of similar changes in several taxa of a related group would likewise be dismissed as not requiring any explanation. I am not very happy with this outlook, but until a better understanding is reached it should perhaps not be wholly rejected.

A possibly useful variant of this approach would be to say that there is a limited number of possibilities for morphologic evolution in any one group, that some of the myriad individual selections for obscure physiological characters will pleiotropically affect these morphological characters, and that a stepwise morphological progression, an apparent evolutionary trend of no functional significance, can therefore be produced by a series of scarcely related selections. The viability of this interpretation cannot be fully tested until we know more about the chemical nature of the individual genes which govern the characters affected by evolutionary trends. If the many mutations which are serially fixed to produce a trend are chemically very different from each other, this outlook has possibilities. If, on the other hand, these mutations are chemically related, then their physiological as well as their morphological effects should fall into patterns, and any progressive morphological change caused by a pleiotropic effect should be associated with a progressive change of some other sort which has survival value. This brings us back to the fact that efforts to find adaptive trends associated pleiotropically with the seemingly nonadaptive trends in angiosperms have not been well rewarded.

The problem of the causes of evolutionary trends has been eased somewhat by the general recognition among evolutionary theorists that mutation is not mathematically at random. It is random in the sense that we cannot predict when and to what a particular gene will mutate, and in the sense that mutations are not specifically directed toward suiting the organism to its environment, but it is not random in the sense that any one mutation is as likely as any other. Different genes mutate at different rates; the kinds of mutation occurring in any one gene occur at different frequencies; and opposite mutations from one allele to another and back again occur at different frequencies. Furthermore, both the rate and the direction of mutation at any one locus are subject to influence by genes at other loci. As long ago as 1941 Dobzhansky[20] summarized this situation by saying that mutability, like other characters, is under partial genetic control. Differential mutation might be more accurately descriptive than the familiar term random mutation.

It is clear that the nature of the supply of mutations imposes limits on the evolutionary possibilities of any group, providing more raw material for selection to act on in some directions than in others. The influence of differential mutation is often so strong that a group appears to be pre-

[20] Th. Dobzhansky, *Genetics and the Origin of Species*, 2nd ed. (New York: Columbia University Press, 1941), p. 36.

disposed to evolve in a certain direction. Wernham's comments on critical tendencies, referred to in Chapter 1, are largely a reflection of these genetic facts, as is also Vavilov's law of homologous series in variation.[21]

Sometimes the direction of predisposed evolution does not appear to correlate with survival value or ecologic niches. Then we come to the question of whether mutation pressure, combined with happenstance fixation in small populations, can govern an evolutionary trend in the absence of any selective impetus. The orthodox answer is that mutation pressure can function only in combination with selection; that selection is a sine qua non of all evolutionary trends. Yet in many instances we are hard put to find even an indirect selective significance in an admitted trend.

Such problems apparently present themselves more often to botanists than to zoologists, possibly because the more tightly integrated morpho-logic-physiologic system of animals is less tolerant of casual variation. Yet it may be instructive to consider the case of the drongos, a group of tropical birds with weird and wonderful tails. Failing to find any ecologic or other selective significance in the trends they discovered, the impeccably selectionist authors of a careful study[22] of this group concluded (p. 246) that "throughout the family there is a genetic predisposition toward a lengthening of the tail, toward a modification of the outermost tail feather, and toward the development of a frontal crest, a disposition that is materialized in a certain number of species," and (p. 264) that "evolution within the family has been restricted to a realization of a limited number of trends." Certainly their interpretation of the nature of evolution in the drongos appears to be equally applicable to much of the evolution in the angiosperms, both collectively and severally as regards the individual families and orders.

## SELECTED REFERENCES

Axelrod, D. I., Origin of deciduous and evergreen habits in temperate forests. Evolution 20: 1–15. 1966.

Bailey, I. W., The development of vessels in angiosperms and its significance in morphological research. Am. Jour. Bot. 31: 421–428. 1944.

———, Evolution of tracheary tissue in land plants. Am. Jour. Bot. 40: 4–8. 1953.

———, Nodal anatomy in retrospect. Jour. Arn. Arb. 37: 269–287. 1956.

[21] N. I. Vavilov, Law of homologous series in variation. Jour. Genetics 12: 47–89. 1922.

[22] Ernst Mayr and C. Vaurie, Evolution in the family Dicruridae (birds). Evolution 2: 238–265. 1948.

————, The potentialities and limitations of wood anatomy in the study of the phylogeny and classification of angiosperms. Jour. Arn. Arb. 38: 243–254. 1957.

Bailey, I. W., and B. G. L. Swamy, The conduplicate carpel of dicotyledons and its initial trends of specialization. Am. Jour. Bot. 38: 373–379. 1951.

Barghoorn, E. S., The ontogenetic development and phylogenetic specialization of rays in the xylem of dicotyledons. I. The primitive ray structure. Am. Jour. Bot. 27: 918–928. 1940. II. Modification of the multiseriate and uniseriate rays. Am. Jour. Bot. 28: 273–282. 1941. III. The elimination of rays. Bull. Torrey Club 68: 317–325. 1941.

Bessey, C. A., The phylogenetic taxonomy of flowering plants. Ann. Mo. Bot. Gard. 2: 109–164. 1915.

Bierhorst, David W., and P. M. Zamora, Primary xylem elements and element associations of angiosperms. Am. Jour. Bot. 52: 657–710. 1966.

Carlquist, Sherwin, *Comparative Plant Anatomy*. New York: Holt, Rinehart and Winston, 1961. (Oriented toward phylogenetic interpretation, and with comprehensive bibliographies.)

————, A theory of paedomorphosis in dicotyledonous woods. Phytomorph. 12: 30–45. 1962.

Carpenter, F. M., The evolution of insects. Am. Scientist 41: 256–270. 1953.

Carr, S. G. M., and D. J. Carr, The functional significance of syncarpy. Phytomorph. 11: 249–256. 1961.

Chalk, L., The phylogenetic value of certain anatomical features of dicotyledonous woods. Ann. Bot. (N.S.) 1: 409–428. 1937.

Cheadle, V. I., Secondary growth by means of a thickening ring in certain monocotyledons. Bot. Gaz. 98: 535–555. 1937.

Corner, E. J. H., Centrifugal stamens. Jour. Arn. Arb. 27: 423–437. 1946.

Cutter, E., Recent experimental studies of the shoot apex and shoot morphogenesis. Bot. Rev. 31: 7–113. 1965. (For phyllotaxy and Fibonacci numbers)

Douglas, G. E., The inferior ovary. II. Bot. Rev. 23: 1–46. 1957.

Ehrlich, P. R., and P. H. Raven, Butterflies and plants: A study in co-evolution. Evolution 18: 586–608. 1964.

Esau, K., V. I. Cheadle, and E. M. Gifford, Comparative structure and possible trends of specialization of the phloem. Am. Jour. Bot. 40: 9–19. 1953.

Faegri, K., and L. van der Pijl, *Principles of Pollination Ecology*. Oxford: Pergamon Press, 1966.

Grant, Verne, The protection of ovules in flowering plants. Evolution 4: 179–201. 1950.

————, *The Origin of Adaptations*. New York: Columbia University Press, 1963. (A comprehensive, lucid synthesis of the current theory of evolutionary mechanics in plants.)

Haccius, B., and E. Fischer, Embryologische und histogenetische Studien an "monokotylen Dikotylen." III. *Anemone appenina* L. Österr. Bot. Zeits. 106: 373–389. 1959.

Hiepko, P., Vergleichend-morphologische und entwicklungsgeschichtliche Untersuchungen über das Perianth bei den Polycarpicae. Bot. Jahrb. 84: 359–508. 1965.

Leppik, E. E., Evolutionary relationship between entomophilous plants and anthophilous insects. Evolution 11: 466–481. 1957.

Mauritzon, J., Die Bedeutung der embryologischen Forschung für das natürliche System der Pflanzen. Lunds Univ. Arrskr. N.F. II 35(15): 1–70. 1939.

Parkin, J., The protrusion of the connective beyond the anther and its bearing on the evolution of the stamen. Phytomorph. 1: 1–18. 1951.

———, A plea for a simpler gynoecium. Phytomorph. 5: 46–57. 1955.

———, The unisexual flower again — A criticism. Phytomorph. 7: 7–9. 1957.

van der Pijl, L., Ecological aspects of flower evolution. I. Phyletic evolution. Evolution 14: 403–416. 1960. II. Zoophilous flower classes. Evolution 15· 44–59. 1961.

———, Sarcotesta, aril, pulpa and the evolution of the angiosperm fruit. Proc. K. Nederl. Akad. Wetens. Amsterdam C58: 154–161; 307–312. 1955.

———, Ecological aspects of fruit evolution. A functional study of dispersal organs. Proc. K. Nederl. Akad. Wetens. Amsterdam. C69: 597–640. 1966.

Puri, V., The role of floral anatomy in the solution of morphological problems. Bot. Rev. 17: 471–553. 1951.

———, Placentation in angiosperms. Bot. Rev. 18: 603–651. 1952.

———, Floral anatomy and the inferior ovary. Phytomorph. 2: 122–129. 1952.

———, The classical concept of the carpel: A reassessment. Jour. Indian Bot. Soc. 40: 511–524. 1961.

Rickett, H. W., The classification of inflorescences. Bot. Rev. 10: 187–231. 1944.

Richards, F., The geometry of phyllotaxis and its origin. Symp. Soc. Exp. Biol. 2: 217–245. 1948.

Solbrig, O., Rol de polinización zoófila en la evolución de las angiospermas. Bol. Soc. Argent. Bot. 11: 1–18. 1966.

Sporne, K. R., Statistics and the evolution of dicotyledons. Evolution 8: 55–64. 1954.

Stebbins, G. L., The probable growth habit of the earliest flowering plants. Ann. Mo. Bot. Gard. 52: 457–468. 1965.

———, Adaptive radiation in trends of evolution in higher plants. Vol. I, pp. 101–142, in Th. Dobzhansky, M. K. Hecht, and W. C. Steere, eds., *Evolutionary Biology*. New York: Appleton-Century-Crofts, 1967.

Thompson, D'Arcy, *On Growth and Form*, 2nd ed. New York: The Macmillan Co., 1942. (For phyllotaxy and Fibonacci numbers.)

Thorne, R. F., Some guiding principles of angiosperm phylogeny. Brittonia 10: 72–77. 1958. (A good, concise statement of well known principles which will bear repetition.)

————, Some problems and guiding principles of angiosperm phylogeny. Am. Nat. 97: 287–305. 1963.

Tippo, O., The role of wood anatomy in phylogeny. Am. Midl. Nat. 36: 362–372. 1946.

## Chapter 4

*Flowering branch of Magnolia grandiflora,
an evergreen species of the southeastern
U. S. Magnolia is the most familiar and
widespread genus among the more primitive
genera of angiosperms. U. S. Forest Service
photo by W. D. Brush.*

# The Subclasses, Orders, and Families of Dicotyledons

## Distinction Between Monocots and Dicots

It has been recognized for more than a century that the angiosperms, or flowering plants, form a natural group which consists of two subgroups. These two subgroups have usually been called monocotyledons and dicotyledons (or equivalent names with Latinized endings), from the most nearly constant of the several differences between them. As formal taxa they are here considered to be classes, called Liliatae and Magnoliatae. The terms monocotyledons and dicotyledons, or monocots and dicots, continue to be useful English names. The dicots are much the larger of the two groups. By rough estimate there are perhaps 165,000 species of dicots, and about 55,000 monocots.

Both groups occur in a wide range of habitats, but the dicotyledons are the more diverse in habit. About half of all the species of dicots are more or less woody-stemmed, and many of them are definitely trees, usually with a deliquescently branched trunk. The monocots on the other hand are predominantly herbaceous. Less than ten percent of all monocots are woody, and most of these belong to a single large family, the Palmae. Woody monocots usually have an unbranched (or sparingly branched) stem with a terminal crown of large leaves, a habit which is rare in the dicots. The difference in habit is partly a reflection of the complete absence of typical cambium in monocots, in contrast to its usual presence in dicots.

There are average differences in the underground as well as the aerial parts of monocots and dicots. In monocots the primary root soon aborts, and the mature root system is wholly adventitious. Many dicots likewise have an adventitious root system, but a primary root system, derived from the radicle, is more common.

All of the differences between monocots and dicots are subject to over-lapping or exception. The most nearly constant difference is the number of cotyledons, but as we have noted there are some dicots which have only one cotyledon. The special features of monocots are further discussed in Chapter 5. The several differences between monocots and dicots are summarized in the following table.

| DICOTS | MONOCOTS |
| --- | --- |
| Cotyledons 2 (seldom 1, 3, or 4) | Cotyledon 1 (or the embryo sometimes undifferentiated) |
| Leaves mostly net-veined | Leaves mostly parallel-veined |
| Intrafascicular cambium usually present | Intrafascicular cambium lacking; usually no cambium of any sort |
| Vascular bundles usually borne in a ring which encloses a pith | Vascular bundles generally scattered, or in 2 or more rings |
| Floral parts, when of definite number, typically borne in sets of 5, less often 4, seldom 3 (carpels often fewer) | Floral parts, when of definite number, typically borne in sets of 3, seldom 4, never 5 (carpels often fewer) |
| Pollen typically triaperturate, or of tri-aperturate-derived type, except in a few of the more primitive families | Pollen of uniaperturate or uniaper-turate-derived type |
| Mature root system either primary or ad-ventitious, or both | Mature root system wholly adventitious |

It is universally agreed that the monocots are derived from primitive dicots, and that the monocots must therefore follow rather than precede the dicots in any proper linear sequence. The solitary cotyledon, parallel-veined leaves, absence of a cambium, the disected steele, and adventitious root system of the monocots are all regarded as secondary rather than primitive characters in the angiosperms as a whole, and any plant which was more primitive than the monocots in these several respects would certainly be a dicot. The monocots are more primitive than the bulk of the dicots in having uniaperturate pollen, but several of the more primitive families of dicots also have uniaperturate pollen, so there is no problem here.

### The Subclasses of Dicots (Magnoliatae)

Takhtajan (Taxon 13: 160–164. 1964) has proposed to arrange the families and orders of angiosperms into ten subclasses, six in the Magnoliatae and four in the Liliatae. His subclasses of Magnoliatae appear to be very largely natural, and they are here adopted, with a few controversial families or orders being shifted from one subclass to another. This grouping into subclasses reflects some ideas of relationship which have been developed over the years by a number of phylogenists, and in my opinion it is a major advance in the conceptual scheme.

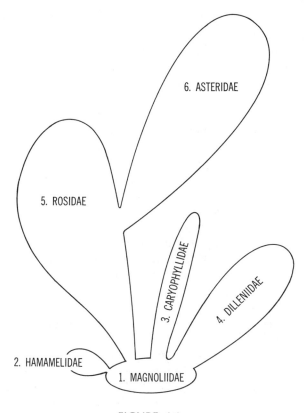

FIGURE 4.1

Probable relationships among the subclasses of Magnoliatae. The size of the balloon for each subclass is proportional to the number of species.

The six subclasses to be recognized in the Magnoliatae are the Magnoliidae, Hamamelidae, Caryophyllidae, Dilleniidae, Rosidae, and Asteridae. These subclasses can be characterized only in generalities and in terms of critical tendencies. They are groups which hang together on the basis of all of our information, but which cannot be precisely defined phenetically.

In general, it may be said that the Magnoliidae are the basal complex, from which the other angiosperms have been derived. The Magnoliidae consist of nearly all of what has been called the Ranalian complex, plus a few other families. The Hamamelidae are a group of mostly wind-pollinated families with reduced, chiefly apetalous flowers that are often borne in catkins. This subclass consists chiefly of the core of the traditional Amentiferae, after some unrelated families such as the Salicaceae have been removed. The Caryophyllidae are the Caryophyllales and their immediate allies. These families show a strong tendency toward free-central or basal placentation, and many of them have betacyanins, so far unknown in the other subclasses. Only a few are sympetalous.

The Rosidae and Dilleniidae are parallel groups which are not well distinguished morphologically. Those members of the Rosidae with numerous stamens have the stamens developing (or initiated) in centripetal sequence, so far as known. More advanced members of the group show a strong tendency toward uniovulate locules and the development of a nectariferous disk which represents a reduced whorl of stamens. The vast majority of the Rosidae are polypetalous; only a few are sympetalous or apetalous. Those of the Dilleniidae which have numerous stamens have the stamens developing centrifugally, so far as known. They usually have more than one ovule in each locule, and seldom have a nectariferous disk of staminoidal origin. Several of the more advanced families of the Dilleniidae are sympetalous, but they are in one way or another less advanced than the Asteridae. The Asteridae are the higher sympetalous families, with the stamens rarely more than as many as the corolla lobes, and with tenuinucellate ovules that have a single massive integument. Preliminary studies suggest the possibility of some fairly consistent chemical differences between the Dilleniidae, on the one hand, and the Rosidae and Asteridae on the other. Not too much weight can be placed on these results, however, because some individual families, such as the Rubiaceae, are heterogeneous in these same features. (See R. D. Gibbs, Biochemistry as an aid in establishing the relationships of some families of dicotyledonous plants. Proc. Linn. Soc. London 169: 216–230. 1958.)

The several subclasses of Magnoliatae might be treated in any of several different linear sequences, so long as the Magnoliidae come first. It is also well for the Asteridae to come last, inasmuch as these are the most highly advanced group. The remaining four subclasses are all derived separately from the Magnoliidae. I treat the Hamamelidae next after the Magnoliidae because the Hamamelidae seem to represent the remnants of a very early line of specialization that waxed and then waned as the other subclasses expanded. This is the only subclass other than the Magnoliidae to have any primitively vesselless members. The Rosidae are treated just before the Asteridae, because of the accumulating evidence that the Asteridae are derived from the Rosidae. That leaves the Caryophyllidae and Dilleniidae to be inserted between the Hamamelidae and the Rosidae. The Caryophyllidae and Dilleniidae collectively differ from the other subclasses in having the stamens, when numerous, initiated in centrifugal sequence. The Caryophyllidae, being the smaller and less highly evolved group, are treated before the Dilleniidae. Thus we come up with a sequence: Magnoliidae – Hamamelidae – Caryophyllidae – Dilleniidae – Rosidae – Asteridae. This sequence is identical to that of Takhatjan (1966).

## SYNOPTICAL ARRANGEMENT OF THE SUBCLASSES OF MAGNOLIATAE[1]

1. Plants relatively primitive, the flowers typically apocarpous, always polypetalous or apetalous and nearly always with an evident perianth, usually with numerous, centripetal stamens, the pollen always binucleate and often uniaperturate; ovules always bitegmic and crassinucellate          **I. Magnoliidae**

1. Plants more advanced in one or more respects than the Magnoliidae, and never with uniaperturate pollen

   2. Flowers more or less strongly reduced and often unisexual, the perianth poorly developed or wanting, often borne in catkins, but never forming bisexual pseudanthia, and never with numerous seeds on parietal placentae
      **II. Hamamelidae**

   2. Flowers usually more or less well developed and with an evident perianth, but if not so, then usually either grouped into bisexual pseudanthia or with numerous seeds on parietal placentae, only rarely with all of the characters of the Hamamelidae as given above

      3. Flowers polypetalous or less often apetalous or sympetalous, if sympetalous then usually either with more stamens than corolla-lobes, or with the stamens opposite the corolla lobes, or with bitegmic or crassinucellate ovules; carpels 1 to many, free or more often united

         4. Stamens, when numerous, developing in centrifugal sequence; placentation various, often parietal or free-central or basal, but often also axile; species with few

[1] It should be clearly understood that the keys presented in this book are intended primarily as conceptual aids rather than as means of identification. Characters which are difficult to observe are frequently used, and many of the numerous exceptions are necessarily ignored or minimized. For a workable but necessarily artificial and repetitive set of keys to the families of angiosperms on a worldwide basis, the student is referred to the second edition (1964) of John Hutchinson's *Families of Flowering Plants.*

stamens and axile placentation usually either bearing several or many ovules per locule, or with a sympetalous corolla, or both

5.  Pollen trinucleate; ovules bitegmic, crassinucellate, most often campylotropous or amphitropous; seeds very often with perisperm; plants usually either with betalains instead of anthocyanins, or with free-central to basal placentation, or both; only a few species are ordinary trees                    **III. Caryophyllidae**

5.  Pollen usually binucleate (notable exception: Cruciferae); ovules various, but seldom campylotropous or amphitropous; seeds seldom with perisperm; no betalains; placentation rarely free-central or basal, except in the Primulales; many species are ordinary trees
                                                        **IV. Dilleniidae**

4.  Stamens, when numerous, developing in centripetal sequence; flowers seldom with parietal placentation (notable exception: many Saxifragaceae) and also seldom (except in parasitic species) with free-central or basal placentation in a unilocular, compound ovary, but very often (especially in species with few stamens) with 2-several locules that have only 1 or 2 ovules each; flowers polypetalous or less often apetalous, only rarely sympetalous                                    **V. Rosidae**

3.  Flowers sympetalous; stamens generally isomerous with the corolla lobes or fewer, never opposite the lobes; ovules unitegmic and tenuinucellate; carpels most commonly 2, occasionally 3-5 or more            **VI. Asteridae**

### Subclass I. Magnoliidae

The subclass Magnoliidae as here defined consists of six orders, 36 families, and more than 11,000 species. About half of the species belong to the single order Magnoliales (5600), and most of the remainder belong to the Rananculales (3200). The Piperales have about 1500 species, the Aristolochiales and Papaverales about 600 each, and the Nymphaeales scarcely 100.

The Magnoliidae are clearly the most primitive subclass of the Magnoliatae and of the angiosperms as a whole, but there is no one character or easy set of characters by which the group can be defined. In general, they have a well developed perianth, which may or may not be differentiated into sepals and petals; they have numerous, centripetal stamens, and they are apocarpous; their ovules are bitegmic and crassinucellate, and their seeds usually have a small embryo and copious endosperm. Many of them have specialized secretory cells called ethereal oil cells, and many of them have uniaperturate pollen. Other highly primitive characters, such as unsealed carpels, vesselless xylem, and two-trace, unilacunar nodes, occur in various members of the subclass. All members of the group which have been investigated have binucleate pollen, but this character is of course widely distributed outside the subclass as well.

All of these characters are subject to exception. The Piperales have vestigial or no perianth; the Lactoridaceae have only six stamens; the Canellaceae are syncarpous; the Lauraceae have no endosperm; etc.

The subclass Magnoliidae conforms in large part to what has often been called the Ranalian complex. Perhaps the most notable exclusion is the family Dilleniaceae, which is often taken as part of the Ranalian complex but which is also well known to be closely allied to the group which has sometimes been called the Guttiferalean complex. On natural (i.e., phyletic plus phenetic) grounds the Dilleniaceae could be referred to either group. As a bridging family, they would be advanced in the (primitive) Ranalian complex, and basal or near-basal in the (more advanced) Guttiferalean complex. On formal morphology they are more at home in the Ranalian complex, but on the basis of transitional forms and close relationships they are better associated with the Guttiferalean complex. I here follow Takhtajan in assigning them to the advanced instead of the primitive group. The opposite course would require the provision of a new subclass name to replace Takhtajan's name Dilleniidae.

The Aristolochiales and Papaverales, which have not generally been included in the Ranalian complex, are here referred to the Magnoliidae, following Takhtajan. These orders are short evolutionary side-branches which have not given rise to other large groups. In a formal classification,

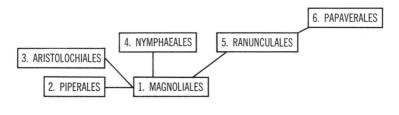

FIGURE 4.2

Probable relationships among the orders of Magnoliidae.

they must either be treated as distinct subclasses, or (as here) included in the Magnoliidae. On formal morphology they are admittedly anomalous in the Magnoliidae.

The monotypic families Tetracentraceae and Trochodendraceae are here excluded from the Magnoliidae and assigned to the Hamamelidae, following Takhtajan. Again, they could on natural grounds be assigned to either group. In the Magnoliidae they would be specialized (in spite of their vesselless wood) and peripheral; in the Hamamelidae they are near-basal. Both families have elongate idioblasts which appear to be homologous with the spherical ethereal oil cells found in many of the Magnoliidae. In *Tetracentron* these idioblasts are secretory, but in *Trochodendron* they are not.

The Sarraceniales, which were assigned to the Magnoliidae by Takhtajan, are here assigned to the Dilleniidae. The reasons for this placement are discussed under the order.

### SYNOPTICAL ARRANGEMENT OF THE ORDERS OF MAGNOLIIDAE

1.  Plants with ethereal oil cells

    2.  Plants woody; flowers normally developed, with an evident perianth of separate tepals which may or may not be differentiated into sepals and petals          **1. Magnoliales**

    2.  Plants differing in one or more respects from the above, most species herbaceous

        3.  Flowers with much reduced or no perianth, tending to be crowded into a spadix; ovules orthotropous; seeds often with perisperm          **2. Piperales**

        3.  Flowers with a well developed perianth typically consisting of a gamosepalous, more or less corolloid calyx, not crowded into a spadix; ovules anatropous; seeds never with perisperm          **3. Aristolochiales**

1.  Plants without ethereal oil cells

    4.  Plants aquatic, lacking vessels; placentation laminar; most species with uniaperturate or uniaperturate-derived pollen          **4. Nymphaeales**

    4.  Plants terrestrial or occasionally aquatic, with vessels; placentation marginal; pollen triaperturate or triaperturate-derived

5. Gynoecium mostly apocarpous or seemingly or actually unicarpellate, seldom evidently syncarpous; sepals usually
more than 2                                              **5. Ranunculales**

5. Gynoecium syncarpous; placentation parietal; sepals 2 or occasionally 3                                             **6. Papaverales**

ORDER 1. **Magnoliales**

Although most of the families grouped into the Magnoliidae are primitive as compared to other angiosperms, no one family combines all the
primitive features. The proper arrangement of the families of Magnoliidae
into orders is a source of major and continuing disagreement among students of phylogeny, although some parts of the pattern are clear enough.
There are certainly some clusters of closely related families whose relationship inter se is admitted by all. There are also a number of more or
less strongly isolated families, whose relationship to one or another of the
obvious groups is controversial.

All of the families which are generally regarded as being among the most
primitive in the Magnoliidae are woody plants with spherical ethereal oil
cells. Within the Magnoliidae, the absence of these cells appears to be
secondary rather than primitive. My nose tells me that the ethereal oils
in the Magnoliidae are all chemically allied, and different from the ethereal
oils found in other groups of flowering plants, but I have not found an
account of the chemistry. The strictly morphological differences between
the ethereal oil cells of the Magnoliidae and those of various more advanced families, such as the Rutaceae and Zingiberaceae, are apparently
minimal.

Thorne has recently[2] proposed to define the order Magnoliales on the
presence of the characteristic ethereal oil cells. This may be defensible on
natural grounds. The families with oil cells form a natural group, in the
sense that no other families have to be included in a chart showing the
relationship of these families to each other. On the other hand, most students have felt that the diversity within this group warrants the recognition
of one or more additional orders. The Piperales and Aristolochiales, in
particular, have usually been treated as distinct orders, even though they
have ethereal oil cells. In my opinion the inclusion of these groups in the
Magnoliales would result in an unnecessarily heterogeneous and conceptually difficult group, united mainly by the single technical character.
They are therefore here retained as separate orders.

The remaining families with ethereal oil cells, after the exclusion of the

[2] Robert F. Thorne. A paper presented orally at the X International Botanical Con-
gress in Edinburgh in 1964.

Piperales and Aristolochiales, have been organized in several different ways, on differing bases, by various authors. It is generally recognized that there is one cluster which includes the Magnoliaceae and some other families, and another cluster which includes the Lauraceae and some other families, but there are also several families which do not fit well into either of these two clusters, or which are placed in one cluster by some authors and in the other by others. As a possible way out of this confusion, the orders Canellales and Laurales, which I attempted to recognize at an earlier time, are here submerged in an expanded order Magnoliales.

The Magnoliales may then be characterized as those families of the Magnoliidae which have ethereal oil cells and a well developed perianth of separate tepals. They are all woody plants. Most of them are apocarpous and polystemonous, with centripetal stamens. The order as here defined consists of some 19 families and 5600 species. More than 4000 of these belong to only two families, the Annonaceae (2100), and the Lauraceae (2200). The Monimiaceae have about 450 species, the Myristicaceae about 250, and the Magnoliaceae a little over 200. None of the other families has more than about a hundred species, and six of them have only one or two species each.

The family Amborellaceae, consisting of a single New Caledonian genus and species, is referred to the Magnoliales in spite of the fact that preliminary studies indicate a possible absence of ethereal oil cells. Its relationships are generally admitted to be here; no other order could be stretched to accommodate it; and it is not sufficiently distinctive, amongst a group of primitive relic families, to warrant separation as a monotypic order.

The Magnoliales have attracted an unusual amount of study during the past several decades because of their position as ancestors or near-ancestors to the rest of the angiosperms. The work of I. W. Bailey and his students and associates has been especially significant. As a result of these studies it has become clear that no simple chart of phylogenetic relationships among the families of the order can be drawn. Several of them do not fit well into any of the clusters that have been perceived, and even within these clusters the families consist mostly of isolated end-lines no one of which can be ancestral to any of the others. We have instead a heterogeneous group of relic families, some of them consisting of only one or a few species. Each of these families has evolved slowly, independently of the others, and now presents a mixture of primitive and more advanced characters. Much of the evolution has evidently been parallel in different groups, reflecting the independent realization of initially similar evolutionary potentialities. The sharing of advanced characters is therefore no guarantee of close relationship.

Primitive characters in the angiosperms are of course not restricted to the Magnoliales, but all of the most primitive characters known in angiosperms do occur somewhere in the order. No family assigned to any other order is known to be more primitive in any respect than all of the Magnoliales. Among the notable primitive characters which occur in one or more families of the order are xylem without vessels, two-trace unilacunar nodes, beetle pollination (associated with an absence of nectar and nectaries), uniaperturate pollen, laminar stamens with embedded microsporangia, and unsealed carpels. Laminar placentation and the presence of more than two cotyledons, as seen in *Degeneria*, are perhaps also primitive.

Within the Magnoliales as here defined, the Magnoliaceae, Winteraceae, Degeneriaceae, Himantandraceae, and Annonaceae form a cluster which has been evident to most recent students of the group, although Hutchinson considers the Annonaceae to stand somewhat apart from the others. These families have hypogynous, nearly always apocarpous flowers with mostly numerous stamens that frequently are laminar and not divided into filament and anther; the pollen is uniaperturate, and the seeds have a copious endosperm. These are all primitive characters in the order; other families, except the Austrobaileyaceae, are more advanced in one feature or another. These families are also alike in having trilacunar (or pentalacunar to multilacunar) nodes. As we have seen, there is some doubt as to whether the trilacunar or the unilacunar node is the more primitive, but the odds seem to be with the two-trace, unilacunar type. It would appear, therefore, that this group of five families should not be regarded as ancestral to the two groups next considered, which have two-trace, unilacunar nodes.

Two small families, the Schisandraceae and Illiciaceae, form an obvious, closely knit group which differs from the magnoliaceous group in having unilacunar nodes and triaperturate or more advanced pollen. They also have typical stamens, with well differentiated filament and anther. If our analysis of the characters is correct, the Schisandraceae and Illiciaceae are more primitive than the magnoliaceous group in nodal anatomy, but more advanced in pollen structure. The Schisandraceae and Illiciaceae might with some reason be united into a single family, although neither is ancestral to the other.

A third group which has been evident to most recent students consists of the Amborellaceae, Trimeniaceae, Monimiaceae, Calycanthaceae, Lauraceae, Gomortegaceae, and Hernandiaceae, although there have been some doubts about the Calycanthaceae. These families have unilacunar nodes (usually with two traces), specialized pollen which is usually neither uniaperturate nor triaperturate, usually more or less perigynous to epigynous flowers, and only one (seldom two) ovules per carpel. Most of the species

lack endosperm. Obviously this group cannot be considered either ancestral to or derived from either of the other two, since it combines primitive
and advanced characters in different organs.

Five families remain unassigned (in this discussion) amongst the three
preceding groups. These are the Austrobaileyaceae, Myristicaceae, Eupomatiaceae, Lactoridaceae, and Canellaceae. None of them appears to be
closely allied to any of the other four nor to the Schisandraceae-Illiciaceae
group, but three of them combine, in one way or another, characters of
the magnoliaceous and lauraceous groups.

The Austrobaileyaceae have been considered by some students to be
allied to the lauraceous cluster of families, but they are more primitive than
that group in having hypogynous flowers with laminar stamens, uniaperturate pollen, and several ovules per carpel, in all of which respects they
resemble the magnoliaceous group. They differ from the magnoliaceous
group, however, in having two-trace, unilacunar nodes, like the lauraceous
group. They cannot be considered ancestral to the other groups because
they are lianas with vessels and closed (rather than merely conduplicate)
carpels. They seem to represent an isolated end-line, parallel to and in
between the magnoliaceous and lauraceous lines.

The Myristicaceae have trilacunar nodes, hypogynous flowers, and uniaperturate pollen, like the magnoliaceous group. They have nevertheless
been associated by several students with the lauraceous group, partly because both the perianth and the gynoecium resemble those of the Lauraceae. In both families the perianth is usually small and inconspicuous,
without a clear differentiation into sepals and petals, and the uniovulate,
seemingly unicarpellate ovary has often been interpreted as pseudomonomerous. The gynoecial structure in both families is still controversial; in
the current Engler Syllabus the Lauraceae are interpreted as being pseudomonomerous, but the Myristicaceae, even though they have two stigmas,
are interpreted as being strictly monomerous (as to the gynoecium). The
Myristicaceae are in any case well isolated from any other family by the
peculiar androecium, in which the filaments of the numerous stamens in
the male flowers are united into a solid column. I here hazard the guess
that the resemblance between the Myristicaceae and Lauraceae is secondary, and that the closer relationship of the Myristicaceae is with the
magnoliaceous group of families. In a paper published after this book was
essentially completed, Wilson and Maculans (cited below) consider the
gynoecium of the Myristicaceae to be truly monomerous, and they suggest
a possible relationship to the Canellaceae.

The Eupomatiaceae, consisting of only two South Pacific species, present
still another combination of the characters of the magnoliaceous and
lauraceous clusters of families. The multilacunar nodes and numerous
ovules would seem to ally them to the magnoliaceous group, but they have

perigynous flowers and biaperturate pollen, suggesting an affinity to the lauraceous group. Most authors have associated the Eupomatiaceae with the Annonaceae (perhaps because of the ruminate endosperm, an otherwise rare character) and thus ultimately with the magnoliaceous group. I am not entirely happy with this assignment. If the Eupomatiaceae are assigned to the magnoliaceous group, then the nodal anatomy is the only remaining character by which the two groups differ. In view of the fact that only a relatively few species in each group have been investigated for nodal anatomy, we should be cautious about assuming that this character will prove to be any more absolute than the others. On the other hand, the evidence for associating the Eupomatiaceae with the lauraceous instead of the magnoliaceous group is certainly not overwhelming; they are anomalous in either place.

The family Lactoridaceae consists of a single species, restricted to the Juan Fernandez Islands, off the western coast of South America. The plant is a shrub with small, simple, alternate, stipulate leaves, two-trace, unilacunar nodes, and small, trimerous, hypogynous, axillary flowers. There are three sepals, no petals, six stamens, and three essentially separate carpels, each with several ovules. The pollen is uniaperturate, and the seeds have a copious endosperm. *Lactoris* clearly belongs to the Magnoliidae, but beyond that its relationships are debatable. In the current Engler Syllabus it is doubtfully referred to the Piperales, and indeed the flower would look much like that of *Saururus* in the Piperales if the sepals aborted. However, the axillary flowers with well developed (if small) sepals and anatropous ovules would make it anomalous at best in that order. Furthermore, it differs from typical members of the Piperales in enough other characters (woody; apocarpous; ovules several; no perisperm) so that one cannot consider that the ensemble of characters calls for it to be placed there in spite of a few key differences. On the other hand, the small, three-membered perianth and few stamens are anomalous in the Magnoliales. In other respects *Lactoris* has much in common with some members of the Magnoliales. All in all, it seems most appropriate to consider the Lactoridaceae as a rather isolated family of the Magnoliales, related eventually to the Magnoliaceae and perhaps also to the ancestors of the Piperales.

The Canellaceae have in the past often been referred to the vicinity of the Violales in the subclass Dilleniidae. It is now generally agreed, however, that on the ensemble of characters they belong with the Magnoliidae, in spite of the syncarpous ovary. The trilacunar nodes, hypogynous flowers, mostly uniaperturate pollen and endospermous seeds of the Canellaceae ally them to the magnoliaceous rather than the lauraceous cluster of families in the Magnoliales. The compound ovary is of course an advanced character, but its significance in setting the Canellaceae apart from the other families is largely vitiated by the existence of two similarly syncarpous genera (*Isoloma, Monodora*) in the Annonaceae.

The three family-clusters which we have noted within the Magnoliales conform fairly well to the orders Magnoliales, Laurales, and Illiciales of Takhtajan (1966). However, he refers the Eupomatiaceae to the Magnoliales, and the Lactoridaceae to the Laurales, in contrast to the assignments here tentatively proposed. If these two families and the three others which we have considered to be of doubtful affinities did not exist, I would not hesitate to use three orders instead of one for the remaining families here assigned to the Magnoliales.

The Magnoliales occupy no particular ecologic niche, in comparison with other angiosperms, or if they do its characters are so recondite that they have so far escaped detection. They occur mostly in tropical and warm-temperate regions with a moist, equable climate, but this is not unusual among angiosperms in general. It may be significant that nine of the 19 families of the order are largely or wholly confined to the southwestern Pacific region, especially the tropical islands. This concentration has led many botanists to assume that the southwestern Pacific is the cradle of the angiosperms, as indeed it may well be. On the other hand, this area may merely have suffered less climatic disturbance and been less subject to invasion by better-adapted competitors than other parts of the tropics.

### SYNOPTICAL ARRANGEMENT OF THE FAMILIES OF MAGNOLIALES

1. Flowers hypogynous; endosperm well developed; most families with several or numerous ovules per carpel; pollen mostly uniaperturate or triaperturate

    2. Pollen uniaperturate; stamens often more or less laminar

       3. Nodes unilacunar, two-trace; stamens 6-9

          4. Stamens laminar; tepals numerous; carpels 6-14; lianas with opposite, exstipulate leaves     **1. Austrobaileyaceae**

          4. Stamens with normal filament and anther; tepals 3; carpels 3; erect shrubs with alternate, stipulate leaves
          **2. Lactoridaceae**

       3. Nodes trilacunar or multilacunar; stamens few to more often numerous

          5. Leaves stipulate     **3. Magnoliaceae**

5.  Leaves exstipulate

    6.  Vessels absent; stamens usually laminar; endosperm not ruminate    **4. Winteraceae**

    6.  Vessels present; stamens and endosperm various

        7.  Stamens laminar; perianth not trimerous

            8.  Carpel solitary, pluriovulate, unsealed; endosperm ruminate    **5. Degeneriaceae**

            8.  Carpels several, uniovulate, largely or wholly closed; endosperm not ruminate    **6. Himantandraceae**

        7.  Stamens not laminar; perianth most commonly trimerous

            9.  Stamens distinct, with short, thick filament and with the connective surpassing the anther and enlarged above it; endosperm ruminate; ovules 1-many    **7. Annonaceae**

            9.  Stamens with the filaments united; endosperm not ruminate

                10.  Carpels separate; ovule solitary; flowers unisexual; filaments united into a solid column    **8. Myristicaceae**

                10.  Carpels united into a compound pistil with parietal placentation; ovules 2-many; flowers perfect; filaments united into a tube    **9. Canellaceae**

2.  Pollen triaperturate; stamens with normal filament and anther; nodes unilacunar; leaves alternate, exstipulate

    11.  Flowers perfect; trees; carpels in one whorl; fruit follicular    **10. Illiciaceae**

    11.  Flowers unisexual; lianas; carpels spirally arranged, becoming fleshy in fruit    **11. Schisandraceae**

1. Flowers perigynous to epigynous (except in Amborellaceae and Trimeniaceae); ovules only 1 or 2 per carpel (except in Eupomatiaceae); pollen various but seldom either uniaperturate or triaperturate; stipules none

  12. Nodes multilacunar; ovules numerous; endosperm ruminate; leaves alternate; pollen biaperturate   **12. Eupomatiaceae**

  12. Nodes unilacunar; ovule 1 (or 2 but only 1 maturing); endosperm, when present, not ruminate

    13. Vessels absent; stamens numerous; leaves alternate
                                                  **13. Amborellaceae**

    13. Vessels present; stamens numerous or often few

      14. Endosperm present; leaves opposite

        15. Ovary superior; flowers mostly unisexual

          16. Flowers hypogynous; carpel 1 (2)   **14. Trimeniaceae**

          16. Flowers perigynous; carpels (1) several to many
                                                  **15. Monimiaceae**

        15. Ovary inferior; flowers perfect; carpels 2-3
                                                  **16. Gomortegaceae**
      14. Endosperm wanting

        17. Carpels numerous; pollen biaperturate or rarely triaperturate; leaves opposite   **17. Calycanthaceae**

        17. Carpel 1 (or seemingly so); pollen nonaperturate or rarely uniaperturate; leaves mostly alternate

          18. Ovary superior                           **18. Lauraceae**

          18. Ovary inferior                        **19. Hernandiaceae**

## SELECTED REFERENCES

Bailey, I. W., and C. G. Nast, The comparative morphology of the Winteraceae. Journ. Arn. Arb. 24: 340–346; 472–481. 1943. 25: 97–103;

215–221; 342–348; 454–466. 1944. 26: 37–47 (summary and conclusions). 1945.

————, Morphology and relationships of *Illicium*, *Schisandra* and *Kadsura*. Jour. Arn. Arb. 29: 77–89. 1948.

Bailey, I. W., C. G. Nast, and A. C. Smith, The family Himantandraceae. Jour. Arn. Arb. 24: 190–206. 1943.

Bailey, I. W., and A. C. Smith, Degeneriaceae, a new family of flowering plants from Fiji. Jour. Arn. Arb. 23: 356–365. 1942.

Bailey, I. W., and B. G. L. Swamy, *Amborella trichopoda*, a new morphological type of vesselless dicotyledon. Jour. Arn. Arb. 29: 245–254. 1948.

————, The morphology and relationships of *Austrobaileya*. Jour. Arn. Arb. 30: 211–226. 1949.

————, The conduplicate carpel of dicotyledons and its initial trends of specialization. Am. Jour. Bot. 38: 373–379. 1951.

Canright, J. E., The comparative morphology and relationships of the Magnoliaceae. I. Trends of specialization in the stamens. Am. Jour. Bot. 39: 484–497. 1952. II. Significance of the pollen. Phytomorph. 3: 355–365. 1953. III. Carpels. Am. Jour. Bot. 47: 145–155. 1960. IV. Wood and nodal anatomy. Jour. Arn. Arb. 36: 119–140. 1955.

————, Contributions of pollen morphology to the phylogeny of some Ranalean families. Grana Palyn. 4: 64–72. 1963.

Carlquist, S., Morphology and relationships of Lactoridaceae. Aliso 5: 421–435. 1964.

Fahn, A., and I. W. Bailey, The nodal anatomy and the primary vascular cylinder of the Calycanthaceae. Jour. Arn. Arb. 38: 107–117. 1957.

Hiepko, P., Vergleichend-morphologische und entwicklungsgeschichtliche Untersuchungen über das Perianth bei den Polycarpicae. Bot. Jahrb. 84: 359–508. 1965.

Hotchkiss, A. T., Pollen and pollination in the Eupomatiaceae. Proc. Linn. Soc. N. S. Wales. 83: 86–91. 1958.

Kasapligil, B., Morphological and ontogenetic studies on *Umbellularia californica* Nutt. and *Laurus nobilis* L. Univ. Calif. Pub. Bot. 25: 115–240. 1951.

Leinfellner, W., Wie sind die Winteraceen-Karpelle tasächlich gebaut? Österr. Bot. Zeits. 112: 554–575. 1965. 113: 84–95; 245–264. 1966.

————, Über die Karpelle verschiedener Magnoliales. Österr. Bot. Zeits. 113: 383–389; 440–458; 563–569. 1966.

Money, L. L., I. W. Bailey, and B. G. L. Swamy, The morphology and relationships of the Monimiaceae. Jour. Arn. Arb. 31: 372–404. 1950.

Nair, N. C., and P. N. Bahl, Vascular anatomy of the flower of *Myristica malabarica* Lamk. Phytomorph. 6: 127–134. 1956.

Periasamy, K., and B. G. L. Swamy, Studies in the Annonaceae. I. Microsporogensis in *Cananga odorata* and *Miliusa wightiana.* Phytomorph. 9: 251–263. 1959.

Sastri, R. L. N., Studies in the Lauraceae. IV. Comparative embryology and phylogeny. Ann. Bot. N. S. 27: 425–433. 1963. V. Comparative morphology of the flower. Ann. Bot. N. S. 29: 39–44. 1965

Schaeppi, H., Morphologische Untersuchungen an der Karpellen der Calycanthaceae. Phytomorph. 3: 112–118. 1953.

Shutts, C. F., Wood anatomy of Hernandiaceae and Gyrocarpaceae. Trop. Woods 113: 85–123. 1960.

Smith, A. C., The families Illiciaceae and Schisandraceae. Sargentia 7: 1–224. 1947.

Stern, W. L., Comparative anatomy of xylem and phylogeny of the Lauraceae. Trop. Woods. 100: 1–72. 1954.

————, Xylem anatomy and relationships of Gomortegaceae. Am. Jour. Bot. 42: 874–885. 1955.

Swamy, B. G. L., Further contributions to the morphology of the Degeneriaceae. Jour. Arn. Arb. 30: 10–38. 1949.

Tucker, S. C., and E. M. Gifford, Jr., Carpel vascularization of *Drimys lanceolata.* Phytomorph. 14: 197–203. 1964.

Van der Wyk, R. W., and J. E. Canright, The anatomy and relationships of the Annonaceae. Trop. Woods. 104: 1–24. 1956.

Wilson, T. K., Comparative morphology of the Canellaceae. IV. Floral morphology and conclusions. Am. Jour. Bot. 53: 336–343. 1966.

Wilson, T. K., and L. M. Maculans, The morphology of the Myristicaceae. I. Flowers of *Myristica fragrans* and *M. malabarica.* Am. Jour. Bot. 54: 214–220. 1967.

Wood, C. E., Jr., The genera of woody Ranales in the southeastern United States. Jour. Arn. Arb. 39: 296–346. 1958.

ORDER 2. **Piperales**

The order Piperales as here defined consists of three families and about 1500 species, the vast majority of them belonging to the single family Piperaceae. The Chloranthaceae have about 70 species, and the Saururaceae only about five.

The Piperales are Magnoliidae with reduced, crowded flowers and orthotropous ovules. Vegetatively they resemble the Magnoliales in having ethereal oil cells, but they differ in that most species are herbaceous or nearly so. Only a few are arborescent. Two of the families (Piperaceae, Saururaceae) have a copious perisperm (which is unknown in the Magnoliales) and only scanty endosperm. The Chloranthaceae stand somewhat apart from the other two families, and might perhaps as well be referred to the Magnoliales. However, their reduced perianth and orthotropous ovules would make them as anomalous in that order as they are (on other grounds) in the Piperales.

The Piperales are clearly derived from the Magnoliales. All of the features by which the Piperales differ from the Magnoliales represent evolutionary advances. A plant more primitive than the Piperales in all the features by which the two orders differ would surely be referred to the Magnoliales. In spite of their advances in other characters, the Piperales retain the primitive uniaperturate pollen of their magnolialean ancestors.

Within the Magnoliales, the Lactoridaceae probably represent the closest approach to the Piperales. *Lactoris* cannot be ancestral to the Piperales, however, because it has well developed vessels, whereas one genus (*Sarcandra*) of the Chloranthaceae is vesselless.

Many of the Piperales are herbs of moist tropical forests, but they are by no means confined to this habitat, nor do they occupy it to the exclusion of other groups.

### SYNOPTICAL ARRANGEMENT OF THE FAMILIES OF PIPERALES

1. Nodes unilacunar; leaves opposite; ovary more or less inferior, unicarpellate, with a single ovule; seed with copious endosperm and no perisperm        **1. Chloranthaceae**

1. Nodes trilacunar or multilacunar; leaves mostly alternate; ovary superior, except in some Saururaceae; seeds with copious perisperm and scanty endosperm

    2. Ovules (1) 2–10 per carpel; carpels distinct or united into a compound ovary with parietal or modified laminar placentation
        **2. Saururaceae**

    2. Ovule solitary in the single locule of the compound or pseudomonomerous pistil        **3. Piperaceae**

### SELECTED REFERENCES

Balfour, E., The development of the vascular system of *Macropiper excelsum* Forst. Phytomorph. 7: 354–364. 1957. 8: 224–233. 1958.

Murty, Y. S., Studies in the order Piperales. II. A contribution to the study of vascular anatomy of the flower of *Peperomia*. Jour. Indian Bot. Soc. 37: 474–491. 1958. III. A contribution to the study of the floral morphology of some species of *Peperomia*. Jour. Indian Bot. Soc. 38: 120–139. 1959. V. A contribution to the study of floral morphology of some species of *Piper*. Proc. Indian Acad. Sci. B49: 52–65. 1959. VII. A contribution to the study of the morphology of *Saururus cernuus*. L. Jour. Indian Bot. Soc. 38: 195–203. 1959.

Raju, M. V. S., Morphology and anatomy of the Saururaceae. I. Floral anatomy and embryology. Ann. Mo. Bot. Gard. 48: 107–124. 1961.

Swamy, B. G. L., The morphology and relationships of the Chloranthaceae. Jour. Arn. Arb. 34: 375–408. 1953.

Swamy, B. G. L., and I. W. Bailey, *Sarcandra*, a vesselless genus of Chloranthaceae. Jour. Arn. Arb. 31: 117–129. 1950.

Vijayaraghavan, M. R., Morphology and embryology of a vesselless dicotyledon — *Sarcandra irvingbaileyi* Swamy, and systematic position of the Chloranthaceae. Phytomorph. 14: 429–441. 1964.

ORDER 3. **Aristolochiales**

The order Aristolochiales consists of a single family, the Aristolochiaceae, with some 600 species. The Hydnoraceae and Rafflesiaceae, which have sometimes been referred to the Aristolochiales, are here referred to the Rafflesiales, and the Nepenthaceae are referred to the Sarraceniales.

The Aristolochiales present an unusual combination of primitive and advanced features. They have uniaperturate or nonaperturate (never triaperturate) pollen. Within the dicotyledons uniaperturate pollen is known only in the subclass Magnoliidae. Their seeds have a small embryo and copious endosperm. Such seeds are not restricted to the Magnoliidae, but they are more common in that subclass than in others, and they do represent the primitive type. The Aristolochiales have ethereal oil cells like those of the Magnoliales, and we have noted that this is also probably a primitive character. On the other hand, they have strongly perigynous to epigynous, usually apetalous flowers, usually with united carpels. Apetalous species often have the sepals joined into a highly irregular, corolloid calyx. They are variously herbaceous to woody (but scarcely arborescent) and often climbing, but both the shrubs and the woody lianas appear to have herbaceous ancestors within the group; i.e., woodiness in this order is secondary rather than primitive. The stamens range from numerous and spiralled to few and unicyclic.

The monotypic genus *Saruma* helps to bridge the otherwise wide gap between typical members of the Aristolochiales and more characteristic members of the subclass Magnoliidae. *Saruma* has perigynous rather than epigynous flowers, with well developed petals as well as sepals, and it has several essentially distinct carpels which ripen into follicles. *Saruma* is perfectly at home in the subclass Magnoliidae, and its affinity to the other members of its family (Aristolochiaceae) is also widely accepted.

In recent years it has been generally agreed that the ancestry of the Aristolochiales must be sought within the Magnoliidae rather than in any more advanced group. Within the Magnoliidae, the Magnoliales, Piperales, and Ranunculales as here defined have been suggested as possible ancestors. The uniaperturate pollen and the presence of ethereal oil cells in the Aristolochiales would seem to rule out the Ranunculales as possible ancestors. A certain habital similarity of *Asarum* (Aristolochiaceae) to species of *Piper* and *Peperomia* (Piperaceae), as well as the similarity of the pollen of *Saruma* to that of some members of the Chloranthaceae (Piperales) lend some substance to the thought of a possible relationship between these two orders. On the other hand, all known members of the Piperales are already too far advanced in their own type of floral reduction to serve as possible ancestors of the Aristolochiales. The Aristolochiales must therefore be taken as derived directly from the Magnoliales, possibly from ancestors similar to those which gave rise to the Piperales.

Although the Aristolochiales do not present the wide range of diverse ecological types found in some of the larger orders, they do not fall into any one type, nor do they as a group occupy any important niche to the exclusion of other kinds of plants. Many of them are tropical woody vines or scandent shrubs, but there are many other tropical woody vines in other orders, and the tropical species of *Aristolochia* have little in common ecologically with the species of *Asarum* which are familiar as low, rhizomatous herbs of the forest floor in temperate regions.

## SELECTED REFERENCES

Daumann, E., Zur Kenntnis der Blütennektarien von *Aristolochia*. Preslia 31: 359–372. 1959.

Gregory, M. P., A phyletic rearrangement in the Aristolochiaceae. Am. Jour. Bot. 43: 110–122. 1956.

Hegnauer, R., Chemotaxonomische Betrachtungen. II. Phytochemische Hinweise für die Stellung der Aristolochiaceae in System der Dicotyledonen. Pharmacie 15: 634–642. 1960.

Johri, B. M., and S. P. Bhatnagar, A contribution to the morphology and life history of *Aristolochia*. Phytomorph. 5: 123–137. 1955.

Lorch, J. W., The perianth of *Aristolochia* — a new interpretation. Evolution 13: 415–416. 1959.

Nair, N. C., and K. R. Narayanan, Studies on the Aristolochiaceae. I. Nodal and floral anatomy. Proc. Nat. Inst. Sci. India B28: 211–227. 1962.

Wyatt, R. L., An embryological study of four species of *Asarum*. Jour. Elisha Mitchell Soc. 71: 64–82. 1955.

ORDER 4. **Nymphaeales**

The order Nymphaeales consists of only three families, the Nymphaeaceae, Nelumbonaceae, and Ceratophyllaceae, with scarcely a hundred species in all. They are all aquatics which lack vessels. They have laminar (rather than marginal) placentation, and all of them except *Nelumbo* have uniaperturate or uniaperturate-derived pollen.

The Nelumbonaceae have traditionally been included in the Nymphaeaceae, but the differences between them seem so fundamental as to call for at least familial segregation. Li (cited below) and subsequently Takhtajan (1959) have even taken the Nelumbonales as a separate order, with one family, one genus, and two species. The Nelumbonaceae are indeed so isolated taxonomically that their affinities are not entirely certain. I choose to follow the traditional concept that they are related to the Nymphaeaceae sens. strict., in preference to leaving them as homeless orphans. The similarities in habitat, habit, aspect of the flowers, and placentation might conceivably reflect mere parallelism, but there is no reason to believe that they must. In any case the ancestors of the two families must have had much in common. The triaperturate pollen of *Nelumbo*, as opposed to the uniaperturate or uniaperturate-derived pollen[3] of the rest of the order, need cause no great concern. *Nelumbo* is advanced in other characters as well, and it is perfectly clear that triaperturate pollen has originated from uniaperturate more than once within the Magnoliidae.

Two other small families, the Cabombaceae (eight spp) and the Barclayaceae (three spp) have sometimes been segregated from the Nymphaeaceae. Here the relationships are seldom disputed, only the rank at which the agreed groups should be received. From a phylogenetic standpoint, either treatment is acceptable.

On purely phenetic grounds, the order Nymphaeales might with some justification be treated as a suborder of the Ranunculales. The Ranunculales as thus expanded would be no more diverse than the Magnoliales, which comfortably accommodate vesseliferous and vesselless types as well as groups with uniaperturate and triaperturate pollen. On phylogenetic grounds, however, such a treatment is unacceptable. The Ranunculales, with triaperature pollen, appear to be related to the families of Magnoliales which also have triaperature pollen. The ancestry of the Nymphaeales, on the other hand, must be sought in the uniaperturate families. The laminar stamens with embedded microsporangia, seen in some of the Nymphaeaceae, are a very primitive feature, found otherwise only in a few families of the Magnoliales. Furthermore, the laminar placentation of the Nymphaeales is very probably a primitive survival, albeit sufficiently specialized in its own way so that the more ordinary marginal placentation of the Ranunculales could hardly have been derived from it. It therefore

---

[3] I use the term uniaperaturate-derived as equivalent to "monocolpate-derived" of much recent literature, for lack of a better alternative. Triaperturate pollen is also derived from uniaperturate pollen, but triaperturate pollen and its derivatives are specifically excluded from the concept intended by the terms uniaperturate-derived and monocolpate-derived.

seems unlikely that the Nymphaeales and Ranunculales have any common ancestor short of some unknown, extinct, highly primitive member of the Magnoliales.

The absence of vessels from the Nymphaeales has been variously interpreted as a primitive survival or a reduction related to the aquatic habitat. In recent years Takhtajan (1959) has taken the former point of view, and Cheadle (Phytomorphology 3: 23-44. 1953) the latter. If the vesselless condition is primitive, it furnishes one more reason to dissociate the Nymphaeales from the Ranunculales.

I do not feel competent to judge the technical anatomical evidence critically, but I incline toward the view that the absence of vessels here is secondary rather than primitive. I suggest that the vesselless and generally primitive xylem structure of the Nymphaeales results from a synchronic disharmony in which maturation of the xylem is progressively postponed, and ultimately postponed out of existence, so that the earliest ontogenetic stages, resembling also early phylogenetic stages, are fixed as the mature type. This type of synchronic disharmony is comparable though not identical to the type of synchronic disharmony known as neoteny. Neoteny is widely believed to be an important aspect of angiosperm evolution, being involved in the change from the woody to the herbaceous habit.

The absence of cambium from the Nympaeales is surely secondary rather than primitive. If the terrestrial ancestors of the Nymphaeales had vessels only in the secondary xylem, as many of the more primitive vesseliferous angiosperms do, then reduction and loss of the cambium would in itself have deprived the Nymphaeales of vessels.

It is clear enough that the aquatic habitat of the Nymphaeales provides no selective pressure toward the retention of vessels in the xylem. Deterioration of complex structures when the selective pressure for them declines is a well known evolutionary phenomenon. A logical next step in the process of reduction of the xylem would be the disappearance of the cells which suggest to anatomists that simplification has occurred. It is interesting that the Nymphaeales are often regarded as near-ancestral to the monocots, and that Cheadle finds no anatomical evidence that primitive, aquatic, vesselless monocots such as some of the Alismatales ever had a vesseliferous ancestry.

The Ceratophyllaceae are obviously more advanced than the Nymphaeaceae and presumably derived from them. Every feature which helps to mark the Ceratophyllaceae as a distinctive group could logically have been derived from a nymphaeaceous ancestry.

Unlike many other orders of angiosperms, the Nymphaeales do have a distinctive ecological niche. They are plants of quiet fresh water, usually with floating or submersed leaves. They do not occupy this niche to the exclusion of other groups, however. The Menyanthaceae, especially the genus *Nymphoides*, in a wholly different order and subclass, occupy similar habitats and are habitally very much like members of the Nymphaeales. Many of the Alismatales are also ecologically similar to the Nymphaeales.

## SYNOPTICAL ARRANGEMENT OF THE FAMILIES OF NYMPHAEALES

1.  Plants with roots; flowers perfect; carpels 1–many; leaves alternate, entire to dissected; integuments 2

    2.  Pollen of uniaperturate or uniaperturate-derived type; seeds with small embryo, some endosperm, and abundant perisperm    **1. Nymphaeaceae**

    2.  Pollen triaperturate; seeds with large embryo and no endosperm or perisperm    **2. Nelumbonaceae**

1.  Plants rootless, free-floating; flowers unisexual; carpel 1; leaves whorled, cleft, with slender segments; integument 1; pollen uniaperturate    **3. Ceratophyllaceae**

## SELECTED REFERENCES

Li, H. L., Classification and phylogeny of the Nymphaeaceae and allied families. Am. Midl. Nat. 54: 33–41. 1955.

Moseley, F. M., Morphological studies of the Nymphaeaceae. I. The nature of the stamens. Phytomorph. 8: 1–29. 1958. II. The flower of *Nymphaea*. Bot. Gaz. 122: 233–259. 1961. III. The floral anatomy of *Nuphar*. Phytomorph. 15: 54–84. 1965.

Wood, C. E., Jr., The genera of the Nymphaeaceae and Ceratophyllaceae in the southeastern United States. Jour. Arn. Arb. 40: 94–112. 1959.

### ORDER 5. **Ranunculales**

The order Ranunculales as here defined consists of eight families with about 3200 species. Nearly 3100 of these belong to only three of the families, the Ranunculaceae (2000), Berberidaceae (650), and Menispermaceae (425). These three families, together with the smaller families Circaeasteraceae (including *Kingdonia*) and Lardizabalaceae (including Sargentodoxaceae), form a group whose mutual affinity seems to be well established. The remaining three families (Coriariaceae, Corynocarpaceae, and Sabiaceae) are only doubtfully referred to here. They might with some reason be referred to the Rosidae, but their affinities within that subclass would be obscure. Their wood anatomy suggests that they may be only secondarily woody, like other woody members of the Ranunculales. It may be noteworthy that the glycosides of the Coriariaceae are very much like some of those in the Menispermaceae. The pollen morphology of the Corynocarpaceae would be compatible with the assignment of this family to the Rosales.

The name Ranunculales is here adopted in preference to the more familiar term Ranales, following the current Engler Syllabus. Inasmuch as the order takes its name eventually from *Ranunculus* rather than from *Rana*, the longer form is etymologically preferable. Furthermore, the name Ranales has often been used in a much broader sense, and the change in the form of the name may serve as a reminder of the change in contents. Alternatively, one might adopt the name Berberidales, but more than half of the species of the order belong to the family Ranunculaceae, and it seems appropriate to base the ordinal name on this large and familiar family.

The Ranunculales are the herbaceous (or secondarily woody) equivalent of the Magnoliales, from which they further differ in lacking ethereal oil cells and in lacking some of the primitive features (unsealed carpels, uniaperturate pollen, beetle pollination, laminar stamens, absence of vessels) found in various members of the Magnoliales. Woody members of the Ranunculales have wide or very wide wood rays and are probably only secondarily woody. Many members of the order contain isoquinoline alkaloids, most notably berberin, which occurs in many of the Berberidaceae and Menispermaceae and some few of the Ranunculaceae (e.g., *Hydrastis*, *Xanthorhiza*). Berberin is also known in the Papaverales, Annonaceae, Lauraceae, and Rutaceae.

The Ranunculales are pretty clearly derived from the Magnoliales. Within the Magnoliales they find their closest allies in the Illiciaceae and Schisandraceae, which have alternate, exstipulate leaves, hypogynous flowers, separate carpels, triaperturate pollen, and mostly numerous stamens with normal filament and anther. Collectively these two families show all of the more primitive characters found in the Ranunculales, but individually each of them is in its own way too advanced in one or more characters to be directly ancestral.

Within the Ranunculales, there is no obvious reason why the Ranunculaceae might not be ancestral to all the other families of the order. All of the features in which the other families differ from the Ranunculaceae appear to represent phyletic advances. The main caveat is that the Ranunculaceae are chiefly boreal (or montane tropical), whereas the Menispermaceae are mostly tropical forest dwellers. However, if the Menispermaceae are only secondarily woody (as seems to be the case), they may also be secondarily tropical.

The Ranunculales show no ecologic or geographic unity, beyond the fact that many of them occur in moist or wet places and relatively few are adapted to arid conditions. The nectariferous petals found in most of the Ranunculaceae, Berberidaceae, Lardizabalaceae, and some of the Sabiaceae are one of the characteristic features of the order, but they do not fit the plants to any one group of pollinators. Indeed different species of the single genus Aquilegia are suited to such different pollinators as humming-birds, bees, and moths. The fact that the haplostemonous members of the order always have the stamens opposite the petals is of doubtful functional significance.

SYNOPTICAL ARRANGEMENT OF THE FAMILIES OF RANUNCULALES

1.  Gynoecium mostly of separate carpels, or seemingly or actually of a single carpel; endosperm copious to scanty or none; herbs and woody vines, seldom shrubs

2.  Leaves mostly alternate; petals not adhering to the carpels; flowers variously organized, often trimerous in part or with numerous stamens and carpels, never pentamerous throughout

3.  Flowers perfect; herbs (including herbaceous v es) or shrubs; endosperm well developed

4.  Leaves with dichotomous, free venation; nodes unilacunar; ovules orthotropous; flowers reduced, without petals or petaloid nectaries          **2. Circaeasteraceae**

4.  Leaves with net venation; nodes trilacunar to multilacunar; ovules anatropous; flowers usually well developed, often with nectariferous petals

5.  Carpels usually 2 or more and distinct, seldom solitary or weakly united; stamens usually numerous, the anthers opening by longitudinal slits; herbs, rarely shrubs          **1. Ranunculaceae**

5.  Carpel seemingly solitary (the pistil probably actually 2–3-carpellate with all but one carpel obsolete); stamens typically in 1–several sets of 3, usually opening by valves; herbs and shrubs          **3. Berberidaceae**

3.  Flowers unisexual; nearly all woody plants, mostly vines

6. Leaves mostly compound; seeds with small embryo and abundant endosperm; petals when present nectariferous; mostly monecious **4. Lardizabalaceae**

6. Leaves mostly simple; seeds mostly with large embryo and scanty endosperm; petals not nectariferous; dioecious **5. Menispermaceae**

2. Leaves opposite or whorled; petals persistent, enlarging in fruit and adhering to the individual carpels to produce a set of pseudodrupes; flowers wholly pentamerous; nodes unilacunar **6. Coriariaceae**

1. Gynoecium obviously of united carpels; endosperm scanty or none; trees and shrubs

7. Flowers diplostemonous, the inner set of stamens modified into staminodes; carpels 2 but only one fertile, this with one ovule **7. Corynocarpaceae**

7. Flowers haplostemonous; carpels 2–3, each with 1 or 2 ovules **8. Sabiaceae**

## SELECTED REFERENCES

Chapman, M., Carpel anatomy of the Berberidaceae. Am. Jour. Bot. 23: 340–348. 1936.

Ernst, W. R., The genera of Berberidaceae, Lardizabalaceae and Menispermaceae in the southeastern United States. Jour. Arn. Arb. 45: 1–35. 1964.

Foster, A. S., The floral morphology and relationships of *Kingdonia uniflora*. Jour. Arn. Arb. 42: 397–415. 1961.

————, The morphology and relationships of *Circaeaster*. Jour. Arn. Arb. 44: 299–327. 1963.

Gregory, W. C., Phylogenetic and cytological studies in the Ranunculaceae. Trans. Am. Phil. Soc. n.s. 36: 443–521. 1941.

Hammond, H. D., Systematic serological studies in Ranunculaceae. Serol. Mus. Bull. 14: 1–3. 1955.

Hegnauer, R., Chemotaxonomische Betrachtungen. V. Die systematische Bedeutung des Alkaloidmerkmales. Planta Medica 6: 1–34. 1958.

Hiepko, P., Vergleichend-morphologische und entwicklungsgeschichtliche Untersuchungen über das Perianth bei den Polycarpicae. Bot. Jahrb. 84: 359–508. 1965.

Jensen, U., Die Verwandtschaftsverhältnisse innerhalb der Ranunculaceae aus serologischer Sicht. Ber. Deuts. Bot. Ges. 79: 407–412. 1966. (1967)

Kaute, U., Beiträge zur Morphologie des Gynoecums der Berberidaceen mit einem Anhang über die Rhizomknospe von *Plagiorhegma dubium*. Inaug. Diss. Frei Univ. Berlin. 81 pp. 1963.

Leinfellner, W., Zur Morphologie des Gynözeums von *Berberis*. Öster. Bot. Zeits. 103: 600–612. 1956.

Leppik, E. E., Floral evolution in the Ranunculaceae. Iowa State Jour. Sci. 39: 1–101. 1964.

Manske, R. H. F., and H. L. Holmes, *The Alkaloids*, Vol. IV (Isoquinolines). New York: Academic Press, 1964.

Ruijgrok, H. W. L., Chemotaxonomische Untersuchungen bei den Ranunculaceae. II. Über Ranunculin und verwandte Stoffe. Planta Medica 11: 338–347. 1963.

Willaman, J. J., and B. G. Schubert, Alkaloid-bearing plants and their contained alkaloids. U. S. D. A. Tech. Bull. 1234. 287 pp. 1961.

ORDER 6. **Papaverales**

The order Papaverales as here defined consists of only two families, of about equal size, the Papaveraceae and Fumariaceae. Between them they have about 600 species. Typical members of the Fumariaceae seem very different from the Papaveraceae, but the two groups are undoubtedly closely allied, and they have often been considered as parts of a single family (Papaveraceae). The genera *Hypecoum* and *Pteridophyllum*, with regular flowers and spurless petals as in the Papaveraceae, but otherwise resembling the Fumariaceae, form a sort of connecting link between the two families. My decision to treat the Fumariaceae as a distinct family instead of as a subfamily of the Papaveraceae is purely arbitrary. In any case there are two groups which collectively form a larger group.

It has been traditional to include the Papaveraceae and Fumariaceae in a larger order, usually called the Rhoeadales, which contains also the families Capparaceae, Cruciferae, Moringaceae, Resedaceae, and Tovariaceae. More recently many authors have preferred to recognize two orders instead of one, establishing an order Papaverales for the Papaveraceae and Fumariaceae, and an order Capparales for the other five families. The name Rhoeadales is then abandoned in order to avoid confusion and also to bring the nomenclature into line with the more customary practice of basing the name of an order on one of the included families.

It is perfectly clear that the Papaverales stand apart as a group from the Capparales. The question has been whether the unifying factors outweigh the dividing ones. The unifying factors are mainly the perfect, hypogynous flowers, compound ovary with parietal placentation and sometimes also a replum, and frequently compound leaves. All of these features except the replum also occur in the Violales, although seldom in combination.

The known differences between the two orders have increased in number and significance as a result of studies during the past several decades, and it now appears that they are phylogenetically rather remote. The Papaverales often have a latex system or elongate individual secretory (latex?) cells, which are wanting in the Capparales. The Capparales, on the other hand, commonly have myrosin cells, which are unknown in the Papaverales. There is no apparent phylogenetic connection between these two types of specialized secretory elements. The Papaverales characteristically contain isoquinoline alkaloids (sometimes including berberin), a feature which links them to the rest of the Magnoliidae. Hegnauer (1956, cited below) has noted that "The Polycarpicae (sensu Wettstein) together with the Papaveraceae form a very natural group characterized by the synthesis of alkaloids of the phenylalanine type." The phenylalanine alkaloids of Hegnauer are apparently equivalent to the isoquinoline alkaloids of Manske and Holmes and some other authors. The Capparales have centrifugal stamens, in contrast to the centripetal stamens of the Papaverales. Careful serological and palynological studies further emphasize the separation between the two orders. All told, it appears that the relationships of the Papaverales are with the Ranunculales in the Magnoliidae, whereas the relationships of the Capparales are with the Theales and Violales in the Dilleniidae.

Although the Papaverales usually have a compound ovary with parietal (sometimes deeply intruded) placentae, the monotypic Californian genus *Platystemon* (Papaveraceae) has 6–25 carpels that are only weakly united when young and become free in fruit. It is not to be supposed that *Platystemon*, a semidesert annual, is much like the ancestral prototype of the Papaverales, but it does seem to have a relatively primitive gynoecium. The inclusion of *Platystemon* in the Papaveraceae is unquestioned, but its similarity to the Ranunculaceae is also obvious.

Most of the characters which mark the Fumariaceae as a separate group within the order represent phylogenetic advances over the characters of the Papaveraceae. On the other hand, the well developed latex system of the Papaveraceae is nearly wanting in the Fumariaceae, being represented only by scattered, elongate secretory cells, with contents of uncertain nature, in certain species of some genera. If the Fumariaceae are to be regarded as derived directly from the Papaveraceae, the latex system must have become vestigial in the process. It may be more prudent to regard the two families as parallel groups which show different individual specializations.

The Papaverales are ecologically less varied than some of the larger orders, but even so they show no particular ecological unity. Most of them are herbs, but a few are shrubs or small trees. They occur from subtropical to arctic regions. Some of them grow in moist woods, others in deserts, and others above timberline in the mountains; some of them are ruderal weeds. All of the habitats in which they occur are occupied also by habitally similar species of diverse other orders.

## SYNOPTICAL ARRANGEMENTS OF THE FAMILIES OF PAPAVERALES

1. Stamens numerous; flowers regular; petals neither spurred nor saccate; latex system more or less well developed
   **1. Papaveraceae**

1. Stamens 4 or 6; flowers usually irregular; usually some of the petals spurred or saccate; latex system wanting or occasionally represented by scattered elongate secretory cells
   **2. Fumariaceae**

## SELECTED REFERENCES

Ernst, W. R., The genera of Papaveraceae and Fumariaceae in the southeastern United States. Jour. Arn. Arb. 43: 315–343. 1962.

Frohne, D., Das Verhältnis von vergleichender Serobotanik zu vergleichender Phytochemie, dargestellt an serologischen Untersuchungen im Bereich der "Rhoeadales." Planta Medica 10: 283–297. 1962.

Hegnauer, R., Chemotaxonomische Betrachtungen. V. Die systematische Bedeutung des Alkaloidmerkmales. Planta Medica 6: 1–34. 1958.

————, Die Gliederung der Rhoeadales sensu Wettstein im Lichte der Inhaltstoffe. Planta Medica 9: 37–46. 1961.

Ryberg, M., A morphological study of the Fumariaceae and the taxonomic significance of the characters examined. Acta Horti Bergianai 19: 122–248. 1960.

## Subclass II. Hamamelidae

As the smallest of the six subclasses of dicots, the subclass Hamamelidae contains only about 3400 species. These are distributed amongst nine orders and 23 families. More than 2800 of the species belong to only three families (Moraceae, Urticaceae, Fagaceae), and nearly 3200 to only three of the orders (Urticales, Fagales, Hamamelidales). Thirteen of the 23 families have less than ten species each, and five are monotypic. Although the subclass is small in terms of number of species, it contains some of the commonest and most important genera of deciduous trees of north temperate regions, e.g., *Carya, Quercus, Fagus, Castanea, Juglans, Platanus,* and *Ulmus*.

The Hamamelidae are a loosely knit group of dicots with more or less strongly reduced flowers. The perianth is poorly developed or wanting, and the flowers are often unisexual. In the more advanced types the flowers of one or both sexes are borne in catkins, and the mature fruit is unilocular and indehiscent, with only a single seed. With the notable exception of some of the Urticales, they are all woody plants.

As here defined, the subclass Hamamelidae includes among others a number of families which have traditionally been placed at the beginning of the dicots in the Englerian system and called the Amentiferae. The name Amentiferae (catkin-bearers) refers to the characteristic inflorescences. Several amentiferous families are excluded from the Hamamelidae, however. Thus the Salicaceae are referred to the Dilleniidae, the Garryaceae to the Rosidae, and the Bataceae to the Caryophyllidae.

There is a growing consensus that most or all of the 23 families here assembled as the Hamamelidae do form a natural group. In addition to the reduced, anemophilous flowers, the group is held together by a complex set of overlapping similarities in wood anatomy, pollen morphology, detailed gynoecial structure, and other features.

The floral reduction of the Hamamelidae is loosely associated with the substitution of wind pollination for insect pollination. It is still an open question, however, which change was the horse and which the cart. *Trochodendron*, one of the most primitive genera of the subclass in a number of respects, is reported to be insect-pollinated in spite of the fact that it is completely without perianth. I have observed insects visiting the large and conspicuous staminate inflorescences of *Castanea*, but I am not sure what this means in terms of pollination. Inasmuch as the pistillate flowers of *Castanea* are relatively inconspicuous and often well removed from the staminate ones, the genus certainly does not seem well adapted to entomophily. Forman (Kew Bull. 17: 381–396. 1964) interprets the Fagaceae as evolving from insect pollination to wind pollination, but floral reduction is already far advanced in the most primitive existing members of the family. Thus it appears that, strange as it may seem, anemophily in the Hamamelidae may prove to be consequent to the floral reduction, rather than antecedent or closely and immediately linked to it. The cause of the floral reduction, in that case, would be wholly unknown. Could mutation pressure, without selective value, cause such critically important changes in floral morphology? The orthodox answer certainly would be that it could not. The possibility that entomophily is secondary within the Hamamelidae instead of primitive in the group should also be considered.

Ecologically, the only thing that most members of the Hamamelidae have in common is wind-pollination. Wind-pollination associated with floral reduction has evolved independently in so many other groups of angiosperms, however, that this feature scarcely makes them distinctive. Even so, none of the other groups of anemophilous angiosperms approaches the Hamamelidaes in diversity or in number of species.

It has been suggested by R. M. Schuster (personal conversation) that wind pollination associated with floral reduction was an early adaptation by angiospermous trees to growth in temperate or cold-temperate regions, at a time when the evolutionary explosion of pollinating insects had not yet occurred. Needing the fullest possible time for maturation of the seed before the onset of winter, such plants flowered early in the spring, when even such potential pollinators as might have been available were not yet active. Even now many of the amentiferous trees are among the first plants to flower in the spring. We should note, however, that there is a perfectly good ecological reason for modern amentiferous trees to bloom very early in the spring. The distribution of pollen by wind is more effective when the trees are bare and the forests open than later in the season when there is a dense canopy of leaves.

Schuster's suggestion is certainly worthy of careful consideration, and it may well prove to be correct. It would help to explain why such a large proportion of the amentiferous woody plants do hang together as a natural group, and why the group now appears to be relictual, with many isolated families.

On first consideration it also appears (teste James Doyle) that the fossil pollen record may be compatible with an early proliferation and dominance of the Hamamelidae in extratropical regions, followed by an evolutionary decline of the group as other kinds of woody angiosperms became adapted to similar conditions. The possible significance of the pollen record merits careful investigation.

It is clear that neither any one family nor any one order of the Hamamelidae can be regarded as ancestral to all the others. The Trochodendrales are certainly the most primitive order, on the basis of their vesselless xylem, perfect flowers, sometimes evident perianth, scarcely sealed and only weakly connate carpels, more or less numerous ovules, and dehiscent, follicular fruits. However, they have already progressed farther in floral reduction than the Hamamelidaceae, which have unmistakable (though somewhat reduced) petals, in contrast to the complete absence of petals in the Trochodendrales. The remaining orders may all have been derived from the Hamamelidales.

A plant which combined all the primitive features found in the various families of the Hamamelidae would be a tree with stipulate, alternate, simple leaves, vesselless xylem, and perfect, hypogynous flowers with a relatively inconspicuous perianth that is differentiated into sepals and petals, an androecium of numerous separate stamens, triaperturate pollen, and a gynoecium of separate, multiovulate, scarcely sealed carpels which ripen into follicles. Such a plant would surely be referred to the Magnoliidae. Within the Magnoliidae, such a composite prehamamelid would have no obvious close relatives.

We must face the prospect that the similarities which lead us to recognize the subclass Hamamelidae as a group reflect the independent realization of similar potentialities by more or less closely allied ancestors. We have already noted that insistence on a strict application of the monophyletic principle in taxonomy would necessitate the dissolution of many apparently natural groups. A loose application of the monophyletic principle, as discussed in Chapter 1, is all that is necessary or even desirable.

The several amentiferous groups that are here excluded from the Hamamelidae are excluded because they do not meet even the loosest test of monophylesis. They reflect an independent realization of similar evolutionary potentialities in method of pollination and gross floral morphology from only distantly related ancestors. Not surprisingly, the convergence in floral structure is not accompanied by convergence in other characters such as wood anatomy and pollen morphology.

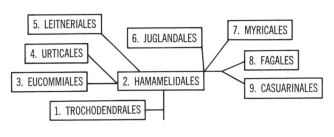

FIGURE 4.3

Probable relationships among the orders of Hamamelidae.

SYNOPTICAL ARRANGEMENT OF THE ORDERS OF HAMAMELIDAE

1. Flowers either with distinct carpels (a single carpel in *Didymeles,* of the Hamamelidales), or with dehiscent fruits, or both, variously perfect or unisexual; ovules 1–many per carpel; seeds endospermous, with small to large embryo

2. Vessels wanting; some or all of the flowers perfect; carpels 4 or more, in a single whorl, laterally coherent but scarcely forming a compound pistil, the group of carpels ripening into a follicetum; ovules several to many in each carpel; stipules wanting or represented only by a pair of flanges on the

petiole; leaves with unique, elongate (often branched) idioblasts                                                **1. Trochodendrales**

2.  Vessels present, flowers perfect or more often unisexual; carpels either 1–several and separate, or 2–4 and united (at least below) to form a compound pistil; ovules 1–many; stipules nearly always present except in *Euptelea* and *Didymeles;* no conspicuous idioblasts                **2. Hamamelidales**

1.  Flowers either with united carpels, or with a pseudomonomerous, seemingly unicarpellate pistil; fruits indehiscent; ovules not more than 2 per carpel; flowers in most families unisexual; seeds with or without endosperm, but the embryo always fairly large

3.  Leaves pinnately compound or trifoliolate; ovary 1-2-locular, with 1–2 ovules per locule, or one locule empty.
                                                                 **6. Juglandales**

3.  Leaves simple (though sometimes deeply lobed or cleft); gynoecium various

4.  Flowers, or at least the male flowers, not in catkins; ovary unilocular or rarely bilocular; calyx often present; plants woody or herbaceous

5.  Ovules 2, unitegmic, intermediate between crassinucellate and tenuinucellate; stipules none; nodes unilacunar; flowers solitary, without a perianth      **3. Eucommiales**

5.  Ovule solitary, bitegmic, crassinucellate; stipules usually present; nodes trilacunar to multilacunar; flowers clustered, usually with a vestigial perianth      **4. Urticales**

4.  Flowers, or at least the male flowers, usually in catkins (except some Fagaceae); ovary with 1–several locules; calyx mostly wanting or very much reduced; plants woody

6.  Pistil pseudomonomerous, seemingly composed of a single carpel, with a single style and locule; ovary superior
                                                                 **5. Leitneriales**

6.  Pistil composed of more than one carpel, as evidenced by

the presence of more than one style and often also more than one locule; ovary inferior or nude

7. Leaves opposite or alternate, seldom whorled, always more or less well developed; ovules each with a single embryo sac

    8. Ovule solitary, orthotropous (except in *Canacomyrica*), unitegmic; plants aromatic, the leaves resinous-dotted     **7. Myricales**

    8. Ovules more than one, anatropous, bitegmic or unitegmic; plants not notably aromatic   **8. Fagales**

7. Leaves whorled, reduced to scales; ovules with multiple embryo-sacs     **9. Casuarinales**

## SELECTED REFERENCES

Hjelmqvist, H., Floral morphology and phylogeny of the Amentiferae. Bot. Notis. Suppl. 2: 78–140. 1948.

Petersen, A. E., A comparison of the secondary xylem elements of certain species of the Amentiferae and Ranales. Bull. Torrey Club 80: 365–384. 1953.

Tippo, O., Comparative anatomy of the Moraceae and their presumed allies. Bot. Gaz. 100: 1–99. 1938.

ORDER 1. **Trochodendrales**

The order Trochodendrales consists of two families, the Trochodendraceae and Tetracentraceae, each with only a single species. The mutual affinity of these two families has been generally recognized for the last two decades, since the studies of Bailey, Nast, and Smith (cited below). It is clear enough that they are among the more primitive families of angiosperms, as shown especially by their conduplicate, scarcely sealed carpels and vesselless xylem. It is also clear that they are derived from the Magnoliales. The peculiar idioblasts (elongate, secretory, and simple or branched in *Tetracentron*; elongate, nonsecretory, and branched in *Trochodendron*) are evidently derived from the spherical ethereal oil cells of the Magnoliales. On the other hand, the obsolete or poorly developed perianth of the Trochodendrales clearly reflects a reduction, and neither the stamens nor the pollen are especially primitive.

Admitting that the Trochodendrales have no really close relatives, one turns to the Cercidiphyllaceae and Eupteliaceae as the least distant families. These, like the Trochodendrales, have a fairly primitive gynoecium combined with a reduced perianth, and they are generally regarded as eventually related to the Magnoliales, without the intervention of any other group. The Cercidiphyllaceae and Eupteliaceae also recall the somewhat more advanced families Platanaceae and Hamamelidaceae.

From a practical standpoint it seems more useful to associate the Trochodendrales with the Hamamelidae, in which they are near-basal, than with the Magnoliidae, in which they are peripheral. There is no order of the Magnoliidae to which they could reasonably be referred, in the scheme here presented. Within the Hamamelidae, a case could be made for including them in the already heterogeneous order Hamamelidales, but I think it is conceptually more satisfactory to maintain them as a distinct order, forming a near-basal side-branch within their subclass.

As so defined, the order Trochodendrales consists of only two families, genera, and species. Neither family, as now represented, can be considered ancestral to the other. *Tetracentron* is the more primitive in having a perianth, in having vestiges of stipules, and in having secretory idioblasts. *Trochodendron* is the more primitive in having numerous stamens, instead of only four. The pinnate venation of *Trochodendron* may also be more primitive than the palmate venation of *Tetracentron*.

## SYNOPTICAL ARRANGEMENT OF THE FAMILIES OF TROCHODENDRALES

1. Perianth present; stipular flanges present; stamens 4, opposite the sepals; leaves palmately veined; idioblasts simple or branched, secretory                            **1. Tetracentraceae**

1. Perianth none; stipules none; stamens numerous; leaves pinnately veined; idioblasts branched, not secretory
                                             **2. Trochodendraceae**

## SELECTED REFERENCES

Bailey, I. W., and C. G. Nast, Morphology and relationships of *Trochodendron* and *Tetracentron*. I. Stem, root, and leaf. Jour. Arn. Arb. 26: 143–154. 1945. II. Inflorescence, flower, and fruit. Jour. Arn. Arb. 26: 267–276. 1945.

Keng, H., Androdioecism in the flowers of *Trochodendron aralioides*. Jour. Arn. Arb. 40: 158–160. 1959.

Smith, A. C., A taxonomic review of *Trochodendron* and *Tetracentron*. Jour. Arn. Arb. 26: 123–142. 1945.

Foster, A. S., The foliar sclereids of *Trochodendron aralioides* Sieb. & Zucc. Jour. Arn. Arb. 26: 155–162. 1945.

ORDER 2. **Hamamelidales**

The order Hamamelidales, as here defined, consists of six families, the Hamamelidaceae, Myrothamnaceae, Platanaceae, Eupteleaceae, Cercidiphyllaceae, and Didymelaceae. Of these, the Hamamelidaceae contain about 100 species, the Platanaceae about six species, and the others only two species each.

The Hamamelidaceae are a diverse and loosely knit family. Thirteen of the 26 genera are monotypic. Two small genera, Altingia and Liquidambar, are sometimes separated collectively as a distinct family Altingiaceae. The monotypic genus Disanthus has at least as much claim to family status as the Altingiaceae.

Disanthus is clearly the most primitive genus in the Hamamelidaceae. It has perfect, nearly hypogynous flowers with both sepals and petals and with five to six ovules in each locule of the ovary. It is interesting and perhaps significant that the leaves of Disanthus are so much like those of Cercidiphyllum that the species is named D. cercidifolius. Not too much should be made of this likeness, however, inasmuch as neither Disanthus nor Cercidiphyllum is alled to Cercis.

The mutual affinity of the Hamamelidaceae, Platanaceae, and Myrothamnaceae is now widely accepted. Their status as distinct families is, however, unquestioned. Each of the three has a combination of primitive and more advanced characters which indicates that it could not be ancestral to either of the others. In addition to the more obvious differences, Myrothamnus has distinctive pollen reminiscent in some respects of that of the Monimiaceae.

The position of the Cercidiphyllaceae, Eupteleaceae, and Didymelaceae, each with a single genus, is more debatable. Each of them is so distinctive that the problem is to find any relatives at all, rather than to choose among different possibilities. On the basis of their wood anatomy as well as floral morphology the Cercidiphyllaceae and Eupteleaceae cannot be very far removed from the Magnoliidae, but this is also true of the Hamamelidaceae cum Platanaceae and Myrothamnaceae. McLaughlin (Tropical Woods 34: 3–37. 1933) found the wood of both Cercidiphyllum and Euptelea to be similar in many respects to that of both Illicium (Magnoliales) and the Hamamelidaceae, with some additional special similarities between Cercidiphyllum and Liquidambar (Hamamelidaceae). The flowers of Cercidiphyllum and Euptelea are more primitive than the remainder of the Hamamelidales in having numerous stamens, but they still fall into the general pattern of reduction associated with anemophily that marks the Hamamelidae as a whole.

The Didymelaceae have the most reduced and specialized flowers and inflorescence of all the families here referred to the Hamamelidales. However, Takhtajan reports (personal correspondence) that both the carpel and the wood are very primitive, the vessels having scalariform perforations with numerous bars. The alternative to including the Didymelaceae in the Hamamelidales is to establish a separate order for them, as indeed Takhtajan has done (p. 66 in *System and Phylogeny of Flowering Plants* [in Russian], 1966).

At the very least, it appears that *Cercidiphyllum, Euptelea, Didymeles,* and the remainder of the Hamamelidales have undergone a series of similar (as well as some individually distinctive) changes from similar ancestors. It therefore seems useful to include them all in the same order, especially inasmuch as this can be done without making the group conceptually difficult.

There is a special problem in the interpretation of the pistillate flowers of *Cercidiphyllum.* The usual and most obvious interpretation is that they have four sepals (like the staminate flowers) and several separate carpels which are unusual in that the seed-bearing suture is dorsal (extrose) rather than ventral. This unusual position of the placenta could be achieved if each carpel were twisted 180 degrees at the base. Finding no anatomical evidence of such twisting, Swamy and Bailey (cited below) interpreted the apparent flower as a pseudanthium composed of several naked, unicarpellate flowers. It is perhaps a measure of the taxonomic isolation of the family that one's choice between the two interpretations of floral morphology makes no difference in its taxonomic position.

Six of the remaining seven orders of the Hamamelidae are probably derived separately and individually from the Hamamelidales. Each of the six is sufficiently distinctive in its own way so that it is not likely to be ancestral to any of the others. The Casuarinales, on the other hand, might be derived from either primitive Fagales or more directly from the Hamamelidales.

At one time I thought it useful to group the Eucommiales, Fagales, Leitneriales, Myricales, and Urticales of the present treatment into a broadly defined order Urticales. Under the relaxed concept of phylogeny in taxonomy that I have expounded, such an arrangement might be tenable, since all the taxa involved probably take their origin in the Hamamelidales. However, I fear that it does more to obscure than to clarify the relationships in this technically very diverse group, and I now prefer to return to a more nearly traditional concept in which several orders are recognized.

SYNOPTICAL ARRANGEMENT OF THE FAMILIES OF HAMAMELIDALES

1. Carpels separate (or solitary)
   2. Ovules numerous; fruit follicular; plants dioecious; leaves stipulate, palmately veined, entire     **1. Cercidiphyllaceae**

2. Ovules 1–3; fruit indehiscent; flowers variously perfect or unisexual; leaves various

   3. Carpels several; stamens 3–many; plants monoecious or with perfect flowers

      4. Stamens numerous; ovules more or less anatropous; flowers perfect; connective of ordinary type, not apically enlarged and peltate; leaves exstipulate, pinnately veined
**2. Eupteleaceae**

      4. Stamens 3–4 (7); ovules more or less orthotropous; plants monoecious; connective apically enlarged and peltate; leaves stipulate, palmately veined   **3. Platanaceae**

   3. Carpel solitary; stamens 2, united by their filaments; plants dioecious   **4. Didymelaceae**

1. Carpels united (at least below) into a compound pistil

   5. Leaves alternate (rarely opposite); carpels 2 (rarely 3); flowers perfect or unisexual, with at least a vestigial calyx and often with a corolla as well; filaments free; ovules 1–many; trees or shrubs   **5. Hamamelidaceae**

   5. Leaves opposite; carpels 3–4; plants dioecious; perianth none; filaments united; ovules many; undershrubs
**6. Myrothamnaceae**

## SELECTED REFERENCES

Boothroyd, L. E., The morphology and anatomy of the inflorescence and flower of the Platanaceae. Am. Jour. Bot. 17: 678–693. 1930.

Ernst, W. R., The genera of Hamamelidaceae and Platanaceae in the southeastern United States. Jour. Arn. Arb. 44: 193–221. 1963.

Leandri, J., Sur l'aire et la position systematique du genre malgache *Didymeles* Tour. Ann. Sci. Nat. Bot. Ser. 10. 19: 309–317. 1937.

Nast, C. G., and I. W. Bailey, Morphology of *Euptelea* and comparison with *Trochodendron*. Jour. Arn. Arb. 27: 186–192. 1946.

Smith, A. C., A taxonomic review of *Euptelea*. Jour. Arn. Arb. 27: 175–185. 1946.

Swamy, B. G. L., and I. W. Bailey, The morphology and relationships of *Cercidiphyllum*. Jour. Arn. Arb. 30: 187–210. 1949.

ORDER 3. **Eucommiales**

The Eucommiaceae, with a single genus and species, have usually been associated with the Urticales or Hamamelidales, or sometimes with the Magnoliales. Morphologically they are most nearly at home in the Urticales, but Eckardt (cited below) has pointed out a series of embryological differences which make them anomalous there also. The Eucommiaceae are more primitive than the Urticales in having two ovules instead of only one, but more advanced in several other features (ovules unitegmic, not fully crassinucellate; stipules none). The pollen suggests that of *Cercidiphyllum*. It seems likely that the Eucommiaceae are derived from the Hamamelidales independently of the Urticales. Although I am not fond of taxonomic monotypes, it seems appropriate to treat the Eucommiales as a distinct order.

SELECTED REFERENCES

Eckardt, T., Zur systematischen Stellung von *Eucommia ulmoides*. Ber. Deuts. Bot. Ges. 69: 487–498. 1956.

————, Some observations on the morphology and embryology of *Eucommia ulmoides*. Jour. Indian Bot. Soc. 42A: 27–34. 1963.

Tippo, O., The comparative anatomy of the secondary xylem and the phylogeny of the Eucommiaceae. Am. Jour. Bot. 27: 832–838. 1940.

ORDER 4. **Urticales**

The order Urticales, as here defined, consists of five families and nearly 2400 species. The Moraceae are by far the largest family, containing some 1500 species. The Urticaceae have 700 or more species, the Ulmaceae about 150, the Cannabaceae only four, and the Barbeyaceae only one.

The naturalness of the order Urticales is not in dispute. The two smallest families, the Barbeyaceae and Cannabaceae, are often submerged in the Ulmaceae and Moraceae, respectively. Their status depends on one's assessment of the significance of the similarities and differences.

The differences between the Moraceae and Urticaceae look good on paper, but they are not wholly constant. One whole subfamily (Conocephaloideae, six genera, 140 species) of the Moraceae is more or less transitional to the Urticaceae and has sometimes been referred to that family. The monotypic herb *Fatoua* is always retained in the Moraceae, but externally it looks very urticaceous except for the inconspicuous, vestigial second style.

It is still an open question whether the Urticaceae should be regarded as derived from the Moraceae or whether the two diverged from a common ancestor that would not have fit well into either family. The herbaceous or softly woody habit, pseudomonomerous gynoecium, and orthotropous, basal ovule of the Urticaceae are clearly advanced, as compared to the Moraceae. The absence of milky juice from the Urticaceae is another matter. Whether the poorly developed latex system of some urticaceous genera such as *Laportea* or *Urera* is vestigial or (in contrast) rudimentary remains to be determined.

### SYNOPTICAL ARRANGEMENT OF THE FAMILIES OF URTICALES

1. Trees, without milky juice; seeds with straight embryo and no endosperm; ovule pendulous, anatropous; anthers erect in bud

    2. Leaves alternate, stipulate; styles 2         **1. Ulmaceae**

    2. Leaves opposite, exstipulate; style 1      **2. Barbeyaceae**

1. Plants either not woody, or with milky juice, or with endospermous seeds, often with 2 or even all 3 of these features; anthers inflexed in bud, or less often erect

    3. Styles or style branches 2, sometimes one of them reduced; ovules mostly anatropous and pendulous; embryo straight or more often curved

        4. Woody plants, seldom herbs, mostly with milky juice; anthers usually inflexed in bud      **3. Moraceae**

        4. Herbs, without milky juice; anthers erect in bud
                                  **4. Cannabaceae**

    3. Style 1, unbranched; ovule orthotropous, basal; embryo straight; herbs, seldom shrubs or soft-wooded trees, without milky juice      **5. Urticaceae**

### SELECTED REFERENCES

Corner, E. J. H., The classification of Moraceae. Gard. Bull. Singapore 19: 187–252. 1962.

Johri, B. M., and R. N. Konar, The floral morphology and embryology of *Ficus religiosa* Linn. Phytomorph. 6: 97–111. 1956.

Ram, H. Y. M., and R. Nath, The morphology and embryology of *Cannabis sativa* Linn. Phytomorph. 14: 414–429. 1964.

Tippo, O., Comparative anatomy of the Moraceae and their presumed allies. Bot. Gaz. 100: 1–99. 1938.

ORDER 5. **Leitneriales**

The order Leiteneriales includes only the family Leitneriaceae, with the single species *Leitneria floridana*, native to the southeastern United States. *Leitneria* is by all accounts highly isolated, and its relationships are uncertain. Hjelmqvist has noted that the gynoecium is pseudo-monomerous rather than strictly unicarpellate, as shown by the occasional occurrence of bicarpellate pistils with two styles. The well developed resin canals in the stem and leaves might be compared with those of the Juglandaceae. The multicellular, clavate glands of the leaves might or might not be homologous with the peltate glands of the Juglandales, Myricales, and Fagales. The origin of *Leitneria* by reduction from a hamamelidalean ancestry has been postulated by several authors, and this possibility seems as likely as any.

## SELECTED REFERENCES

Abbe, E. C., and T. T. Earle, Inflorescence, floral anatomy, and morphology of *Leitneria floridana*. Bull. Torrey Club 67: 173–193. 1940.

Channell, R. B., and C. E. Wood, Jr., The Leitneriaceae in the southeastern United States. Jour. Arn. Arb. 43: 435–438. 1962.

Hjelmqvist, H., Floral morphology and phylogeny of the Amentiferae. Bot. Notis. Suppl. 2: 78–140. 1948.

ORDER 6. **Juglandales**

The order Juglandales as here defined consists of three families. Of these, the Juglandaceae have a little over 50 species, the Picrodendraceae have three species, and the Rhoipteleaceae only one. Traditionally the Juglandaceae have usually been the only family of the order. *Rhoiptelea chiliantha*, described from China in 1902, is now usually taken to be another family to be associated with the Juglandaceae. The affinities of *Picrodendron* are still obscure, but in our present state of ignorance it seems best associated with the Juglandaceae. All three of these families have pinnately compound or trifoliolate leaves, a convenient character which sets them apart from the rest of the Hamamelidae.

The Juglandaceae have sometimes been thought to be related to the Anacardiaceae rather than to the orders with which they are here associated. However, their wood is distinctly more primitive than that of the Anacardiaceae, their pollen is similar to that of several families of the Hamamelidae, and they share with the Myricaceae and Fagales the possession of peculiar peltate glands on the leaves. Furthermore, the apparently related but obviously more primitive family Rhoipteleaceae forms a much better connecting link to the Urticales than to anything near the Anacardiaceae.

Anatomically the Juglandaceae are much like the Myricaceae, and the two families have sometimes been associated in a single order. However, an order consisting of these two families, or of these plus the Rhoipteleaceae and/or Picrodendraceae, would be difficult to characterize. It seems more useful to admit a probable relationship without trying to establish a conceptually difficult order. Here, as in many other groups, it is not easy to determine how many of the similarities reflect direct in-. heritance and how many of them result from independent realization of similar evolutionary potentialities.

## SYNOPTICAL ARRANGEMENT OF THE FAMILIES OF JUGLANDALES

1. Flowers in triplets, the middle one perfect and fertile, with 4 sepals, 6 stamens, and a superior, bilocular ovary, one locule empty, the other with a single ovule attached to the partition; lateral flowers female but sterile; leaves pinnately compound, stipulate; fruit samaroid **1. Rhoipteleaceae**

1. Flowers unisexual, the staminate ones with 3–many stamens, borne in catkins, the calyx wanting or adnate to the bract; female flowers solitary or spicate, with 1–4 sepals or calyx-teeth; fruit drupaceous or nutlike

    2. Ovules 4 (2 in each of 2 locules), apical; ovary superior; leaves trifoliolate, minutely stipulate **2. Picrodendraceae**

    2. Ovule solitary, basal, in a unilocular, inferior ovary; leaves pinnately compound, exstipulate **3. Juglandaceae**

## SELECTED REFERENCES

Heimsch, C. J., and R. W. Wetmore, The significance of wood anatomy in the Juglandaceae. Am. Jour. Bot. 25: 651–660. 1939.

Leroy, J. F., Etude sur les Juglandaceae. Mem. Mus. Nat. Hist. Bot. Paris 6: 1–246. 1955.

Manning, W. E., The morphology of the flowers of the Juglandaceae. I. The inflorescence. Am. Jour. Bot. 25: 407–419. 1938. II. The pistillate flowers and fruit. Am. Jour. Bot. 27: 839–852. 1940. III. The staminate flowers. Am. Jour. Bot. 35: 606–621. 1948.

Stone, D. E., J. Reich, and S. Whitfield, Fine structure of the walls of *Juglans* and *Carya*. Pollen et Spores 6: 379–392. 1964.

Whitehead, D. R., Pollen morphology in the Juglandaceae. I. Pollen size and pore number variation. Jour. Arn. Arb. 44: 101–110. 1963. II. Survey of the family. Jour. Arn. Arb. 46: 369–410. 1965.

Withner, C. L., Stem anatomy and phylogeny of the Rhoipteleaceae. Am. Jour. Bot. 28: 872–878. 1941.

ORDER 7. **Myricales**

The order Myricales contains only the family Myricaceae, with about 50 species. All but one of these species belong to the widespread, chiefly subtropical genus Myrica, or to Myrica and two small segregate genera that are recognized by some botanists. Sharply distinct from Myrica is the monotypic New Caledonian genus Canacomyrica, which has a pendulous, anatropous ovule instead of the erect, orthotropous, basal ovule of Myrica. In this respect Canacomyrica is suggestive of the Fagaceae and Betulaceae, but in other respects it seems properly referable to the Myricaceae.

The conspicuous, peltate, resinous glands of the leaves of the Myricaceae, from which the characteristic odor emanates, are evidently homologous with similar peltate glands in the Juglandales, and Fagales. The Myricaceae appear to be allied to both the Fagales and Juglandales, but they would be peripheral and aberrant in either order. At the present state of our knowledge it seems most useful to maintain the Myricaceae in an order of their own.

ORDER 8. **Fagales**

The order Fagales as here defined consists of three families and a little over 700 species. The Fagaceae (600) are by far the largest family, followed by the Betulaceae (100) and Balanopaceae (ten).

It is widely agreed that the Fagaceae and Betulaceae are fairly closely related and that they are derived by floral reduction from ancestors similar to the Hamamelidaceae. The Betulaceae are in some respects more advanced than the Fagaceae, but the pistillate inflorescence of the two families is so different that neither is likely to be directly ancestral to the other. The two families must be regarded as derived from a common ancestor which would not fit well into either of them.

The position of the Balanopaceae is more controversial. Traditionally they have been referred to a distinct order, the Balanopales, variously thought to be related to the Juglandales, Euphorbiales, or Fagales. The pistillate flower of Balanops (the only genus of the family) is solitary, as in many Fagaceae, and the ripe fruit is very like an acorn in appearance, even to being subtended by a similar hull. On the other hand, Balanops presents its own combination of primitive and advanced characters, as compared to the Fagaceae. It is primitive in having endosperm and in having the ripe fruit 2–3-locular with as many seeds, but it is advanced in being dioecious and exstipulate. I agree with Takhtajan (1959) that the closest relationship of *Balanops* is with the Fagaceae, and that these two groups probably had a common ancestor which in turn was derived from the Hamamelidales. In my opinion the taxonomic system is best served by treating the Balanopaceae as a family of the Fagales.

## SYNOPTICAL ARRANGEMENT OF THE FAMILIES OF FAGALES

1. Seeds with endosperm; leaves exstipulate; ovules erect, basal or nearly so, unitegmic; plants dioecious; pistillate flower solitary in a multibracteate involucre which ripens into a hull subtending the fruit                              **1. Balanopaceae**

1. Seeds without endosperm; leaves stipulate; ovules pendulous, apical or nearly so; plants monoecious or seldom dioecious, or the flowers rarely perfect

  2. Styles and locules 3 (seldom 6); integuments 2; female flowers mostly not in catkins, subtended individually or in small groups by an involucre of many bracts which are concrescent at maturity into an indurated hull    **2. Fagaceae**

  2. Styles and locules 2 (seldom 3); integument 1; female flowers mostly borne in catkins; fruits without a hull of bracts, or sometimes with a foliaceous hull derived from 2 or a few bracts                                       **3. Betulaceae**

## SELECTED REFERENCES

Abbe, E. C., Studies in the phylogeny of the Betulaceae. I. Floral and inflorescence anatomy and morphology. Bot. Gaz. 97: 1–67. 1935. II. Extremes in the range of variation of floral and inflorescence morphology. Bot. Gaz. 99: 431–469. 1938.

Brett, D. W., The inflorescence of *Fagus* and *Castanea* and the evolution of the cupules of the Fagaceae. New Phytol. 63: 96–117. 1964.

Forman, L. L., *Trigonobalanus*, a new genus of Fagaceae, with notes on the classification of the family. Kew Bull. 17: 381–396. 1964.

Hall, J. W., The comparative anatomy and phylogeny of the Betulaceae. Bot. Gaz. 113: 225–270. 1952.

Hjelmqvist, H., Floral morphology and phylogeny of the Amentiferae. Bot. Notis. Suppl. 2: 78–140. 1948.

————, Some notes on the endosperm and embryo development in Fagales and related orders. Bot. Notis. 110: 173–195. 1957.

Langdon, L. M., The comparative morphology of the Fagaceae. I. The genus *Nothofagus*. Bot. Gaz. 108: 350–371. 1947.

Poole, A. L., The development of *Nothofagus* seed. Trans. Roy. Soc. New Zealand 78: 363–380; 502–508. 1952.

ORDER 9. **Casuarinales**

The Casuarinales consist of a single family and genus, with perhaps as many as 50 species. Native to the southwestern Pacific region, especially Australia, *Casuarina* is also familiar as a street tree in other tropical and subtropical regions. The equisetoid appearance of the branchlets is striking, but the other differences are so numerous and fundamental that no thought of a relationship between *Casuarina* and *Equisetum* can be sustained.

By all accounts *Casuarina* is a highly isolated genus. In the Englerian system it was taken as the most primitive genus of dicots, but that idea has been generally abandoned along with the change in concepts of evolutionary trends in the angiosperms. It is now clear enough that *Casuarina* is a reduced rather than a primitively simple type.

Most modern workers are agreed that *Casuarina* is eventually related to some of the families here grouped as the Hamamelidae. Its pollen is very like that of *Betula* and resembles to a lesser degree that of *Myrica* and some of the Juglandaceae. There is nothing obvious in the floral or vegetative morphology that could not reasonably have been derived from a generalized fagalean ancestor, but the possibility must also be entertained that the similarities reflect parallel changes from a common ancestry in the Hamamelidales.

## SELECTED REFERENCES

Moseley, M. F., Comparative anatomy and phylogeny of the Casuarinaceae. Bot. Gaz. 110: 232–280. 1948.

Swamy, B. G. L., A contribution to the life history of *Casuarina*. Proc. Am. Acad. 77: 1–32. 1948.

Ueno, J., On the fine structure of the pollen walls of Angiospermae. III. *Casuarina*. Grana palyn. 4: 189–193. 1963.

## Subclass III. Caryophyllidae

The subclass Caryophyllidae as here defined consists of four orders, 14 families, and about 11,000 species. Eleven of the families and some 10,000 of the species belong to the order Caryophyllales. The other three orders consist of single families, the Polygonales with about 800 species, the Plumbaginales with about 350, and the Batales with only two. Each of the three smaller orders has by one author or another been included in the Caryophyllales, but no author (so far as I know) has included them all at once. It is purely a matter of opinion whether one should lump or split in each case. One of the families here included in the Caryophyllales, the Cactaceae, has often been treated as a separate order, and again either course could be defended. Still another family, the Didiereaceae, has only recently been recognized as being of this affinity. The definition of the Caryophyllidae here adopted conforms to that of Takhtajan except in that the Theligonaceae are here excluded and referred somewhat doubtfully to the order Haloragales, in the Rosidae.

The Caryophyllales have trinucleate pollen, with the single known exception of *Batis*, which may not properly belong to the subclass. This is the only subclass, and the largest coherent group other than the Compositae and the Cyperales, to have trinucleate pollen.

In formal terms, the Caryophyllidae may be thought of as dicots with trinucleate pollen, bitegmic, crassinucellate ovules, and either with betalains (betacyanins and betaxanthins) instead of anthocyanins, or with free-central to basal placentation, or both. On the basis of tests which have been conducted so far, it may be projected that more than two-thirds of the species of the subclass will prove to have betalains. Betalains are, so far as known, wholly restricted to the order Caryophyllales. They are further discussed under that order.

The corolla, or apparent corolla, of various members of the Caryophyllidae is of diverse origins. It seems probable that within the subclass the perianth primitively consists of five sepals (or five sepaloid tepals, if you will), as in the Phytolaccaceae. From this base, different groups have by various routes developed a seemingly or actually dichlamydeous perianth with apparent sepals and petals.

In the Cactaceae and Aizoaceae the number of tepals is greatly increased, in association with an increase in the number of stamens. The inner tepals become petaloid, whereas the outer ones remain sepaloid. The result is a perianth that is more or less differentiated into sepals and petals, but without any clear arrangement into alternating cycles.

In the Nyctaginaceae, on the other hand, the entire perianth becomes petaloid, simulating a sympetalous corolla. Various members of the family show intermediate stages in the transformation of some of the bracts of the inflorescence into a false calyx under this petaloid perianth. In the common four o'clock (*Mirabilis*) the transformation is complete, and the flower gives every appearance of having a synsepalous calyx and a sympetalous corolla. Were it not for the transitional forms still in existence, including some in which two or more flowers are subtended by a single false calyx of sepaloid bracts, it would surely be so interpreted.

The Basellaceae and Portulacaceae have usually been interpreted as

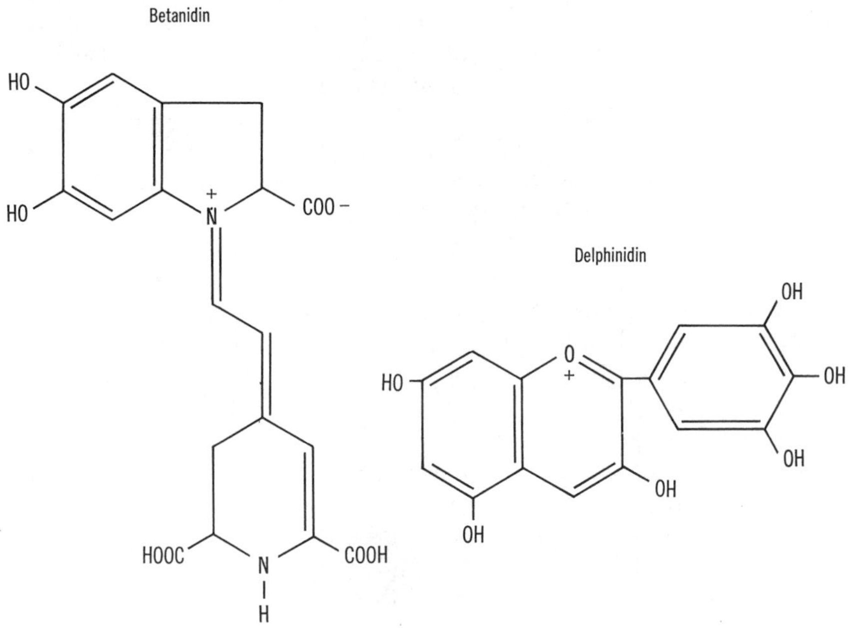

Betanidin

Delphinidin

FIGURE 4.4

Structural formula (left) of betadinin, a typical betacyanin, and (right) of delphinidin, a typical anthocyanin.

having a well defined calyx and corolla, the calyx usually consisting of two sepals. More recently it has been suggested that these two "sepals" are actually bracts, and that the "true" perianth consists only of the apparent petals, just as in the Nyctaginaceae.

The petals of the Caryophyllaceae, on the other hand, have been interpreted as modified stamens or parts of stamens. The resemblance of the bifid petals of species of *Stellaria* to stamens with a pair of "stipular" appendages in the same flower certainly suggests a possible homology. A similar interpretation has been advanced for the corolla of the Plumbaginaceae.

The Amaranthaceae show a strong tendency toward the development of a filament tube, and in extreme types such as *Froelichia* the androecium has exactly the appearance of a sympetalous corolla with epipetalous stamens attached at the throat and alternating with the corolla lobes. Here again it is only the existence of transitional types which permits a more sophisticated interpretation.

In the Polygonaceae, the genus *Polygonum* typically has a perianth of five basally united, quincuncial sepals, but the three outer ones are often larger than the two inner. In *Rumex* there are six tepals in two whorls, the two sets somewhat different, but both sepaloid. In *Eriogonum* there are two sets of three each, arising from a common stipelike base. Both sets are usually petaloid, but the outer and inner sets are not always alike.

We have noted in an earlier chapter that sepals and petals are necessarily defined more by their present position and appearance than by evolutionary homologies. It does seem necessary, however, in the interests of avoiding confusion, to insist on the evolutionary homology of parts given the same name in the same family. *Mirabilis* would acquire an aura of being much more different from other Nyctaginaceae than it really is, if its flowers were described as having a synsepalous calyx and a sympetalous corolla. Whether it is useful to speak of bracts and sepals instead of sepals and petals in the Portulacaceae is perhaps a matter of opinion.

The Caryophyllales are clearly the most primitive order in the subclass, as well as being by far the largest. Each of the other three orders may be regarded as being separately derived from the Caryophyllales, with the reservation that the Batales may not be of this alliance at all.

The Caryophyllales, in turn, take their origin in the order Magnoliales of the subclass Magnoliidae. Within the Magnoliales, the family Illiciaceae (with a single genus *Illicium*) has the general characters which would be required for a hypothetical common ancestor of the Caryophyllales. The relationship is presumably not very close, however, inasmuch as the specialized embryological and chemical features of the Caryophyllales find no obvious antecedents in *Illicium* or any other group of Magnoliidae.

Those families of the order Caryophyllales which have numerous stamens have the stamens initiated in centrifugal rather than centripetal sequence. In this respect the Caryophyllidae resemble the Dilleniidae. Outside of these two subclasses, centrifugal stamens are known only in the Alismatales, of the class Liliatae. The report of centrifugal stamens in *Drimys*, of the Winteraceae, was based on sequence of maturation rather than initiation of the stamens, and it is only the sequence of *initiation* which has turned out to be of critical taxonomic importance among the angiosperms in general. This character is further discussed under the order Dilleniales.

Any possible common ancestry of the Caryophyllidae with the Dilleniidae must be at or near the Magnolialean level, inasmuch as both subclasses contain apocarpous members. The embryological features which mark even the more primitive members of the Caryophylliidae are not known to be matched by any of the Dilleniidae.

The adaptive significance of the characters which mark the Caryophylliidae is wholly obscure. The group has an unusually high proportion of succulents and halophytes (amounting to probably more than half of

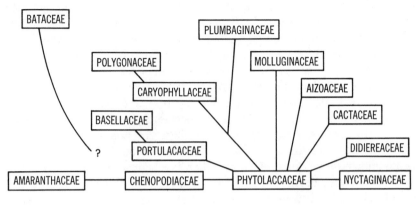

FIGURE 4.5

Probable relationships among the families of Caryophyllidae. The Bataceae, Plumbaginaceae, and Polygonaceae represent separate, unifamilial orders. The other families all belong to the order Caryophyllales.

the species), but similar ecological types are known in other subclasses, and various members of the group are perfectly ordinary trees, or shrubs, or herbs. There is no obvious relationship between the technical characters of the subclass and the resistance to desiccation shown by many of its members.

SYNOPTICAL ARRANGEMENT OF THE ORDERS OF CARYOPHYLLIDAE

1.  Seeds with very scanty or no true endosperm, but very often with an evident perisperm; betalains often present; plants often with anomalous secondary thickening

2.  Ovules campylotropous or amphitropous, seldom anatropous; seeds with a curved embryo that is typically peripheral to a well defined perisperm; pollen trinucleate; flowers seldom very much reduced or borne in catkins, but if so (some Chenopodiaceae) then with unilocular, uniovulate ovary                                                    **1. Caryophyllales**

2.  Ovules anatropous; seeds with nearly or quite straight embryo and neither endosperm nor perisperm; pollen binucleate; flowers very much reduced, borne in unisexual catkins; ovary 4–locular, with one ovule per locule; plants lacking both anthocyanins and betalains                        **2. Batales**

1. Seeds with more or less copious endosperm, but without perisperm; embryo straight or curved, peripheral or often embedded in the endosperm; ovary unilocular, with a single
   basal ovule; no betalains; no anomalous secondary thickening

3. Perianth undifferentiated, or with 2 sets of more or less similar
   tepals, not dichlamydeous and sympetalous; carpels (2)
   3 (4); stamens mostly 6–9, in 2 or 3 cycles; plants often
   with sheathing stipules                    **3. Polygonales**

3. Perianth differentiated into calyx and corolla, the corolla sympetalous; carpels 5; stamens 5, opposite the petals; leaves
   mostly exstipulate                        **4. Plumbaginales**

ORDER 1. **Caryophyllales**

The Caryophyllales are perhaps better known as the Centrospermae, in
reference to the free-central or basal placentation of some of the families.
Descriptive names for orders are still permitted under the rules of nomenclature, but they are going out of style. The name Caryophyllales has a
long history of usage more or less parallel to the name Centrospermae
and is here adopted. Formal priority does not enter into consideration,
because it does not apply to names above the rank of a family.

About four-fifths of the 10,000 odd species of Caryophyllales belong to
only four of the 11 families, the Aizoacease (2500), Cactaceae (2000),
Caryophyllaceae (2000), and Chenopodiaceae (1500). Of the remaining
2000 species, some 900 belong to the Amaranthaceae, 500 to the Portu-
lacacaea, 300 to the Nyctaginaceae, 150 to the Phytolaccaceae, 100 to the
Molluginaceae, and only 20 to the Basellaceae and nine to the Didiereaceae.

The Caryophyllales are embryologically very distinctive. They characteristically have campylotropous or amphitropous (rarely anatropous), bitegmic, crassinucellate ovules, commonly with the inner integument longer
than the outer, and very often with a space between the integuments
toward the chalazal end. The seed has a curved, peripheral embryo more
or less surrounding the food storage tissue, which consists mainly of perisperm, with little or no endosperm. Sometimes both endosperm and perisperm are wanting. No plants outside the order are known to present this
combination of characters.

Five of the 11 families of the Caryophyllales, embracing about half of
the species, have free-central or basal placentation in a compound ovary,
a condition known in relatively few groups (notably the Primulales) outside the subclass. These five families (Amaranthaceae, Basellaceae, Caryophyllaceae, Chenopodiaceae, and Portulacaceae) are sometimes regarded
as the core of the order, to which the others have been added on the basis

of similarities in various other respects. It should be noted, however, that the Caryophyllaceae differ from the four other "core" families, and from most of the order as a whole, in having anthocyanins and lacking betalains.

Eight of the 11 families of the order have anomalous secondary thickening in at least some species, usually resulting in the formation of concentric rings of vascular bundles. In three of these families, the Aizoaceae, Amaranthaceae, and Chenopodiaceae, this feature is so consistent as to be almost a family character, and it also turns up very often in the Nyctaginaceae. In the other four families of the eight it is more sporadic. It is clear that although the tendency to develop anomalous secondary thickening permeates these eight families, the realization of the tendency has occurred independently many times. The Basellaceae, Didiereaceae, and Portulacaceae are not reported to have anomalous secondary thickening, but we should not be surprised to see it turn up somewhere when more genera and species are studied.

Three of the families of the order, the Aizoaceae, Cactaceae, and Didiereaceae, are characteristically succulent, and a fourth family, the Portulacaceae, has many succulent members. Succulents are also known in several other families of the order, notaby the Basellaceae, Chenopodiaceae, and Phytoloaccaceae. Many of the Chenopodiaceae are halophytes (with or without succulence), and it may well be that succulence and halophytism are merely two expressions of the same physiological evolutionary potentialities.

Nine of the 11 families of the order have betalains instead of anthocyanins, according to the most recent reports. Two families (Caryophyllaceae and Molluginaceae), on the other hand, are reported to have anthocyanins and lack betalains. If the Molluginaceae are (as here) held as a separate family instead of being submerged in the Aizoaceae, no family is known to have both anthocyanins and betalains, except as noted in the next paragraph.

The single recorded exception to the mutual antipathy of betalains and anthocyanins has been overlooked or ignored by most later authors. Gascoigne, Ritchie, and White (cited below) found both betacyanin and anthocyanin in the flowers of a species of *Mesembryanthemum*. Obviously their finding should be rechecked instead of ignored.

The betalains are a unique group of vascular pigments, resembling the flavonoids in certain physical properties, but belonging to an entirely different class of chemical compounds. Betalains in the blue-violet to red series are called betacyanins, and those in the yellow to orange-red series are called betaxanthins, the terms being deliberately parallel to anthocyanin and anthoxanthin. The betaxanthins have only recently been recognized as a distinct group closely allied to the betacyanins; earlier work on the betalains was concentrated entirely on the betacyanins, and indeed the term betalains was only introduced in 1967.

Unlike the anthocyanins, the betalains contain nitrogen. Before their structure was understood, the betacyanins were often called nitrogenous anthocyanins. The red pigment of garden beets is a betacyanin, and this class of pigments is in fact named for the genus *Beta*.

Except as noted with regard to *Mesembryanthemum*, no individual plant, and indeed no genus or even family of plants, has yet been found to contain both betacyanins and anthocyanins. It would therefore appear that the two groups of pigments may be functionally equivalent in spite of their chemical differences. The actual function is in both cases wholly unknown, save for the fact that they often serve as flower pigments. It should be noted, however, that plants which have betalains instead of anthocyanins commonly have various other flavonoid compounds, and specifically that anthoxanthins and betaxanthins often occur together.[4] Therefore the betalain taxa evidently have most of the genetic mechanism necessary for the production of anthocyanins.

The circumstances which would lead to the substitution of betalains for anthocyanins are obscure, especially inasmuch as the two are chemically unrelated. It has been surmised that the substitution may have happened only once, and that therefore it must be of great phyletic and taxonomic significance. Some authors have gone so far as to propose that the Centrospermae be defined on this character alone, with the Caryophyllaceae and Molluginaceae being consigned to an undefined limbo outside the order. Such a definition goes counter to the great weight of other evidence, however, and taxonomists have learned through painful experience to be suspicious of groups arbitrarily defined by a single character. Parallelism is so rampant among angiosperms in general that one need not expect this character to be a notable exception. The Caryophyllaceae and Molluginaceae may have retained the ancestral pigment-complex while other members of the subclass were acquiring betalains, or they may have lost their betalains and reverted to the production of anthocyanins.

The reader may have noted that the traditional Centrospermae might be defined in any of several ways, according to which character is emphasized. Among the 11 families here referred to the group, only the Amaranthaceae and Chenopodiaceae have all the features we have discussed as characterizing the order — i.e., the full list of embryological features, plus betalains, anomalous secondary thickening, and basal or free-central placentation. Among the five core families mentioned in a previous paragraph, the Car-

[4] R. Alston, "Biochemical Systematics," in Th. Dobzhansky, M. K. Hecht, and W. C. Steere, eds., *Evolutionary Biology* (New York: Appleton-Century-Crofts, 1967), pp. 197–305 (see p. 251).

yophyllaceae would be excluded by emphasis of the pigment character, and the Portulacaceae and Basellaceae would be excluded by emphasis of the nature of secondary thickening.

Nevertheless, the mutual affinity among all 11 of the families here referred to the Caryophyllales now appears to be well established. The Cactaceae and Didiereaceae might be excluded as separate, monotypic orders if one wished, but this would only add to the number of monotypic satellite orders without greatly increasing the homogeneity of the restricted Caryophyllales. The fact that *Didierea* has been successfully cross-grafted to a cactus may be immaterial, but the strong serological correspondence of *Allaudia* (Didiereaceae) with tested members of the Caryophyllales, plus the presence of betalains, would seem to be definitive.

The Phytolaccaceae are clearly the most primitive family of the Caryophyllaceae. All of the other families of the order might well have been derived directly or indirectly from the Phytolaccaceae (i.e., from plants which, if we had them, would be referred to the Phytolaccaceae). The Amaranthaceae are probably derived directly from the Chenopodiaceae, being more advanced in both of the critical characters (scarious perianth and connate filaments) which mark them as a group. The Basellaceae may well have been derived from the Portulacaceae, being more advanced in all of the obvious respects in which the two families differ. The remaining families of the order are probably all derived directly from the Phytolaccaceae.

Among the families which have traditionally been referred to the Caryophyllales, only the Theligonaceae (Cynocrambaceae) are here wholly excluded from this alliance. In the Caryophyllales they would be unusual in their basal style, unitegmic, tenuinucellate ovules, in having endosperm instead of perisperm, in the apparent absence of both anthocyanins and betalains, in the absence of anomalous secondary thickening, and in the unique and distinctive sculpture of the pollen. This is too long a list to ignore. In the present treatment they are referred somewhat doubtfully to the Haloragales, following the relationship suggested by Melchior in the 12th Engler Syllabus. The other alternative would be to treat them as a monotypic order of the Caryophyllidae, as proposed by Takhtajan (1966).

## SYNOPTICAL ARRANGEMENT OF THE FAMILIES OF CARYOPHYLLALES

1. Gynoecium apocarpous, monocarpous, or sometimes syncarpous, if syncarpous then usually with as many locules as carpels; ovules 1 per locule, or some locules empty

    2. Locules all ovuliferous (though not always fertile); perisperm

well developed; flowers perfect or less often unisexual; plants not cactuslike

3. Sepals mostly free or nearly so, seldom petaloid; carpels mostly 2–several, occasionally only one; inflorescence usually racemose or spicate; leaves alternate; stem seldom with anomalous secondary thickening  **1. Phytolaccaceae**

3. Sepals joined into a tube which resembles a sympetalous corolla, and often subtended by sepaloid bracts; carpel 1; inflorescence mostly cymose; leaves opposite or sometimes alternate; stem very often with anomalous secondary thickening                    **2. Nyctaginaceae**

2. Two locules empty, the third with a single ovule; perisperm apparently wanting; flowers unisexual; inflorescence cymose; spiny, cactuslike trees            **3. Didiereaceae**

1. Gynoecium mostly syncarpous, either unilocular or with the ovules more numerous than the locules, or both

4. Tepals and stamens usually more or less numerous; plants succulent; ovary very often inferior

5. Ovary unilocular, nearly always inferior, with parietal (rarely basal) placentation; mostly spiny stem-succulents with much reduced leaves; New World            **4. Cactaceae**

5. Ovary plurilocular, or occasionally unilocular with a free-central placenta; ovary superior to inferior; leaf-succulents, not spiny; chiefly Old World    **5. Aizoaceae**

4. Tepals and stamens mostly few and cyclic; plants sometimes with succulent leaves, but not spiny or cactuslike; ovary superior

6. Perianth usually dichlamydeous or seemingly so (except notably the Molluginaceae, with axile placentation)

7. Sepals 4 or 5; plants without betalains, but usually with anthocyanins; stamens not at once of the same number as and opposite the petals; plants sometimes with anomalous secondary thickening

8. Ovary 2–5-locular, with axile placentation; petals much reduced or commonly wanting; leaves opposite, alternate or whorled    **6. Molluginaceae**

8. Ovary unilocular (sometimes partitioned toward the base), with free-central or basal placentation; petals usually more or less well developed; leaves opposite
**7. Caryophyllaceae**

7. Sepals (or sepaloid bracts) mostly 2, seldom more; plants with betalains, lacking anthocyanins; stamens usually as many as and opposite the petals (or petaloid sepals), or sometimes more numerous; no anomalous secondary thickening

9. Ovules (1) 2–many; plants not climbing; internal phloem usually wanting; fruit usually capsular
**8. Portulacaceae**

9. Ovule solitary; climbers; internal phloem present; fruit indehiscent    **9. Basellaceae**

6. Perianth monochlamydeous, seldom petaloid, often small or sometimes even obsolete; ovules mostly solitary (rarely several) on a basal placenta; stem nearly always with anomalous secondary thickening

10. Perianth not scarious, often green; filaments mostly free; flowers perfect or not infrequently unisexual
**10. Chenopodiaceae**

10. Perianth more or less scarious; filaments mostly connate below (the androecial tube sometimes even simulating a sympetalous corolla); flowers mostly perfect, seldom unisexual    **11. Amaranthaceae**

### SELECTED REFERENCES

Bailey, I. W., Comparative anatomy of the leaf-bearing cacti. Jour. Arn. Arb. 41: 341–356. 1960. 42: 144–156; 334–346. 1961. 43: 187–202; 234–278; 376–388. 1962. 44: 127–137; 222–231; 390–401. 1963. 45: 140–157; 374–389. 1964. 46: 74–85; 445–461. 1965. 47: 273–287. 1966.

————, The significance of the reduction of vessels in the Cactaceae. Jour. Arn. Arb. 47: 288–292. 1966.

Bisalputra, T., Anatomical and morphological studies in the Chenopodiaceae. Austr. Jour. Bot. 8: 226–242. 1960. 9: 1–19. 1961. 10: 13–24. 1962.

Bocquet, G., The structure of the placental column in the genus *Melandrium* (Caryophyllaceae). Phytomorph. 9: 217–221. 1959.

Boke, N., The cactus gynoecium: a new interpretation. Am. Jour. Bot. 51: 598–610. 1964.

————, Ontogeny and structure of the flower and fruit of *Pereskia aculeata*. Am. Jour. Bot. 53: 534–542. 1966.

Buxbaum, F., Zur Klärung der phylogenetischen Stellung der Aizoaceae und Cactaceae im Pflanzenreich. Jahrb. Schweiz. Kakt.-Ges. 1948: 3–16.

————, Vorläufer des Kakteen-Habitus bei den Phytolaccaceen. Österr. Bot. Zeits. 96: 5–14. 1949.

————, Vorläufige Untersuchungen über Umfang, systematische Stellung und Gliederung der Caryophyllales (Centrospermae). Beitr. Biol. Pfl. 36: 3–56. 1961.

Eckardt, T., Morphologische und systematische Auswertung der Placentation von Phytolaccaceen. Ber. Deuts. Bot. Ges. 67: 113–128. 1954.

————, Blütenbau und Blütenentwicklung von *Dysphania myriocephala* Benth. Bot. Jahrb. 86: 20–37. 1967.

————, Vergleich von *Dysphania* mit *Chenopodium* und mit Illecebraceae. Bauhinia 3: 327–344. 1967.

Fahn, A., and T. Arzee, Vascularization of articulated Chenopodiaceae and the nature of their fleshy cortex. Am. Jour. Bot. 46: 330–338. 1959.

Friedrich, H. C., Studien über natürliche Verwandtschaft der Plumbaginales und Centrospermae. Phyton 6: 220–263. 1956.

Gascoigne, R. M., E. Ritchie, and D. E. White, A survey of anthocyanins in the Australian flora. Jour. Proc. Roy. Soc. N. S. Wales 82: 44–70. 1948 (1949).

Jensen, U., Serologische Untersuchungen zur Frage der systematischen Einordnung der Didiereaceae. Bot. Jahrb. 84: 233–253. 1964.

Kajale, L. B., A contribution to the embryology of the Phytolaccaceae. Jour. Indian Bot. Soc. 33: 206–225. 1954.

Mabry, T., "The Betacyanins and Betaxanthins," in T. Swain, ed., *Comparative Phytochemistry* (London and New York: Academic Press, 1966), pp. 231–244.

Mabry, T., and A. S. Dreiding, The betalains. Recent advances in phytochemistry, in press, 1967. (Manuscript available in advance of publication)

Rauh, W., and H. Reznik, Zur Frage der systematischen Stellung der Didiereaceen. Bot. Jahrb. 81: 94–105. 1961.

Rohweder, O., Centrospermen-Studien 2. Entwicklung und morphologische Deutung das Gynöciums bei *Phytolacca*. Bot. Jahrb. 84: 509–526. 1965.

Roth, I., Histogenese und morphologische Deutung der basalen Plazenta von *Herniaria*. Flora 152: 179–195. 1962.

————, Histogenese und morphologische Deutung der Zentralplazenta von *Cerastium*. Bot. Jahrb. 82: 100–118. 1963.

Schölch, H. F., Die systematische Stellung der Didieraceen im Lichte neuer Untersuchung über ihren Blütenbereich. Ber. Deuts. Bot. Ges. 76: (49)– (55). 1963.

Sharma, H. P., Studies in the order Centrospermales. I. Vascular anatomy of the flower of certain species of the Portulacaceae. Jour. Indian Bot. Soc. 33: 98–111. 1954. II. Vascular anatomy of the flower of certain species of the Molluginaceae. Jour. Indian Bot. Soc. 42: 19–32. 1963. III. Vascular anatomy of the flower of some species of the family Ficoidaceae. Proc. Indian Acad. Sci. 56B: 269–285. 1962. IV. Pollen morphology of some species of families Ficoidaceae, Molluginaceae, Nyctaginaceae, and Portulacaceae. Jour. Indian Bot. Soc. 42: 637–645. 1963.

Vishnu-Mittre, Pollen morphology of Indian Amaranthaceae. Jour. Indian Bot. Soc. 42: 86–101. 1963.

Vishnu-Mittre and H. P. Gupta, Studies of Indian pollen grains. III. Caryophyllaceae. Pollen et Spores 6: 99–111. 1964.

ORDER 2. **Batales**

The Bataceae, with only the small genus *Batis*, may or may not be closely allied to the Caryophyllales. The habital resemblance of *Batis* to *Sarcobatus*, of the Chenopodiaceae, can scarcely be overlooked, and the maritime habitat of *Batis* is in accord with the alkaline, inland habitat of *Sarcobatus* and many other Chenopodiaceae. Both genera have the staminate flowers in catkins, both have the pistillate flowers wholly without perianth, and both lack perisperm as well as endosperm.

There are also notable differences, however. *Batis* has a 4-locular ovary, with one ovule per locule, wholly unlike the ovary of anything in the Chenopodiaceae, and indeed unlike anything in the Caryophyllidae except to some extent some of the Phytolaccaceae. The embryo is straight instead of curved, in contrast to the Caryophyllales in general. *Batis* lacks both anthocyanins and betalains, so that no strong evidence is available from this source. The mere absence of a structure or chemical substance is, as we have noted, a poor guide to affinity. The absence of both perisperm and endosperm from the seed again provides no help. The facts that the female flowers are in catkins (instead of solitary as in *Sarcobatus*) and that the staminate flowers have a peculiar, tiny perianth are perhaps not incompatible with a Caryophyllalean relationship, but they do nothing to bolster the thought of an affinity with *Sarcobatus*. The binucleate pollen of *Batis* presents another difficulty. Inasmuch as *Batis* is clearly an advanced, reduced type, its assignment to the Caryophyllidae would imply an evolutionary reversal from a trinucleate ancestry. Such a reversal is perhaps not impossible, but there is no other place in the whole system where the evidence seems to require it.

Altogether, *Batis* would be highly anomalous in the Caryophyllales, and also somewhat anomalous in the Caryophyllidae. On the other hand, no other likely allies present themselves to mind. On purely formal morphology *Batis* could be referred to the Hamamelidae, but it has no obvious allies there and would still have to be treated as a separate order. At the present state of our knowledge, it seems best to maintain *Batis* as a separate order, and to refer this order somewhat doubtfully to the Caryophyllidae. Serological studies might be useful in resolving the present uncertainties.

## SELECTED REFERENCES

Eckardt, T., Das Blutendiagramm von *Batis*. Ber. Deuts. Bot. Ges. 72: 411–418. 1960.

Johnson, D. S., The development of the shoot, male flower and seedling of *Batis maritima*. Bull. Torrey Club 62: 19–31. 1935.

Mabry, T. J., and B. L. Turner, Chemical investigations of the Batidaceae. Betaxanthins and their systematic implications. Taxon 13: 197–200. 1964.

ORDERS 3 AND 4. **Polygonales** and **Plumbaginales**

Both the Polygonaceae and Plumbaginaceae are pretty clearly related to the Caryophyllales. Both have a single, basal, bitegmic, crassinucellate ovule in a compound, unilocular ovary, and both have trinucleate pollen. These characters are not known to occur in combination outside the subclass Caryophyllidae. The Polygonaceae are further bound to the Caryophyllales by similarities in the pollen and by a more or less transitional group of genera which are variously referred to the Caryophyllaceae or treated as a separate family Illecebraceae. The Plumbaginaceae are somewhat more isolated but may also be derived from the Caryophyllaceae.

It may be noted that the Caryophyllaceae resemble the Polygonaceae and Plumbaginaceae in having anthocyanins, not betacyanins. At the present state of our knowledge one would be uneasy about a scheme which required the repeated renascence of anthocyanins to replace the chemically very dissimilar betacyanins. Although these three families are here linked chiefly on the basis of other evidence, it is comforting to think that the maximum number of reversals in pigment type that might be required is only one.

The putative relationship of the Polygonaceae and Plumbaginaceae to the Caryophyllaceae implies an evolutionary reversal in the nature of the food storage tissue of the seed. I accept this thought with a reasonable degree of equanimity. Inasmuch as a more or less vestigial endosperm occurs in many of the Caryophyllales, and inasmuch as endosperm is presumably initiated (via a triple fusion nucleus) even in those ovules which are wholly without endosperm at maturity, there is no necessity to re-evolve a structure which has been completely lost. All that is required is a shift in the proportions and a delay in the time at which the endosperm degenerates. The observed fact that many evolutionary trends appear to continue indefinitely without reversals should not blind us to other possibilities.

In spite of their probable relationship, it is expedient to exclude both the Polygonaceae and the Plumbaginaceae from the Caryophyllales to keep the order from being too heterogeneous. We have noted that both families have endosperm instead of perisperm, and that both have anthocyanins instead of betacyanins. The embryo in the Plumbaginaceae is straight, and that in the Polygonaceae is straight or variously curved. Neither family has campylotropous or amphitropous ovules. In the Plumbaginaceae they are anatropous, and in the Polygonaceae usually orthotropous. The strongly sympetalous corolla of the Plumbaginaceae is an additional distinguishing feature of this family.

## SELECTED REFERENCES

Channell, R. B., and C. E. Wood, Jr., The genera of Plumbaginaceae in the southeastern United States. Jour. Arn. Arb. 40: 391–397. 1959.

Friedrich, H. C., Studien über der natürliche Verwandtschaft der Plumbaginales und Centrospermae. Phyton 6: 220–263. 1956.

Graham, S., and C. E. Wood, Jr., The genera of Polygonaceae in the southeastern United States. Jour. Arn. Arb. 46: 91–121. 1965.

Roth, I., Sobre el desarrollo del obturator en el gineco de "Armeria." Acta Cient. Venez. 12: 172–174. 1961.

————, Histogenese und morphologische Deutung der basilären plazenta von Armeria. Osterr. Bot. Zeits. 109: 18–40. 1962.

————, Histogenese und morphologische Deutung der Kronblätter von Armeria. Portug. Acta Biol. A6: 211–250. 1962.

Vautier, S., La vascularization florale chez les Polygonacées. Candollea 12: 219–343. 1949.

Veillet-Bartoszewska, M., Embryogénie des Plombagacées. Developpement de l'embryon chez le Plumbago europaea L. Compt. Rend. Acad. Sci. Fr. 247: 2178–2181. 1958.

## Subclass IV. Dilleniidae

The subclass Dilleniidae as here defined consists of 12 orders, 69 families, and nearly 24,000 species. More than three-fourths of the species belong to only five orders, the Violales (5200), Capparales (4000), Malvales (3500), Theales (3000), and Ericales (3000). The Primulales (1900) and Ebenales (1700) are also fairly large orders, whilst the remaining five orders have only about 1400 species amongst them.

The 12 orders which make up the Dilleniidae evidently hang together as a natural group, but this group cannot be fully characterized morphologically. Like the Rosidae, the Dilleniidae are more advanced than the Magnoliidae in one or another respect, but less advanced than the Asteridae. Except for the rather small (400 spp) order Dilleniales, the Dilleniidae are sharply set off from typical members of the Magnoliidae by being syncarpous. The Dilleniales are undoubtedly closely allied to the Theales, however, and they differ from at least most of the Magnoliidae in having centrifugal stamens. Insofar as they have been investigated, the species of Dilleniidae with numerous stamens have the stamens initiated in centrifugal sequence; in this respect they differ sharply from the subclass Rosidae, in which the species with numerous stamens are always, so far as known, centripetal. More than a third of the species of Dilleniidae have parietal placentation, in contrast to the relative rarity of this type in the subclass Rosidae. About a third of the species (not the same third) are sympetalous, but only a very few of these have isomerous stamens alternate with the corolla lobes and also unitegmic, tenuinucellate ovules as in the subclass Asteridae. The ovules in the Dilleniidae as a whole are bitegmic or less commonly unitegmic, with various transitional types, and they range from crassinucellate to tenuinucellate. The pollen is chiefly binucleate, but some groups, notably the large family Cruciferae (3000 spp), are trinucleate. Only a single family and order (Salicales) is amentiferous. This order differs from the subclass Hamamelidae in its parietal placentation and numerous seeds.

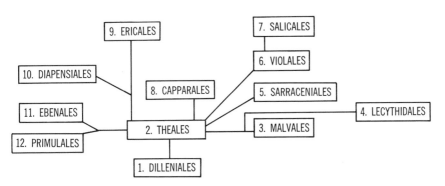

FIGURE 4.6

Probable relationships among the orders of Dilleniidae.

It is perfectly clear that the Dilleniidae take their origin in the Magnoliidae. The apocarpous order Dilleniales, especially the family Dilleniaceae, forms the connecting link between the two subclasses. If the rest of the subclass Dilleniidae did not exist, the order Dilleniales could easily be accommodated in the Magnoliidae. Within the Magnoliidae, the family Illiciaceae in the order Magnoliales may be somewhere near to the ancestry of the Dilleniaceae, but the relationship between the two families is still not particularly close.

The Dilleniales appear to be directly ancestral to the closely related order Theales, which differs mainly in being syncarpous. All the other orders of the subclass are derived directly or indirectly from the Theales. Sympetaly, isomerous stamens, parietal placentation, and unitegmic, tenuinucellate ovules have evolved independently several times among the orders derived from the Theales.

The characters which mark the Dilleniidae as a group have no obvious ecologic or adaptive importance. The biological significance of the sequence of initiation of stamens is obscure, as is also that of the kind of placentation. The number of ovules in a locule would seem a priori to be of some possible significance in relation to seed dispersal, but a broad-scale evaluation of that significance remains to be presented.

### SYNOPTICAL ARRANGEMENT OF THE ORDERS OF DILLENIIDAE

1. Carpels separate; stamens numerous; ovules bitegmic, crassinucellate; seeds arillate and with a well developed endosperm
   **1. Dilleniales**

1. Carpels united to form a compound pistil; other characters various, the stamens numerous to few, the ovules bitegmic or unitegmic, crassinucellate to tenuinucellate, the seeds with or without an aril and with or without endosperm

   2. Flowers mostly polypetalous or apetalous, seldom (mainly some Violales) sympetalous; stamens numerous to few; ovules chiefly bitegmic and crassinucellate, but sometimes unitegmic, or tenuinucellate, or both

      3. Plants insectivorous herbs or shrubs    **5. Sarraceniales**

      3. Plants not insectivorous; habit various

         4. Placentation mostly axile, seldom parietal

5. Sepals imbricate; filaments free or connate in groups; mucilage cells and sacs mostly wanting **2. Theales**

5. Sepals valvate; filaments often connate into a tube or column; mucilage cells or sacs very often present

6. Ovary superior; pubescence very often stellate or lepidote **3. Malvales**

6. Ovary inferior; no stellate or lepidote hairs **4. Lecythidales**

4. Placentation mostly parietal, seldom axile

7. Flowers normally developed, with an evident perianth, perfect or less often unisexual, not borne in catkins

8. Plants usually without myrosin cells, very often woody; flowers hypogynous to often perigynous; or epigynous; leaves mostly simple; carpels mostly 3 or more, seldom only 2; perianth seldom tetramerous **6. Violales**

8. Plants commonly with myrosin cells, herbaceous or less often woody; flowers hypogynous; leaves simple to more often compound or variously dissected; carpels most commonly 2; perianth often tetramerous **8. Capparales**

7. Flowers much reduced, unisexual, without an evident perianth, borne in catkins; woody plants **7. Salicales**

2. Flowers mostly sympetalous (except mainly some Ericales, these with unitegmic, tenuinucellate ovules); stamens seldom more than 2 or 3 times as many as the corolla lobes

9. Placentation axile or less often parietal; stamens various, often more numerous than or alternate with the corolla lobes

10. Plants with the pollen in tetrads (sometimes only one grain of the tetrad maturing), or with poricidal anthers,

or lacking chlorophyll, often with more than one of
these characters; ovules unitegmic, tenuinucellate
**9. Ericales**

10. Plants otherwise, the pollen in monads, the anthers opening by longitudinal slits, the shoot always chlorophyllous; ovules bitegmic or unitegmic, crassinucellate or
tenuinucellate

11. Plants dwarf (commonly prostrate) shrubs; functional
stamens as many as and alternate with the corolla
lobes; ovules unitegmic　　**10. Diapensiales**

11. Plants shrubs or trees; stamens usually either more numerous than or opposite the corolla lobes; ovules
bitegmic or unitegmic　　**11. Ebenales**

9. Placentation free-central or basal; functional stamens as many
as and opposite to the corolla lobes; ovules chiefly bitegmic　　**12. Primulales**

## ORDER 1. Dilleniales

The order Dilleniales as here defined consists of three families, the Dilleniaceae, with about 350 species, the Paeoniaceae with a single genus of
some 33 species, and the Crossosomataceae with a single genus of only
three or four species. The Dilleniaceae may be regarded as the core of the
order, to which the other two families have been added by accretion.

Each of the three families is sharply set off from the others, and they
have not always been considered to be closely related. Nonetheless, they
have a considerable list of features in common. They all have imbricate,
persistent sepals, numerous, centrifugal stamens, binucleate pollen, an apocarpous gynoecium, and bitegmic, crassinucellate ovules which mature into
arillate seeds with well developed endosperm. No other family or order
has this combination of characters, and indeed the combination of centrifugal stamens and separate carpels would by itself distinguish the order
from all other groups of dicots except some of the Phytolaccaceae.

The Dilleniales appear to be derived from the Magnoliales and ancestral
to the Theales and the remainder of the subclass Dilleniidae. It is an open
question whether the Dilleniales should be maintained as a distinct order or submerged in an expanded order Theales. The wholly apocarpous
gynoecium of the Dilleniales, as contrasted to the more or less syncarpous
gynoecium of the Theales, furnishes a convenient character for those who
maintain the distinction into two orders, but the close relationship between

the two groups is widely admitted. On the other hand, authors are agreed that although the Dilleniales are related to the Magnoliales, these two orders should not be combined.

The only character which clearly distinguishes the Dilleniales from the Magnoliidae is the sequence of initiation of the stamens. *Drimys*, in the Winteraceae (Magnoliales) has been reported to have centrifugal stamens, but the report was based on the sequence of formation of the pollen tetrads, rather than on the staminal primordia, which develop in the normal, centripetal sequence. It is now clear that the sequence of maturation of pollen in various genera may or may not conform to the sequence of development of the staminal primordia, and that it is only the primordial sequence which is of major taxonomic importance.[5]

Some investigators of *Paeonia*, *Hypericum*, and the Caryophyllales have maintained that the number of stamens in these groups has been secondarily increased after an earlier reduction to a single whorl. As noted in Chapter 3, I consider this unlikely. So long as the stamens are of indefinite number, the number may increase or decrease, and it seems likely that in some of the Caryophyllales there has actually been such an increase. The possible origin of multistaminate, acyclic androecia from paucistaminate, cyclic ones is quite another matter. In my opinion the multistaminate, centrifugal androecium is derived from a multistaminate, centripetal one, without the intervention of a paucistaminate, cyclic stage.

*Paeonia* has until recent years usually been referred to the Ranunculaceae, where it is anomalous in its persistent sepals, in its sepal-derived rather than staminodial petals, in its perigynous disk, in its ariliferous seeds, and in anatomical, cytological, palynological, and serological characters. The idea that *Paeonia* is closely related to the Ranunculaceae can no longer be seriously entertained.

Beginning with Camp[6] in 1950, most authors have treated *Paeonia* as a separate family to be associated with the Dilleniaceae. This seems reasonable, and no other obvious possibility presents itself to mind. Even so, there are some notable differences between the two families, and it is a matter of opinion whether taxonomic needs are best met by putting them into the same order or by keeping the Paeoniaceae in an order of their own. The report that *Paeonia* has a free-nuclear stage in the development of the embryo, unlike other angiosperms, has by later authors been denied, and then reaffirmed. I do not feel competent to hold an independent opinion on this point.

*Crossosoma* has attracted less attention than *Paeonia*, and its affinities

[5] P. Hiepko, Das zentrifugale Androecium der Paeoniaceae. Ber. Deuts. Bot. Ges. 77: 427–435. 1965.

[6] W. H. Camp, in A. Gundersen, Families of dicotyledons. Chronica Botanica Co. Waltham, Mass. 1950.

are correspondingly less clear. It has usually been treated as a distinct family allied to the Rosaceae, and there is nothing in the gross morphology to negate this disposition. However, it is reported to have centrifugal stamens, and its floral anatomy is, according to Eames,[7] more compatible with that of the Paeoniaceae and Dilleniaceae than with that of the Rosaceae. Both the Crossosomataceae and the Paeoniaceae have more advanced, specialized xylem than the Dilleniaceae, but the specializations in the two differ. At the present state of our knowledge it seems best to follow Camp in associating both the Paeoniaceae and the Crossosomataceae with the Dilleniaceae.

The adaptive significance of the characters which mark the Dilleniales is uncertain. The aril on the seed would seem to be of some importance in distribution or germination or both, but it does not obviously tend to fit the order to any particular ecologic niche to the exclusion of other groups. The ecological significance of centrifugal stamens is even more obscure. Most of the characters which distinguish the three families of the order from each other are also of dubious biological importance, although the xeromorphic aspect of *Crossosoma* is doubtless adaptive. The significance of the crystals in the Dilleniaceae is wholly obscure.

### SYNOPTICAL ARRANGEMENT OF THE FAMILIES OF DILLENIALES

1. Flowers hypogynous, with neither a hypanthium nor an intrastaminal disk; raphides and crystal sand present, often in long sacs or tubes; leaves mostly entire; trees, woody vines, and shrubs, seldom herbs, of tropical and subtropical regions, especially Australia                    **1. Dilleniaceae**

1. Flowers perigynous, or with an intrastaminal disk; raphides and crystal sand absent; North Temperate regions

    2. Seed with copious endosperm and small embryo; flowers with an intrastaminal disk, but without an evident hypanthium; mesophytic shrubs or herbs with large, usually cleft leaves
    **2. Paeoniaceae**

    2. Seed with thin endosperm and large embryo; flowers with an evident hypanthium, but without an intrastaminal disk; xerophytic shrubs with small, entire leaves
    **3. Crossosomataceae**

[7] A. Eames, Floral anatomy of an aid in generic limitation. Chron. Bot. 14: 126–132. 1953.

# SELECTED REFERENCES

Camp, W. H., and M. M. Hubbard, Vascular supply and structure of the ovule and its aril in peony and of the aril in nutmeg. Am. Jour. Bot. 50: 174–178. 1963.

Carniel, K., Über die Embryologie in der Gattung *Paeonia*. Österr. Bot. Zeits. 114: 4–19. 1967.

Cave, M. S., H. J. Arnott, and S. A. Cook, Embryogeny in the California peonies with reference to their taxonomic position. Am. Jour. Bot. 48: 397–404. 1961.

Dickison, W. C., Comparative morphological studies in Dilleniaceae. I. Wood anatomy. Jour. Arn. Arb. 48: 1–29. 1967. II. The pollen. Jour. Arn. Arb. 48: 231–240. 1967.

Gregory, W. C., Phylogenetic and cytological studies in Ranunculaceae. Trans. Am. Phil. Soc. II. 31: 443–521. 1941. (For cytological isolation of *Paeonia* from the Ranunculaceae)

Hammond, H. D., Systematic serological studies in Ranunculaceae. Serol. Mus. Bull. 14: 1–3. 1955. (For serological isolation of *Paeonia* from the Ranunculaceae)

Hiepko, P., Vergleichend-morphologische und entwicklungsgechichtliche Untersuchungen über das Perianth bei den Polycarpicae. Bot. Jahrb. 84: 359–508. 1965. (For nature of the perianth in *Paeonia*)

―――――, Das zentrifugale Androecium der Paeoniaceae. Ber. Deuts. Bot. Ges. 77: 427–435. 1965.

―――――, Zur Morphologie, Anatomie und Function des Diskus der Paeoniaceae. Ber. Deuts. Bot. Ges. 79: 233–245. 1966.

Kapil, R. N., and R. S. Vani, Embryology and systematic position of *Crossosoma californica* Nutt. Curr. Sci. (India) 32: 493. 1963.

Lemesle, R., Position phylogenetique de l'*Hydrastis canadensis* L. et du *Crossosoma californicum* Nutt., d'après les particularities histologiques du xylem. Compt. Rend. Acad. Sci. Paris 227: 221–223. 1948.

Richardson, P., The morphology of the Crossosomataceae: leaf, stem, and node. (Abstract) Am. Jour. Bot. 54: 659. 1967.

Sastri, R. L. N., Floral morphology and embryology of some Dilleniaceae. Bot. Notis. 111: 495–511. 1958.

Wilson, C. L., The floral anatomy of the Dilleniaceae. I. *Hibbertia* Andr. Phytomorph. 15: 248–274. 1965.

Worsdell, W. C., The affinities of *Paeonia*. Jour. Bot. 46: 114–116. 1908.

Wunderlich, R., Zur Deutung der eigenartigen Embryoentwicklung von *Paeonia*. Österr. Bot. Zeits. 113: 395–407. 1966.

Yakolev, M. S., and M. D. Yoffe, On some peculiar features in the embryogeny of *Paeonia* L. Phytomorph. 7: 74–82. 1957.

ORDER 2. **Theales**

The order Theales (Guttiferales) as here defined consists of 13 families and nearly 3000 species. More than 2600 of the species belong to only five of the families: the Guttiferae (900), Theaceae (600), Dipterocarpaceae (400), Ochnaceae (400), and Actinidiaceae (330). The Guttiferae are here taken in the broad sense, to include the Hypericaceae. The remaining families of the order are the Marcgraviaceae (120), Elatinaceae (45), Quiinaceae (37), Sarcolaenaceae (33), Caryocaraceae (25), Rhopalocarpaceae (14), Stachyuraceae (five), and Medusaginaceae (one). Pentaphylax is here referred to the Theaceae, and Strasburgeria to the Ochnaceae.

The Stachyuraceae and Elatinaceae have sometimes been referred to the Violales, and the Sarcolaenaceae (Chlaenaceae) to the Malvales. These differences in formal taxonomy do not reflect any great difference of opinion about relationships, inasmuch as the orders themselves are none too well defined. Takhtajan has pointed out a series of similarities between the Elatinaceae and the genus Hypericum, in the Guttiferae. A direct relationship is unlikely, however, inasmuch as the Elatinaceae have stipules and the Guttiferae, though more primitive in some other respects, do not.

The Rhopalocarpaceae (Sphaerosepalaceae), a small (14 spp) family of Madagascaran trees, have been variously referred to the Theales, Malvales, and Violales. The family is taxonomically isolated, but it seems best accommodated in the Theales near the Sarcolaenaceae.

The Actinidiaceae (Saurauiaceae) are especially interesting because they show affinities with the Dilleniales and Ericales as well as the Theales. Some authors have even submerged the Actinidiaceae in the Dilleniaceae, whereas Takhtajan, in contrast, refers them to the Ericales. The Actinidiaceae are clearly out of place in the Dilleniales, because of their syncarpous ovary with unitegmic, tenuinucellate ovules, in contrast of the apocarpous gynoecium and bitegmic, crassinucellate ovules of the Dilleniales. The Theales are also chiefly bitegmic and crassinucellate, but there are some exceptions and transitional forms. Both Marcgravia and Hypericum, for example, are essentially tenuinucellate, and Marcgravia also shows stages in the fusion of the integuments. It is but a short further step to the single integument of the Actinidiaceae. The ovular structure of the Actinidiaceae would be perfectly at home in the Ericales, as would the frequently poricidal dehiscence of the anthers. The large number of stamens and often wholly separate styles would be out of place in the Ericales, however. Altogether the Actinidiaceae seem to be most at home in the Theales, where they may be regarded as fairly closely related to the Theaceae and near-ancestral to the Ericales.

The Theales are mostly woody plants, less often herbaceous, with simple or occasionally compound leaves. Their flowers are hypogynous or nearly so, with imbricate calyx and mostly polypetalous corolla. The stamens are numerous and centrifugal, or less often few and cyclic, and the pollen is nearly always binucleate. The gynoecium consists of two to many carpels, these more or less united and with mostly axile, seldom parietal placentation. The ovules are bitegmic, except in the Actinidiaceae, and the endosperm varies from copious to more often thin or none.

The Dilleniales differ from the Theales in their apocarpous gynoecium, the Violales differ in their parietal placentation, the Malvales differ in their valvate caylx, and the Ericales differ in their unitegmic ovules and often sympetalous corolla.

The Theales are a prolific source of other orders, comparable to the Rosales in the subclass Rosidae. The Capparales, Ebenales, Ericales, Malvales, Primulales, Sarraceniales, Violales, and possibly the Lecythidales all appear to have been derived directly from the Theales. Extrapolating backwards from any of these derived orders, one conceives a plant which would be most at home in the Theales.

The adaptive significance of the characters which mark the Theales and their included families is obscure. We have noted that syncarpy may have some advantage over apocarpy, and the trailing or climbing habit of two of the families is obviously adaptive, as is the winged fruiting calyx of the Dipterocarpaceae. Beyond that there are few features of obviously adaptive importance in the list of characters which distinguish the order and its included families. There may well be hidden correlations awaiting discovery, but at present the prospect is bleak.

### SYNOPTICAL ARRANGEMENT OF THE FAMILIES OF THEALES

1. Leaves mostly alternate; endosperm present or absent

    2. Leaves mostly stipulate

        3. Leaves simple, or seldom pinnately compound; embryo of normal proportions

            4. Connective not exserted; calyx not winged; endosperm present in most spp.

            4. Style mostly gynobasic; anthers mostly opening by terminal pores; plants lacking mucilage cells; mostly not of Madagascar **1. Ochnaceae**

            5. Style not gynobasic (except in one sp. of Rhopalocar-

paceae); anthers opening by longitudinal slits; plants with mucilage cells; confined to Madagascar

6. Flowers with a prominent intrastaminal disk; filaments irregularly connate toward the base; pollen in monads                                                    **2. Rhopalocarpaceae**

6. Flowers without an intrastaminal disk, the stamens inserted within a ring of staminodes; filaments free; pollen in tetrads                                      **3. Sarcolaenaceae**

4. Connective prominently exserted; calyx mostly winged in fruit; endosperm mostly wanting
**4. Dipterocarpaceae**

3. Leaves palmately compound; embryo with enlarged, thickened hypocotyl and reduced cotyledons
**5. Caryocaraceae**

2. Leaves exstipulate, except in some Theaceae

7. Erect trees or shrubs

8. Petals usually 5; stamens mostly numerous, seldom only one or 2 cycles; stigmas (often also the styles) distinct
**6. Theaceae**

8. Petals 4; stamens 8; style with a capitate, merely lobed stigma                                                    **7. Stachyuraceae**

7. Trailing or climbing shrubs or vines

9. Endosperm copious; integument 1; bracts not highly modified                                                    **8. Actinidiaceae**

9. Endosperm scanty or none; integuments 2; bracts highly modified, becoming pitcherlike, saccate, or spurred
**9. Marcgraviaceae**

1. Leaves mostly opposite or whorled; endosperm wanting
  10. Leaves stipulate

11. Trees; stamens 15 or more; fruit a berry     **10. Quiinaceae**

11. Herbs or low shrubs; stamens 10 or fewer; fruit capsular
**11. Elatinaceae**

10. Leaves exstipulate

12. Ovary 17–25-celled; ovules 2 per cell; no secretory canals
**12. Medusagynaceae**

12. Ovary (1) 3–5 (15)-celled; ovules 1–many per cell; conspicuous secretory canals or cavities in all organs
**13. Guttiferae**

## SELECTED REFERENCES

Capuron, R., Révision des Rhopalocarpacées. Adansonia II. 2: 228–267. 1962.

Carlquist, S., Pollen morphology and evolution of Sarcolaenaceae (Chlaenaceae). Brittonia 16: 231–254. 1964.

Gottwald, H., and N. Parareswaran, Beiträge zur Anatomie und Systematik der Quiinaceae. Bot. Jahrb. 87: 253–303. 1967.

Hunter, G. E., Revision of Mexican and Central American *Saurauia* (Dilleniaceae). Ann. Mo. Bot. Gard. 53: 47–89. 1966.

Keng, H., Comparative morphological studies in Theaceae. U. Calif. Pub. Bot. 33: 269–384. 1962.

Leins, P., Die Frühe Blütenentwicklung von *Hypericum hookerianum* Wight et Arn. und *H. aegypticum* L. Der. Deuts. Bot. Ges. 77: 112–123. 1964.

Swamy, B. G. L., A contribution to the embryology of the Marcgraviaceae. Am. Jour. Bot. 35: 628–633. 1948.

Vestal, P. A., The significance of comparative anatomy in establishing the relationship of the Hypericaceae to the Guttiferae and their allies. Phil. Jour. Sci. 64: 199–252. 1937.

Vijayaraghavan, M. R., Morphology and embryology of *Actinidia polygama* Franch. & Sav. and systematic position of the family Actinidiaceae. Phytomorph. 15: 224–235. 1965.

Wood, C. E., Jr., The genera of Theaceae of the southeastern United States. Jour. Arn. Arb. 40: 413–419. 1959.

## ORDER 3. **Malvales**

The Malvales, as here defined, consist of six families and more than 3500 species. The Malvaceae, with more than 1500 species, are the largest family of the order, followed by the Sterculiaceae with about 1000. The Tiliaceae and Elaeocarpaceae have about 400 species each, the Bombacaceae nearly 200, and the Scytopetalaceae only about 30.

The Malvales have hypogynous flowers with valvate calyx, mostly separate petals that are often convolute in bud, usually numerous and centrifugal stamens that are often connate by their filaments, and basically syncarpous pistil with axile (rarely parietal) placentation and often rather numerous carpels. Most of the families (and species) have stellate or lepidote hairs and scattered internal cells, sacs, or canals that contain mucilage.

Most authors have agreed that all of the families here referred to the Malvales are allied. The Elaeocarpaceae and Scytopetalaceae stand somewhat apart from the others and from each other, but even so the relationship is so close that the Elaeocarpaceae have by some authors been submerged in the Tiliaceae. The remaining four families are even more closely allied, and there has been some controversy about their definition. Individual genera have been shifted from one family to another by various authors; the whole tribe Hibiscieae is referable to the Bombacaceae if the type of dehiscence of the fruit is taken as the critical character, and to the Malvaceae if (as here) the ornamentation of the pollen is stressed instead.

A seventh family, the Sarcolaenaceae (Chlaenaceae), is often also referred to the Malvales. The Sarcolaenaceae do resemble the Malvales in having mucilage cells and in having the phloem stratified into alternating bands of fibrous and nonfibrous tissue. (Students may recall slides of cross sections of *Tilia* twigs.) The pollen, however, is unlike that of anything else in the order, although there are some similarities to *Tilia* in fine structure. The calyx of the Sarcolaenaceae is imbricate, rather than valvate like the Malvales. On balance, the Sarcolaenaceae seem more at home in the Theales than in the Malvales, partly because the Malvales are otherwise relatively homogeneous, whereas the Theales are more heterogeneous.

The monotypic African genus *Hua* is a tree with alternate, entire leaves, valvate sepals, induplicate-valvate, long-clawed petals, ten stamens, and a unilocular ovary with a single basal ovule, the fruit dehiscent into five vales. It has variously been associated with the Malvales, Ebenales, and Linales, recently often as a distinct family. Its affinities are still wholly obscure.

The nature of the nectaries in the Malvales is evidently correlated with the polypetalous condition of the flowers.

> In this order, nectary glands are characteristic, multicellular, glandular hairs which are usually packed close together to form cushion-like growths. In the Tiliaceae, these nectaries are found in various places, including the sepals, petals and androgynophore. In the Malvaceae, Bombacaceae and Sterculiaceae they have become localized and are found in the sepals. . . . The honey is made

available by means of openings between the overlapping bases of the petals. The lack of gamopetalous corollas in these families is probably connected with the occurrence of nectaries in the sepals and the necessity for slits between the petals to make the nectar available.[8]

There has been continuing controversy about whether the large number of stamens and/or carpels found in some or all members of most of the families of the order reflects a primitive survival or a secondary increase. The monothecal stamens of the Malvaceae and Bombacaceae are widely admitted to represent longitudinal halves of ancestrally dithecal stamens, but there the agreement stops. Polyandrous types commonly have one or more sets of five stamen traces, with each trace repeatedly forked out so that eventually there is one bundle for each stamen. Multicarpellate types may show a similar branching of five basic carpel traces.

Some students of floral anatomy have interpreted this pattern to indicate an increase in number of stamens and carpels, based on an originally pentamerous flower. The contrary view is here advanced that the high numbers are inherited directly from a Thealean ancestry, and that consolidation of the vascular supply is merely the first step in a phyletic reduction from numerous stamens and carpels to few. The characteristic compound stamen-trunks which have been interpreted to mean a secondary increase in number are not confined to the Malvales, but are widespread among the multistaminate members of the Dilleniidae in general, and they have also been found in the Myrtales and Alismatales. My interpretation of compound stamen trunks as a stage in reduction rather than multiplication is in accord with that of Eames, as noted in Chapter 3.

The Malvales are probably derived from the less modified members of the Theales, from which they differ in the valvate rather than imbricate calyx, and in a series of less consistently expressed tendencies such as the production of mucilage cells or cavities and stellate hairs. The only other obvious relationship of the Malvales is with the Violales. These two orders are here regarded as groups which have taken different evolutionary paths from a common ancestry in the Theales. The Elaeocarpaceae are one of the more primitive families of the Malvales, as judged by the general floral morphology, wood anatomy, and lack of the specialized type of nectary found in other families of the order. Some of the Elaeocarpaceae are very similar to some of the Flacourtiaceae, and this similarity between the families extends even to pollen morphology. The Flacourtiaceae are one of the less modified families of the Violales. Any common ancestor of the Flacourtiaceae and Eleaocarpaceae would presumably have an imbricate

[8] W. H. Brown, Proc. Am. Phil. Soc. 79: 557–558. 1938.

calyx and axile placentation, and thus would fall within the bounds of the Theales.

Aside from the correlation of the type of nectary with the structure of the corolla, the ecological significance of the characters which mark the Malvales is obscure. If the valvate arrangement of the sepals has any importance to the plant, the fact is not obvious. If the mucilage cells convey any advantage or have any functional significance, it remains to be discovered. If stellate hairs have any advantage over simple hairs, it has not been demonstrated. The burden of proof will rest with anyone who suggests a correlation.

The ecological significance of the families which make up the order is scarcely more evident than that of the order as a whole. The Malvaceae are mostly herbaceous, whereas the other families are mostly woody, but beyond that the problems are mostly unresolved. The spines on the pollen of the Malvaceae, and the tendency in several families for the filaments to be connate, might relate to the kind of pollinators, or they might not. It is hard to see how the number of pollen sacs per anther might affect the competitive ability of the plant, but here as always the possibility must be entertained that there is a hidden significance which might be discovered by a persistent and perceptive investigator.

SYNOPTICAL ARRANGEMENT OF THE FAMILIES OF MALVALES

1. Petals valvate; plants mostly without mucilage cells and always without mucilage sacs or canals; pubescence not stellate or lepidote

    2. Leaves stipulate; sepals separate to the base or nearly so; petals mostly separate; integuments 2     **1. Elaeocarpaceae**

    2. Leaves exstipulate; sepals connate into a cup or tube; petals often connate below; integument 1     **2. Scytopetalaceae**

1. Petals imbricate or contorted; plants mostly with mucilage cells or often with mucilage sacs or canals; pubescence mostly stellate or lepidote

    3. Anthers bilocular; filaments free or connate; mostly trees and shrubs, a few herbs

    4. Stamens numerous, mostly free, arranged in a single whorl
        **3. Tiliaceae**

4. Stamens (including staminodes) 10–numerous, in two or more whorls, mostly connate by their filaments, seldom only 5 and then always connate **4. Sterculiaceae**

3. Anthers mostly unilocular; filaments connate

5. Pollen smooth, triporate; fruit loculicidally dehiscent, or indehiscent; mostly trees and shrubs, a few herbs
**5. Bombacaceae**

5. Pollen minutely spiny, mostly multiporate; fruit typically septicidally dehiscent or schizocarpic, seldom loculicidally dehiscent; plants herbaceous, or occasionally softly woody **6. Malvaceae**

## SELECTED REFERENCES

Brizicky, G. K., The genera of Tiliaceae and Elaeocarpaceae in the southeastern United States. Jour. Arn. Arb. 46: 286–307. 1965.

————, The genera of Sterculiaceae in the southeastern United States. Jour. Arn. Arb. 47: 60–74. 1966.

Carlquist, S., Pollen morphology and evolution of Sarcolaenaceae (Chlaenaceae). Brittonia 16: 231–254. 1964.

Edlin, H. L., A critical revision of certain taxonomic groups of the Malvales. New Phytol. 34: 1–20; 122–143. 1935.

Mauritzon, J., Zur Embryologie der Elaeocarpaceae. Ark. Bot. 26A (10): 1–8. 1934.

Venkata Rao, C., Floral anatomy of some Sterculiaceae with special reference to the position of the stamens. Jour. Indian Bot. Soc. 28: 237–244. 1949.

————, Pollen grains of the Sterculiaceae. Jour. Indian Bot. Soc. 29: 130–137. 1950.

————, Contributions to the embryology of Sterculiaceae. III. *Melochia corchorifolia* L. Jour. Indian Bot. Soc. 30: 122–131. 1951.

————, Floral anatomy of some Malvales and its bearing on the affinities of families included in the order. Jour. Indian Bot. Soc. 31: 171–203. 1952.

## ORDER 4. **Lecythidales**

The order Lecythidales consists of the single wholly tropical family Lecythidaceae, with about 450 species. The Lecythidaceae have traditionally been referred to the Myrtales because of their combination of mostly separate petals, numerous stamens, and syncarpous, inferior ovary with axile placentation. They differ from characteristic members of the Myrtales, however, in their alternate leaves, centrifugal stamens, lack of internal phloem, and a series of embryological features which have been elucidated by Mauritzon. This is too long a list to ignore, and the Lecythidaceae must be removed from the Myrtales. No other order can accommodate the Lecythidaceae without undue strain, and it therefore becomes necessary to maintain them in an order by themselves.

The Lecythidales can readily be accommodated to the subclass Dilleniidae, but would be anomalous in the Rosidae because of their centrifugal stamens. I hesitate to put too much weight on a single character, especially one which has been carefully observed in only a relatively small number of species, but it does provide a lead to possible relatives. Within the Dilleniidae, the Lecythidales are readily distinguished from other polyandrous groups by their combination of epigynous flowers and axile placentation. The Lecythidales resemble the Malvales in their valvate calyx and connate filaments, and in having the secondary phloem usually stratified into fibrous and soft portions. It may also be of interest that mucilage canals, comparable to those of the Malvales, have been observed in several genera of the Lecythidales. The Lecythidales lack the stellate pubescence found in so many of the Malvales, however, and of course the inferior ovary would be quite out of place in the latter order. A study of the nectaries might be instructive.

On the basis of presently available evidence it appears that the Lecythidales are related to the Malvales, either as direct descendants of primitive, now extinct members of that order, or possibly through a common ancestry in the Theales.

### SELECTED REFERENCE

Mauritzon, J., Contributions to the embryology of the orders Rosales and Myrtales. Lunds Univ. Årsskr. N. F. Avd. 2. 35 (2): 1–120. 1939.

Order 5. **Sarraceniales**

The order Sarraceniales as here defined consists of three well marked small families, scarcely 200 species in all. The Droseraceae have about 90 species, the Nepenthaceae about 80, and the Sarraceniaceae only about 16. A fourth family, the Byblidaceae (including Roridulaceae), has often been referred to this order, but is now generally associated with the Pittosporaceae in the Rosales.

The Sarraceniales are herbs or shrubs with alternate, simple leaves that are modified for catching insects. They have regular, hypogynous, polypetalous or apetalous, syncarpous flowers with few to many stamens. The fruit is a loculicidal capsule, and the seeds have a well developed endosperm. The sequence of initiation of stamens has yet to be determined.

The mutual affinity of the three families has been affirmed and denied by different authors. I am more impressed by the similarities than by the admitted differences. In addition to the obvious exomorphic characters, the Droseraceae and Nepenthaceae have very similar pollen. Markgraf has concluded that the insect-catching leaves in all three families are homologous, in spite of the obvious difference between the Droseraceae and the other two families.

None of the three families of the order can be considered ancestral to any of the others. They represent three distinct lines which have undergone more or less similar modifications from a similar ancestry.

The ancestry of the Sarraceniales is to be sought in the Theales. Except for their insectivorous habit, the Sarraceniales would fit very well into the Theales (assuming that the stamens turn out to be centrifugal). Inasmuch as two of the families have axile placentation, the Violales do not seem very likely ancestors. The similarities of *Ancistrocladus* and *Dioncophyllum*, in the Violales, to members of the Sarraceniales (especially the Nepenthaceae) are here regarded as reflecting a common ancestry in the Theales rather than a direct relationship.

The insectivorous habit of the Sarraceniales may be presumed to be an evolutionary response to their growth in habitats deficient in available nitrogen. The Sarraceniaceae and Droseraceae commonly grow in waterlogged soils containing little or no soluble nitrate. The Nepenthaceae occur in wet, tropical forests, which characteristically have nutrient-poor soils. Many other groups of plants have faced similar problems, but very few have learned to meet them by trapping insects.

Aside from the insect-catching apparatus, the characters which mark the order and the individual families are of doubtful ecological significance.

## SYNOPTICAL ARRANGEMENT OF THE FAMILIES OF SARRACENIALES

1. Leaves, or some of them, modified to form pitchers; placentation axile; style solitary, or sometimes very short or none

    2. Flowers perfect; filaments distinct; pollen grains in monads; ovules unitegmic, tenuinucellate; herbs, not climbing; New World      **1. Sarraceniaceae**

    2. Flowers unisexual; filaments united into a column; pollen grains in tetrads; ovules bitegmic, crassinucellate; herbs or shrubs, often climbing; Old World      **2. Nepenthaceae**

1. Leaves not forming pitchers; placentation parietal; styles several      **3. Droseraceae**

## SELECTED REFERENCES

Chanda, S., The pollen morphology of Droseraceae with special reference to taxonomy. Pollen et Spores 7: 509–528. 1965.

Markgraf, F., Über Laubblat-Homologien und verwandtschaftliche Zusammenhänge bei Sarraceniales. Planta 46: 414–446. 1955.

Wood, C. E. Jr., The genera of Sarraceniaceae and Droseraceae in the southeastern United States. Jour. Arn. Arb. 41: 152–163. 1960.

### ORDER 6. **Violales**

The order Violales as here defined consists of some 21 families and more than 5200 species. The names Bixales, Cistales, and Parietales have also been used for this order, and the names Cucurbitales, Datiscales, Passiflorales, and Tamaricales have been applied to certain families or groups of families here included in the Violales. The adoption of the name Violales in the present treatment follows the usage in the 12th edition of the Engler Syllabus and is also influenced by personal conversations with Arman Takhtajan. Inasmuch as the nomenclatural principle of priority does not apply to groups above the rank of family, uniformity in such matters can only be achieved by consensus and accommodation.

The Violales are a rather heterogeneous assemblage, but any consistent attempt to divide the order into smaller, more homogeneous ones leads into a morass in which perhaps as many as a dozen orders, most of them with only one or two families each, might have to be recognized. In spite of its heterogeneity, the order appears to be a natural group, at least to the extent that all of its families take their origin from the related, more primitive order Theales. The seeming contrasts between the treatment here presented, that of Melchior (in the Engler Syllabus), and that of Takhtajan reflect relatively small differences in concepts of relationship, magnified by the exigencies of a formalized system.

Almost nine-tenths of the species of Violales belong to only six of the families: The Flacourtiaceae (1300), Violaceae (850), Cucurbitaceae (850), Begoniaceae (800), Passifloraceae (600) and Loasaceae (250). It is now widely agreed that all of these large families belong to the same general circle of affinity, although some authors distribute them among two or more orders which are themselves considered to be allied.

The remaining 15 families have fewer than 600 species amongst them. Some of these smaller families, notably the Ancistrocladaceae, Dioncophyllaceae, Fouquieriaceae, Hoplestigmataceae, and Scyphostegiaceae, have often been referred to other orders, some of which are not closely allied to the Violaceae. These disputed families encompass fewer than 30 species in all.

The Ancistrocladaceae (16 spp) and Dioncophyllaceae (3 spp) have often been referred to the related order Theales. They have no very close relatives in either order, although they very likely belong in this general alliance. Recent studies by Schmid (cited below) indicate that their closer affinity may be with the Violales, and it is also more comfortable to refer them to the Violales because of their unilocular ovary.

The Scyphostegiaceae (one sp) have unusual flowers which have been variously interpreted. I here accept the interpretation of Swamy (cited below) rather than that of Hutchinson (Fam. Fl. Pl. 1: 326–329. 1959). Under this interpretation the family fits best into the Violales. This position is in accord with the anatomical evidence, which tends to ally the Scyphostegiaceae with the Flacourtiaceae.

The Fouquieriaceae (five spp) are well isolated, without evident close relatives. They have sometimes been associated with the Polemoniaceae, but again the relationship can scarcely be very close. Pollen morphology provides no obvious clue to their relationship, nor does the wood anatomy, although it may be noted that the wood is relatively unspecialized. The Fouquieriaceae have bitegmic ovules, as do other members of the Violales, rather than unitegmic ovules like the Polemoniales and the Asteridae in general. The relatively numerous (10–17) stamens in the Fouquieriaceae would also be anomalous in the Asteridae. At least until contrary evidence is forthcoming, I prefer to retain the Fouquieriaceae in the Violales.

The Hoplestigmataceae (two spp) are an isolated group of uncertain affinities. They have a sympetalous, irregularly 11–14-lobed corolla, numerous, irregularly three-cyclic, epipetalous stamens, and a unilocular ovary with two unitegmic ovules hanging from each of two intruded parietal placentae. They have variously been associated with the Boraginaceae (Takhtajan), Symplocaceae (Wagenitz, in the 12th Engler Syllabus), and Bixaceae (Hutchinson). Their tentative assignment to the Violales in the present system is purely arbitrary.

The most characteristic feature of the Violales is the unilocular, compound ovary with mostly parietal placentation. When the stamens are numerous they are always, so far as known, centrifugal. The ovules are crassinucellate or less often tenuinucellate, and with the single known exception of the family Loasaceae they are bitegmic. The only character which will distinguish the Violales as a whole from the Theales is the placentation (which is axile in the Theales), and we have noted that there are some exceptions in both orders.

The more primitive members of the Violales are trees with alternate, stipulate leaves, perfect, hypogynous, polypetalous flowers with numerous centrifugal stamens, a compound pistil with free styles and parietal placentation, and seeds with a well developed endosperm. Such a combination of characters immediately suggests the Flacourtiaceae, which are usually considered to be the most primitive family of the order. It may also be interesting to note that a few of the Flacourtiaceae have a plurilocular ovary with axile placentation; on formal morphological characters these would be perfectly at home in the Theales.

Tendencies toward perigyny and epigyny, unisexuality, reduction in the number of stamens, fusion of filaments, the development of a corona, reduction in the number of carpels, fusion of styles, and loss of endosperm from the seed can all be seen in the family Flacourtiaceae. These are some of the more prominent characters which are used in combination to define many of the other families of the order. Although the Flacourtiaceae have retained more of the primitive characters than any other family of the order, they are not to be taken as directly ancestral to all the other families. Instead we have the familiar pattern of a group of related taxa undergoing a series of parallel evolutionary changes. Some few of the Cucurbitaceae (*Fevillea*), one of the most advanced families in the order, still retain typical axile placentation.

The Violales and Capparales may be considered as parallel offshoots from the Theales, each having mostly parietal instead of axile placentation. The Violales differ from the Capparales in usually lacking myrosin cells, in having a much higher proportion of woody species, in often having perigynous to epigynous flowers, in seldom having compound leaves, and in very often having three carpels (a rare number in the Capparales), only seldom two (the commonest number in the Capparales).

The adaptive significance of the characters which distinguish the Violales as a group is obscure. Placentation, ovular structure, and sequence of initiation of the stamens are difficult to relate to survival value. Some of the families do show some ecological correlation, however. The Fouquieriaceae are thorny xerophytes which have small leaves that fall off as the soil dries out. The Tamaricaceae, Frankeniaceae, and many of the Cistaceae meet problems of water stress by having small, firm, persistent leaves that can survive desiccation. The Passifloraceae and Cucurbitaceae are chiefly tendriliferous vines, and the Dioncophyllaceae and Ancistrocladaceae are woody vines which climb by means of stout hooks from the branch tips (Ancistrocladaceae) or leaf tips (Dioncophyllaceae). Aside from these instances, one sees the familiar pattern of families defined by characters of little or no obvious biological importance, with the larger families embracing a wide range of form and occurring in diverse habitats.

## SYNOPTICAL ARRANGEMENT OF THE FAMILIES OF VIOLALES

1. Ovary usually superior; plants of various habit, most families with a well developed endosperm

2. Flowers polypetalous or sometimes apetalous; endosperm mostly well developed

  3. Endosperm oily; many species cyanogenetic

    4. Stamens mostly numerous, seldom only 5 or fewer; trees or shrubs, seldom climbing

      5. Ovules mostly on parietal placentae; petals usually present (sometimes scarcely distinguishable from the sepals); carpels 2–10, with free or united styles
        **1. Flacourtiaceae**

      5. Ovules hanging from an apical placenta; petals none; carpels 3–4; styles free    **2. Peridiscaceae**

    4. Stamens mostly numerous, seldom only 5 or fewer; trees or times climbing

      6. Carpels 9–12; stamens 3; flowers unisexual; ovules numerous on a broad, basal placenta; woody plants
        **3. Scyphostegiaceae**

      6. Carpels 2–5; stamens 5; flowers mostly perfect; ovules on parietal placentae

        7. Flowers without a corona; trees, shrubs, or herbs, not climbing

          8. Flowers hypogynous, with a single style, often irregular    **4. Violaceae**

          8. Flowers perigynous, with separate styles, regular
           **5. Turneraceae**

        7. Flowers with a corona; herbs and shrubs, often climbing

          9. Seeds arillate; plants often climbing by tendrils; flowers hypogynous to somewhat perigynous; sepals imbricate; petals imbricate or seldom wanting    **6. Passifloraceae**

9.  Seeds exarillate; undershrubs or herbs, not climbing; flowers strongly perigynous; sepals and petals valvate                                            **7. Malesherbiaceae**

3.  Endosperm starchy; few species cyanogenetic

10.  Plants variously woody or herbaceous, but not climbing; ovules several or many, included in the fruit

11.  Stamens mostly numerous; style 1; leaves small or more often of ordinary size

12.  Leaves alternate, stipulate, palmately lobed or veined; ovules anatropous; upright shrubs or small trees
**8. Bixaceae**

12.  Leaves opposite, exstipulate, entire, not palmate; ovules orthotropous or seldom anatropous; herbs or low shrubs                                **9. Cistaceae**

11.  Stamens mostly 4–10, seldom numerous; styles 1–several; leaves small, often ericoid or reduced to scales; herbs or more often shrubs or trees

13.  Leaves alternate; sepals connate below; styles 3–4
**10. Tamaricaceae**

13.  Leaves opposite; sepals free; style 1 **11. Frankeniaceae**

10.  Plants woody climbers with stout, hooked branch tips or leaf tips

14.  Ovules numerous, on 2–5 parietal placentae, conspicuously exserted from the capsule at maturity; styles several; plants climbing by hooked leaf-tips
**12. Dioncophyllaceae**

14.  Ovule solitary on a basal placenta, included in the mature nut; style solitary; plants climbing by hooked branch-tips                    **13. Ancistrocladaceae**

2.  Flowers sympetalous; endosperm scanty or none except in the Achariaceae

15. Flowers perfect; leaves simple, entire or nearly so; stamens 10 or more

    16. Thorny, xerophytic shrubs with small leaves; carpels 3; fruit capsular; stamens 10–17    **14. Fouquieriaceae**

    16. Unarmed, large-leaved trees; carpels 2; fruit drupaceous; stamens numerous    **15. Hoplestigmataceae**

15. Flowers unisexual; leaves simple or often lobed or compound; stamens 10 or less

    17. Endosperm well developed; plants herbaceous or nearly so and without a latex system; stamens 3–5; style 1
        **16. Achariaceae**

    17. Endosperm scanty or none; plants woody and with a well developed latex system; stamens 10 (5); styles free
        **17. Caricaceae**

1. Ovary mostly inferior; plants mostly herbs or herbaceous vines, a few trees and shrubs; endosperm scanty or none, except in some Loasaceae

    18. Flowers perfect; stamens mostly numerous; style 1; integument 1    **18. Loasaceae**

    18. Flowers mostly unisexual; stamens few or less often numerous; styles 1–several; integuments 2

        19. Stamens 4–many, all with bilocular anthers; plants without tendrils; styles free except in some Begoniaceae; petals separate or none

            20. Sepals 2 (–5); leaves stipulate; carpels 2–5  **19. Begoniaceae**

            20. Sepals 3–9; leaves exstipulate; carpels 3–8  **20. Datiscaceae**

        19. Stamens 1–5, typically 3 with one unilocular and two bilocular anthers; mostly tendriliferous vines; style mostly single; corolla mostly sympetalous    **21. Cucurbitaceae**

## SELECTED REFERENCES

Ayensu, E. S., and W. L. Stern, Systematic anatomy and ontogeny of the stem in Passifloraceae. Contr. U. S. Nat. Herb. 34: 45–73. 1964.

Airy Shaw, H. K., On the Dioncophyllaceae, a remarkable new family of flowering plants. Kew Bull. 1951: 327–347.

Brizicky, G. K., The genera of Turneraceae and Passifloraceae in the southeastern United States. Jour. Arn. Arb. 42: 204–218. 1961.

————, The genera of Violaceae in the southeastern United States. Jour. Arn. Arb. 42: 321–333. 1961.

————, The genera of Cistaceae in the southeastern United States. Jour. Arn. Arb. 45: 346–357. 1964.

Chakravarty, H. L., Morphology of the staminate flowers in the Cucurbitaceae with special reference to the evolution of the stamen. Lloydia 21: 49–87. 1958.

Chopra, R. N., Some observations on endosperm development in the Cucurbitaceae. Phytomorph. 5: 219–230. 1955.

Chopra, R. N., and H. Kaur, Embryology of Bixa orellana Linn. Phytomorph. 15: 211–214. 1965.

Ernst, W. R., and H. J. Thompson, The Loasaceae in the southeastern United States. Jour. Arn. Arb. 44: 138–142. 1963.

Gauthier, R., and J. Arros, L'anatomie de la fleur staminée de l'Hillebrandia sandwicensis Oliver et la vascularization de l'étamine. Phytomorph. 13: 115–127. 1963.

Gibbs, R. D., A classical taxonomist's view of chemistry in taxonomy of higher plants. Lloydia 28: 279–299. 1965.

Hagerup, O., Vergleichende morphologische und systematische Studien über die Ranken und andre vegetative Organe der Cucurbitaceen und Passifloraceen. Dansk. Bot. Ark. 6 (8): 1–103. 1930.

Jeffrey, C., Notes on Cucurbitaceae, including a proposed new classification of the family. Kew Bull. 15: 337–371. 1962.

Johri, B. M., and D. Kak, The embryology of Tamarix Linn. Phytomorph. 4: 230–247. 1954.

Mauritzon, J., Über die Embryologie der Turneraceae und Frankeniaceae. Bot. Notis. 1933: 543–554.

————, Zur Embryologie einiger Parietales-Familien. Svensk. Bot. Tidsk. 30: 79–113. 1936.

Metcalfe, C. R., The anatomical structure of the Dioncophyllaceae in relation to the taxonomic affinities of the family. Kew Bull. 1951: 351–368.

————, Scyphostegia borneensis Stapf. Anatomy of the stem and leaf in relation to its taxonomic position. Reinwardtia 4: 99–104. 1956.

————, Notes on the systematic anatomy of *Whittonia* and *Peridiscus*. Kew Bull. 15: 472–475. 1962.

Murty, Y. S., Studies in the order Parietales. IV. Vascular anatomy of the flower of Tamaricaceae. Jour. Indian Bot. Soc. 33: 226–238. 1954.

Puri, V., Studies in floral anatomy. IV. The vascular anatomy of the flower in certain species of the Passifloraceae. Am. Jour. Bot. 34: 562–573. 1947. V. On the structure and nature of the corona in certain species of the Passifloraceae. Jour. Indian Bot. Soc. 27: 130–149. 1948. VII. On placentation in the Cucurbitaceae. Phytomorph. 4: 278–299. 1954.

Schmid, R., Die systematische Stellung der Dioncophyllaceen. Bot. Jahrb. 83: 1–56. 1964.

Singh, B., Studies on the structure and development of seeds of Cucurbitaceae. Phytomorph. 3: 224–239. 1953.

Swamy, B. G. L., On the floral structure of *Scyphostegia*. Proc. Nat. Inst. Sci. India 19: 127–142. 1953.

Venkatesh, C. S., The curious anther of *Bixa* — its structure and dehiscence. Am. Midl. Nat. 55: 473–476. 1956.

Walia, K., and R. N. Kapil, Embryology of *Frankenia* Linn. with some comments on the systematic position of the Frankeniaceae. Bot. Notis. 118: 412–429. 1965.

ORDER 7. **Salicales**

The order Salicales consists of the single family Salicaceae, with about 350 species. There are only two genera, *Salix* and *Populus*, the former by far the larger. Taxonomists are agreed that no other family should be referred to this order.

The Salicales are dioecious woody plants with much reduced flowers aggregated into catkins. They differ from other such plants in their numerous ovules on parietal placentae in a compound, unilocular pistil which ripens into a two- (to four)-valved capsule. They have alternate, simple, stipulate leaves. The mature seeds are plumose-hairy and lack endosperm.

The Salicales are taxonomically isolated. Traditionally they have been referred to the Amentiferae, because of their unisexual flowers which lack a perianth and are borne in catkins. Most of the traditional Amentiferae are here referred to the subclass Hamamelidae, but some are distributed elsewhere in the system. The gynoecium of the Salicales would be anomalous in the Hamamelidae, in which the pistil, when compound, tends to be few-seeded or one-seeded and indehiscent. It might be possible, however, to see a forerunner of the salicalean gynoecium in some members of the Hamamelidaceae which have capsular fruits with parietal placentae.

A more likely suggestion is that the Salicales are florally reduced derivatives of the Violales. The gynoecium is of course highly compatible with this interpretation. Brown (cited below) has also noted some similarities in nectary structure which suggest a relationship among the Violales, Salicales, and the families here referred to the Capparales. Anatomically the Salicales are advanced and specialized, but their peculiarities find no close parallel in any of the groups which have been suggested as possible relatives. The pollen morphology, on the other hand, is ambiguous because it resembles that of too many other things. Pollen much like that of the Salicaceae occurs in some of the Flacourtiaceae and Tamaricaceae, but also in the Platanaceae and some families of the Rosidae and Asteridae.

Ecologically, the Salicales are characterized by their wind-distributed, tiny, plumose-hairy seeds. They tend to occur in moist places, such as along streams and in bottomlands, but they are not restricted to such habitats. One might suppose that the flowers would be consistently anemophilous, but the supposition would be unfounded. *Salix* is very largely entomophilous, in spite of its seeming adaptation to anemophily. It is not yet certain whether *Salix* has retained the entomophilous habit while undergoing floral reduction, or whether it has reverted to entomophily after an anemophilous evolutionary stage.

Neither wind-distribution of the seeds nor growth in moist habitats is unique to the Salicales, nor is there any structural or ecological need for these two features to be associated. It should also be noted that adoption of wind as the mechanism for distributing the seeds of the Salicales is no hindrance at all to possible similar adaptation by other groups of plants. The Salicales do not and cannot preempt this niche, nor can any other group of plants.

### SELECTED REFERENCES

Brown, W. H., The bearing of nectaries on the phylogeny of flowering plants. Proc. Am. Phil. Soc. 79: 549–595. 1938.

Fisher, M. J., The morphology and anatomy of the flowers of the Salicaceae. Am. Jour. Bot. 15: 307–326; 372–394. 1928.

Hjelmqvist, H., Studies on the floral morphology and phylogeny of the Amentiferae. Bot. Notis. Suppl. 2 (1): 1–171. 1948.

Nagaraj, M., Floral morphology of *Populus deltoides* and *P. tremuloides*. Bot. Gaz. 114: 222–243. 1952.

ORDER 8.  **Capparales**

The order Capparales as here defined consists of five families and approaching 4000 species. More than three-fourths of the species belong to the single family Cruciferae (3000), and most of the rest to the Capparidaceae (800)). The Resedaceae (70), Moringaceae (ten), and Tovariaceae (two) have fewer than 100 species amongst them. The Papaveraceae and

Fumariaceae, often associated with the Capparales in a collective order Rhoeadales, are here treated as a separate order Papaverales in the subclass Magnoliidae.

The spelling Capparaceae and Capparales is here used, following the spelling in the list of conserved family names in the current edition of the Rules. These names have usually been spelled Capparidales and Capparidaceae. Crosswhite and Iltis (cited below) have vigorously expounded the view that the "corrected" spelling used in the Rules is unauthorized and incorrect, and they may well be right. Consistency would require that if I reject the change in this instance, I should investigate the several other similar changes in the list, and this I do not feel properly prepared to do. I am therefore somewhat reluctantly following the "corrected" spelling used in the Rules.

The five families here referred to the Capparales are generally admitted to be closely related, but it is difficult to find a set of distinctive characters which apply to all the families. The Cruciferae and Capparaceae form the core of the order and have many features in common. They have alternate, usually compound or dissected leaves, and hypogynous flowers with mostly four sepals, four petals, centrifugal stamens, and (at least apparently) two carpels which form a compound ovary with parietal placentation. The ovules are often campylotropous and the seeds have little or no endosperm. Nearly all of the Cruciferae and many of the Capparaceae have specialized myrosin cells, which are chiefly though not entirely restricted to this order. Myrosin is an enzyme involved in the formation of mustard oil.

The other families of the order differ in one or another respect from this core group. The Moringaceae have perigynous flowers with five sepals, five petals, five functional stamens, and three carpels. The Resedaceae have evidently irregular flowers, often with more than four sepals and more than two carpels, and are unique among the angiosperms in often having the syncarpous ovary open at the top. Both of these families have myrosin cells and parietal placentae.

The Tovariaceae are the smallest and most distinctive family of the group. They are not reported to have myrosin cells. They have seemingly or actually axile placentation, and their seeds have well developed endosperm. They commonly have eight sepals, eight petals, eight stamens, and five to eight carpels. In spite of the impressive list of differences, most authors are agreed in associating the Tovariaceae with the other families of the group. It may be noted that the embryology and pollen morphology, at least, do suggest a relationship to the rest of the order, and the elongate, true racemes are very suggestive of those of the Cruciferae and many of the Capparaceae. The endospermous seeds and axile placentation of the Tovariaceae appear to be primitive, but the number of sepals, petals, stamens, and carpels may well reflect a secondary increase.

The gynoecium of the Cruciferae is unique. It is bilocular, but the ovules are attached to the *margins of the partition*. At maturity the two valves of the fruit commonly fall off, leaving the ovules, or their stalks, still attached to the margins of the more or less persistent replum (partition). The morphology of the gynoecium has occasioned much controversy and is still not settled to the satisfaction of all concerned. I prefer the two-carpellary interpretation of Puri to the four-carpellary interpretation of Eames. In spite of the unique gynoecium, most authors agree that the Cruciferae are derived directly from the Capparaceae, and there are in fact some members of the Capparaceae which have been interpreted as showing an approach to the peculiar partition of the cruciferous gynoecium.

All of the characters of the Capparales, including the centrifugal stamens, are compatible with an ancestry in the Theales, and no other order provides an obvious starting point for the group. No one family of the Theales is suggested as a direct ancestor, however. As noted in the discussion of the Violales, the Capparales and Violales may be regarded as offshoots of the Theales which have independently evolved parietal placentation. Differences between the Violales and Capparales are discussed under the former order.

The Capparaceae have all the characters that would be required in a hypothetical common ancestor of the Cruciferae, Moringaceae, and Resedaceae. It is not here suggested that any living genus of the Capparaceae is ancestral to the other three families, but only that the ancestors, if we had them, would be referred to the Capparaceae. The Tovariaceae are, as we have noted, more primitive than the other families in having a well developed endosperm. On the other hand, the Tovariaceae cannot be regarded as ancestral to the rest of the order, because they have only as many stamens as petals, whereas many of the Capparaceae have numerous stamens.

The adaptive significance of most of the characters which mark the order and the families within it is obscure. The importance of mustard oil and myrosin cells in the economy of the plant is wholly speculative. The ecological importance of the cruciferous gynoecium is if anything even less clear. Some few of the differences, however, are of biologic as well as taxonomic importance. The importance of growth habit needs no elucidation. We have seen that the number of nuclei in the pollen grain is related to the mechanism of incompatibility. The irregular flowers of the Moringaceae may well relate to the mechanism of pollination.

SYNOPTICAL ARRANGEMENT OF THE FAMILIES OF CAPPARALES

1. Placentation axile or seemingly so; sepals, petals, and stamens 8 each; ovary closed, with 5–8 carpels; endosperm present; coarse herbs or soft shrubs with trifoliate leaves

   **1. Tovariaceae**

1. Placentation nearly always parietal (sometimes with intruded placentae) or seemingly so; sepals, petals, and carpels usually fewer; seeds with little or no endosperm

  2. Flowers hypogynous

    3. Flowers mostly regular or only slightly irregular; sepals mostly 4: carpels mostly 2 (or seemingly so); style short or elongate, with capitate or slightly lobed stigma

      4. Ovary mostly 1-celled; stamens 4–many, never tetradynamous as in the next family; flowers almost always with an evident gynophore or androgynophore; pollen binucleate; leaves simple to trifoliolate or palmately compound, not much dissected; herbs, shrubs, or trees, chiefly tropical and subtropical        **2. Capparaceae**

      4. Ovary mostly 2-celled, with the ovules attached to the margins of the partition; stamens typically 6 and tetradynamous (the 2 outer ones shorter than the 4 inner ones), seldom fewer; flowers rarely with a gynophore or androgynophore; pollen trinucleate; leaves often more or less dissected in pinnate fashion, but without distinct, articulated leaflets; herbs, seldom shrubs, chiefly of temperate regions        **3. Cruciferae**

    3. Flowers usually evidently irregular; sepals (4) 5–6 (–8); carpels 2–6; stigmas sessile, separate, as many as the carpels; ovary often open at the top; herbs        **4. Resedaceae**

  2. Flowers perigynous; sepals, petals and functional stamens each 5; carpels 3; trees with pinnately decompound leaves
        **5. Moringaceae**

## SELECTED REFERENCES

Arber, A., Studies in flower structure. VII. On the gynaeceum of *Reseda*, with a consideration of paracarpy. Ann. Bot. Ser. II. 6: 43–48. 1942.

Eames, A. J., and C. L. Wilson, Crucifer carpels. Am. Jour. Bot. 17: 638–656. 1930.

Crosswhite, F. S., and H. H. Iltis, Studies in the Capparidaceae. X. Orthography and conservation: Capparidaceae vs. Capparaceae. Taxon 15: 205–214. 1966.

Erdtman, G., Pollen morphology and plant taxonomy in some African plants. Webbia 11: 405–412. 1955.

Ernst, W. R., The genera of Capparaceae and Moringaceae in the southeastern United States. Jour. Arn. Arb. 44: 81–95. 1963.

Frohne, D., Das Verhältnis von vergleichender Serobotanik zu vergleichender Phytochemie, dargestellt an serologischen Untersuchungen im Bereich der "Rhoeadales." Planta Medica 10: 283–297. 1962.

Mauritzon, J., Die Embryologie einiger Capparidaceen sowie von *Tovaria pendula*. Ark. Bot. 26A (15): 1–14. 1935.

Narayana, H. S., Studies in the Capparidaceae. I. The embryology of *Capparis decidua* (Forsk.) Pax Phytomorph. 12: 167–177. 1962. II. Floral morphology and embryology of *Cadaba indica* Lamk. and *Crataeva nurvala* Buch.-Ham. Phytomorph. 15: 158–175. 1965.

Puri, V., The life history of *Moringa oleifera* Lamk. Jour. Indian Bot. Soc. 20: 263–284. 1941.

————, Studies in floral anatomy. VI. Vascular anatomy of the flower of *Crataeva religiosa* Forst., with special reference to the nature of the carpels in the Capparidaceae. Am. Jour. Bot. 37: 363–370. 1950.

Stoudt, H. N., The floral morphology of some of the Capparidaceae. Am. Jour. Bot. 28: 664–675. 1941.

### ORDER 9. **Ericales**

The order Ericales as here defined consists of seven families and a little more than 3000 species. The Ericaceae, with more than 2500 species, are by far the largest family, followed by the Epacridaceae with about 400. The Pyrolaceae (45), Monotropaceae (30), Clethraceae (30), Cyrillaceae (13), and Empetraceae (nine) have fewer than 150 species in all.

The Ericales are sympetalous or less often polypetalous plants with unitegmic, tenuinucellate ovules, with the stamens typically twice as many as the petals and usually attached directly to the receptacle, with axile (seldom parietal) placentation, and usually with more or less numerous ovules in each locule. The anthers very often have prominent appendages (like horns or tails) and open by terminal pores. Most members of the group share a long series of embryological features in common, including notably the production of pollen in tetrads. A large proportion of the species are strongly mycorhizal, and some of them are so dependent on their fungal symbiont that they have lost the ability to make their own food. In at least some of these nonchlorophyllous species the fungal symbiont is also symbiotic with coniferous trees, forming a sort of bridge by which nutrients pass from the conifer root to the angiospermous plant. The precise physiological relationships are still unclear, but it is clear enough that the customary designation of these plants as saprophytes is misleading.

The Ericales are evidently related to and derived from the Theales, from which they differ in their consistently unitegmic, tenuinucellate ovules and usually sympetalous corolla, and in having no more than twice as many stamens as petals. The more or less transitional status of the Actinidiaceae has been discussed under the Theales. The Cyrillaceae and Clethraceae may also be regarded as somewhat transitional between the two orders, but they are better accommodated in the Ericales.

The mutual relationship among the Ericaceae, Epacridaceae, Monotropaceae, Pyrolaceae, and Clethraceae has been recognized for many years, and the Pyrolaceae and Monotropaceae have often been submerged in the Ericaceae. The Empetraceae have often in the past been referred to the Sapindales, but their relationship to the Ericales instead has been widely accepted in recent years. This is another of the families which has been moved to a new position in the latest (12th) edition of the Engler Syllabus. A similar consensus about the Cyrillaceae is only now developing. It may be noted that in certain chemical features tested by Gibbs, the Cyrillaceae fit with the Dilleniidae rather than with the Rosidae.

Ecologically, the most distinctive thing about the Ericales is the obligately mycorhizal habit of a large proportion of the species. In order to grow these plants successfully, one must first provide the proper conditions for the mycorhizal fungus. The well known requirement of many Ericaceae for acid soil is one expression of this fact. Mycorhizae are more widespread among seed plants than has been generally realized, but the Ericales are more closely dependent on the mycorhizal relationship than are most other groups. The Monotropaceae have become so dependent on their fungal symbiont that they have lost the ability to make their own food. Stages in the progressive dependence on the mycorhizal fungus are shown by the related family Pyrolaceae, in which the same species sometimes includes "normal" and leafless forms.

The adaptive significance of the other features which characterize the order and its several families is much less clear. It may well be that tetradinous pollen and appendiculate, poricidal anthers, for example, adapt the plants to particular kinds of pollinators, but the fact remains to be demonstrated. The tiny, scarcely differentiated, acotyledonous embryo of the Monotropaceae and Pyrolaceae probably reflects the mycotrophic habit. Similar reduction of the embryo occurs in a number of other mycotrophic groups.

### SYNOPTICAL ARRANGEMENT OF THE FAMILIES OF ERICALES

1. Embryo normally developed, with 2 cotyledons; plants more or less woody, always chlorophyllous
    2. Pollen grains borne singly; stamens mostly twice as many as the petals, commonly opening by terminal pores

3. Ovules 1 (–3) per locule; petals commonly joined at the base; seed without a seed coat; carpels 2–5        **1. Cyrillaceae**

3. Ovules numerous; petals free; seed with a seed coat; carpels 3
**2. Clethraceae**

2. Pollen grains nearly always borne in tetrads (in some Epacridaceae only one grain of the tetrad matures)

4. Sepals and petals each 4–7; corolla mostly sympetalous; habit various

5. Stamens mostly twice as many as the corolla lobes; leaves rarely palmately veined; anthers mostly opening by terminal pores, often appendiculate; widespread
**3. Ericaceae**

5. Stamens mostly as many as the corolla lobes; leaves mostly palmately veined; anthers opening by longitudinal slits, not appendaged; chiefly Australian
**4. Epacridaceae**

4. Stamens and petals each 1–3, the petals separate; stamens opening by longitudinal slits, not appendaged; dwarf, often prostate, evergreen shrubs        **5. Empetraceae**

1. Embryo very small and scarcely differentiated, without cotyledons; herbs or half-shrubs, often without chlorophyll

6. Plants usually with green leaves; anthers opening by pores; pollen usually in tetrads; petals separate; placentation axile
**6. Pyrolaceae**

6. Plants without cholorophyll, the leaves reduced to mere scales; anthers opening by longitudinal slits; pollen grains borne singly; petals separate or united; placentation axile or parietal        **7. Monotropaceae**

## SELECTED REFERENCES

Copeland, H. F., Observations on the Cyrillaceae particularly on the reproduction structures of the North American species. Phytomorph. 3: 405–411. 1953.

————, Observations on certain Epacridaceae. Am. Jour. Bot. 41: 215–222. 1954.

Cox, H. T., Studies in the comparative anatomy of the Ericales. Am. Midl. Nat. 39: 220–245. 1948. 40: 493–516. 1948.

Franks, J. W., and K. Watson, The pollen morphology of some critical Ericales. Pollen et Spores 5: 51–68. 1963.

Ganapathy, P. S., and B. Palser, Studies of floral morphology in the Ericales. VII. Embryology in the Phyllodoceae. Bot. Gaz. 125: 280–297. 1964.

Gibbs, R. D., Biochemistry as an aid in establishing the relationships of some families of dicotyledons. Proc. Linn. Soc. Lond. 169: 216–230. 1958.

Hagerup, O., Studies on the Empetraceae. Danske Vid. Selsk. Biol. Meddel. 20: 1–49. 1946.

Kavaljian, L. G., The floral morphology of *Clethra alnifolia* with some notes on *C. acuminata* and *C. arborea*. Bot. Gaz. 113: 392–413. 1952.

Leins, P., Entwicklungsgeschichtliche Studien an Ericales-Blüten. Bot. Jahrb. 83: 57–88. 1964.

Nowicke, J. W., Pollen morphology and classification of the Pyrolaceae and Monotropaceae. Ann. Mo. Bot. Gard. 53: 213–219. 1966.

Palser, B., "Some Aspects of Embryology in the Ericales," in *Recent Advances in Botany* (Toronto: Univ. Toronto Press, 1961), pp. 685–689.

————, Studies of floral morphology in the Ericales. V. Organography and vascular anatomy in several United States species of the Vacciniaceae. Bot. Gaz. 123: 79–111. 1961.

Palser, B., and Y. S. Murty, Studies of floral morphology in the Ericales. VIII. Organography and vascular anatomy in *Erica*. Bull. Torrey Club 94: 243–320. 1967.

Paterson, B. R., Studies of floral morphology in the Epacridaceae. Bot. Gaz. 122: 259–279. 1961.

Samuelsson, G., Studien über die Entwicklungsgeschichte der Blüten einiger Bicornes-Typen. Svensk. Bot. Tidsk. 7: 97–188. 1913.

Thomas, J. L., A monographic study of the Cyrillaceae. Contr. Gray Herb. 186: 1–114. 1960.

————, The genera of the Cyrillaceae and Clethraceae of the southeastern United States. Jour. Arn. Arb. 42: 96–106. 1961.

Veillet-Bartoszewska, M., Embryogénie des Cléthracées. Développement de l'embryon chez le *Clethra alnifolia* L. Compt. Rend. Acad. Sci. Paris 251: 2572–2574. 1960.

————, Embryogénie des Epacridacées. Développement de l'embryon chez le *Dracophyllum secundum* R. Br. Compt. Rend. Acad. Sci. Paris 253: 1000–1002. 1961.

————, Recherches embryogéniques sur les Ericales comparaison avec les Primulales. Rev. Gen. Bot. 70: 141–230. 1963.

Venkata Rao, C., Pollen types in the Epacridaceae. Jour. Indian Bot. Soc. 40: 409–425. 1961.

Wood, C. E., Jr., The genera of Ericaceae in the southeastern United States. Jour. Arn. Arb. 42: 10–80. 1961.

Wood, C. E., Jr., and R. B. Channell, The Empetraceae and Diapensiaceae of the southeastern United States. Jour. Arn. Arb. 40: 161–171. 1959.

ORDER 10. **Diapensiales**

The order Diapensiales consists of the single family Diapensiaceae, with about 18 species in temperate and arctic regions of the Northern Hemisphere. The family has often been included in the Ericales, but it differs from typical members of that order in a rather long list of embryological and androecial characters. There is a single set of functional stamens, attached well up in the corolla tube, with or without an alternating set of staminodes. The anthers mostly lack appendages and open by longitudinal slits, and the pollen grains are borne singly. The characterisic nectariferous disk about the base of the ovary, seen in most of the Ericales, is wanting or poorly developed in the Diapensiaceae. The Diapensiaceae are also without endosperm haustoria, whereas the Ericales have endosperm haustoria at both ends of the embryo sac. These and other differences have led modern students of the group, such as Palser, to exclude the Diapensiaceae from the Ericales.

In spite of the differences, the Diapensiales may well be related to the Ericales. The unitegmic, tenuinucellate ovules with glandular tapetum suggest that the Diapensiales belong either with the Asteridae or with the more advanced orders of the Dilleniidae. The combination of three carpels, axile placentation, numerous ovules, and frequently a set of staminodes in addition to the functional stamens suggests that the Diapensiales would be more at home among the advanced Dilleniidae than among the Asteridae. This leads us back to the Ericales as the most likely allies, and it may not be without significance that two genera of the Diapensiales have appendages on the anthers which resemble those of some of the Ericales. The wood anatomy is perfectly compatible with a relationship between the two orders, and there are some similarities in floral anatomy as well. Any relationship must of course involve the more primitive members of both groups, rather than the more advanced ones. It seems probable that the two orders have a common ancestry in the Theales.

The adaptive significance of the characters which mark the Diapensiales is dubious.

### SELECTED REFERENCE

Palser, B., Studies of floral morphology in the Ericales. VI. The Diapensiaceae. Bot. Gaz. 124: 200–219. 1963.

ORDER 11. **Ebenales**

The order Ebenales as here defined consists of five families and about 1700 species. The Sapotaceae, with about 800 species, make up nearly half of the order, and the Ebenaceae, with about 450, make up another quarter. The Symplocaceae have about 300 species, the Styracaceae about 150, and the Lissocarpaceae only two.

The Ebenales are woody, sympetalous plants with axile placentation and with usually twice as many stamens (including staminodes) as corolla lobes. When there is only a single set of stamens, these are usually opposite the corolla lobes. Usually there are only one or two (rarely four to many) ovules in each locule. The Ebenales are unusual among the chiefly sympetalous orders in having many members (Ebenaceae, some Styraceae) with bitegmic ovules.

Most authors are agreed that the Ebenales as here constituted form a natural group, and there is a developing consensus that the group takes its origin in the Theales. The Ebenales also resemble the Ericales in some respects, but the relationship is collateral rather than ancestral. The Ebenales are more advanced than the bulk of the Ericales in the reduced number of ovules and in having the stamens attached to the corolla tube, and in not having any polypetalous members, but they lack the set of specialized embryological features which characterize the Ericales, and the bitegmic ovules of the Ebenaceae could scarcely have been derived from the unitegmic ovules of the Ericales. The two orders may be regarded as having undergone certain parallel and other divergent changes from a common ancestry in the Theales.

It seems likely that some of the Ebenales have undergone an increase in the number of floral parts of a kind, in contrast to the general trend toward reduction among the angiosperms as a whole. In the Sapotaceae, some species of *Pouteria* have as many as 12 sepals, *Mimusops* has eight corolla lobes, and some species of *Manilkara* have as many as 14 locules in the ovary. *Symplocos* (Symplocaceae) usually has numerous stamens, which are often grouped into five bundles alternate with the petals. Here it might perhaps be possible to read the series either way, as in *Hypericum* and some members of the Theales and Malvales, but the Symplocaceae do not seem particularly primitive in other respects.

There is nothing ecologically outstanding about the Ebenales as a group, nor about any of the included families. They are all woody and chiefly tropical, but these are typical angiosperm features. The differences in fruit type which characterize some of the families are obviously related to seed dispersal, but none of the families has any special adaptation that is not well known in other groups as well. The biological significance of the latex system in the Sapotaceae is wholly unknown; it does not appear to fit them to any particular habitat. The importance of the differences in pubescence and nodal anatomy is equally obscure, and if the genes which govern these characters are important for other effects, the nature of those other effects is still wholly unknown.

### SYNOPTICAL ARRANGEMENT OF THE FAMILIES OF EBENALES

1. Plants with a well developed latex system; pubescence of 2-armed hairs, one arm sometimes reduced or obsolete; nodes mostly trilacunar **1. Sapotaceae**

1. Plants without a latex system; pubescence not of 2-armed hairs except in some Ebenaceae; nodes unilacunar

   2. Flowers mostly unisexual; styles separate at least distally **2. Ebenaceae**

   2. Flowers perfect; style solitary, with a capitate or slightly lobed stigma

      3. Pubescence of stellate or peltate hairs; fruit mostly dry; ovary superior to less often inferior; anthers more or less linear **3. Styracaceae**

      3. Pubescence of simple hairs (or none); fruit more or less fleshy; ovary inferior (sometimes only half inferior)

         4. Anthers linear; flowers with a corona; stamens 8 (twice as many as the corolla lobes) **4. Lissocarpaceae**

         4. Anthers ovate or rotund; flowers without a corona; stamens usually more than 8 **5. Symplocaceae**

### SELECTED REFERENCES

Copeland, H. F., The Styrax of northern California and the relationships of the Styracaceae. Am. Jour. Bot. 25: 771–780. 1938.

Lam, H. J., On the system of the Sapotaceae, with some remarks on taxonomical methods. Rec. Trav. Bot. Neerl. 36: 509–525. 1939.

Veillet-Bartoszewska, M., Embryogénie des Styracacées. Développement de l'embryon chez le Styrax officinalis L. Compt. Rend. Acad. Sci. Fr. 250: 905–907. 1960.

Wood, C. E., Jr., and R. B. Channell, The genera of the Ebenales in the southeastern United States. Jour. Arn. Arb. 41: 1–35. 1960.

ORDER 12. **Primulales**

The order Primulales as here defined consists of three families and some 1900 species. The Myrsinaceae have about 1000 species, the Primulaceae about 800, and the Theophrastaceae a little more than 100. Most taxonomists agree that these three families are closely allied. The Theophrastaceae have often been submerged in the Myrsinaceae. *Lysimachia* is often cited as a primitive member of the Primulaceae which suggests a connection to the Myrsinaceae.

The Primulales are sympetalous plants with the functional stamens opposite the corolla lobes (with or without an alternating set of staminodes), and with a compound ovary that has a single style and two to numerous usually bitegmic ovules on a free-central or basal placenta. No other group has this combination of characters. The Plumbaginaceae, which have often been referred to the Primulales, differ among other respects in their solitary ovules and partly or wholly distinct styles, and are here treated as an order of the subclass Caryophyllidae.

The Primulales are probably related to the Ebenales. One of the unusual features found in both orders is the combination of sympetalous flowers and bitegmic ovules. Neither of the two orders could be ancestral to the other, but they might have a common ancestry just short of the Theales. Such an ancestor would be a tropical tree with hypogynous, sympetalous flowers that have two or three sets of epipetalous stamens, a compound ovary, separate styles, and numerous bitegmic ovules on axile placentae. Only the sympetalous condition would be at odds with the characters of the Theales. One could also assume that the Ebenales and Primulales achieved the sympetalous condition independently, in which case the nearest common ancestor would be referable to the Theales.

Botanists once entertained the thought that the Primulales might be derived from the Caryophyllales, but further study has made such a relationship highly unlikely. The Primulales have none of the special features of the Caryophyllales other than free-central placentation, and it is hard to see how any possible ancestor of the Primulales could be included in the Caryophyllidae. In seeking an ancestor to the Primulales one must look for something which could give rise to the Theophrastaceae, not something which suggests the obviously advanced family Primulaceae.

The Primulales as a whole do not appear to be adapted to any particular ecological niche, nor are the characters which mark the order of any obvious selective significance. The characters of growth habit and fruit type which distinguish the Primulaceae from the other two families are evidently adaptive, but there is nothing distinctive about them as compared to other angiosperms. The biological significance of the staminodes of the Theophrastaceae and the secretory system of the Myrsinaceae is obscure.

SYNOPTICAL ARRANGEMENT OF THE FAMILIES OF PRIMULALES

1. Plants woody, mostly arborescent, largely tropical; fruits mostly fleshy, often 1-seeded (even though there are several or many ovules in the ovary)

  2. Flowers with staminodes alternate with the corolla-lobes; plants without an evident secretory system, the leaves not gland-dotted                    **1. Theophrastaceae**

  2. Flowers without staminodes; plants with a schizogenous secretory system and gland-dotted leaves      **2. Myrsinaceae**

1. Plants herbaceous or occasionally half-shrubby, chiefly of temperate regions or altitudes; fruit capsular, with (2–) many seeds
                                                      **3. Primulaceae**

SELECTED REFERENCES

Channell, R. B., and C. E. Wood, Jr., The genera of the Primulales of the southeastern United States. Jour. Arn. Arb. 40: 268–288. 1959.

Dickson, J., Studies in floral anatomy. III. An interpretation of the gynaecium in the Primulaceae. Am. Jour. Bot. 23: 385–393. 1936.

Douglas, G. E., Studies in the vascular anatomy of the Primulaceae. Am. Jour. Bot. 23: 199–212. 1936.

Roth, I., Histogenese und morphologische Deutung der Plazenta von *Primula*. Flora 148: 129–152. 1959.

Sattler, R., Zur frühen Inflorescenz- und Blütenentwicklung der Primulales . . . Bot. Jahrb. 81: 358–396. 1962.

Veillet-Bartoszewska, M., Recherches embryogéniques sur les Ericales comparaison avec les Primulales. Rev. Gen. Bot. 70: 141–230. 1963.

## Subclass V. Rosidae

The subclass Rosidae as here defined consists of 16 orders, 108 families, and about 60,000 species. More than a third of all the species of dicotyledons belong to this one subclass. A third of the species in the subclass belong to the single order Rosales (about 20,000), and somewhat more than another third belong to only three additional orders, the Myrtales (9000), Euphorbiales (7600), and Sapindales (6700). The remaining 12 orders have about 16,000 species amongst them.

The 16 orders which make up the Rosidae evidently hang together as a natural group, but this group cannot be fully characterized morphologically. In general, the Rosidae are more advanced than the Magnoliidae in one or another respect, but less advanced than the Asteridae. Most of them have fairly well developed flowers with a polypetalous corolla, but apetalous types and a few sympetalous types are also known in the group. Only a few are amentiferous. Insofar as they have been investigated, the forms with numerous stamens have the stamens developing in centripetal sequence. The pollen is most commonly binucleate, but trinucleate forms have repeatedly been developed within the group. The ovules are bitegmic or less commonly unitegmic, with various transitional types, or sometimes even without integument in parasitic families. The ovules are much more often crassinucellate than tenuinucellate, and there are some transitional types as well. The tenuinucellate ovule with a single massive integument, as seen in the Asteridae, is seldom precisely duplicated in the Rosidae. Typical parietal placentation is rather uncommon in the group, and a great many of the species, especially in the more advanced families, have only one or two ovules per locule. A well developed nectary disk is very often present.

The Rosidae are evidently derived from the Magnoliidae. Every significant respect in which any member of the Rosidae differs from the more primitive members of the Magnoliidae represents a phyletic advance. None of the especially primitive characters which we have noted in some of the Magnoliidae is known in the Rosidae.

The subclass most likely to be confused with the Rosidae is the Dilleniidae. These likewise are more advanced than the Magnoliidae and less advanced than the Asteridae. Most of them are polypetalous, but a considerable number are sympetalous, and a few are amentiferous. In contrast to the Rosidae, the Dilleniidae have centrifugal stamens (when the stamens are numerous), and only a few families (notably the Cruciferae) have trinucleate pollen. Parietal placentation is common in the group, but other types are also well represented. Uniovulate or biovulate locules are much less common in the Dilleniidae than in the Rosidae. Not many of the Dilleniidae have a typical nectary disk, although other types of nectaries are common.

In the last analysis the Rosidae and Dilleniidae are kept apart as subclasses because each seems to constitute a natural group separately derived from the ancestral Magnoliidae, rather than because of any definitive distinguishing characters. The same sorts of evolutionary advances have occurred in both groups, but with differing frequencies. In spite of the lack of solid distinguishing criteria, I believe that it is conceptually more useful to hold the two as separate subclasses than to combine them into one or to abandon any attempt at organization of the Magnoliidae into subclasses.

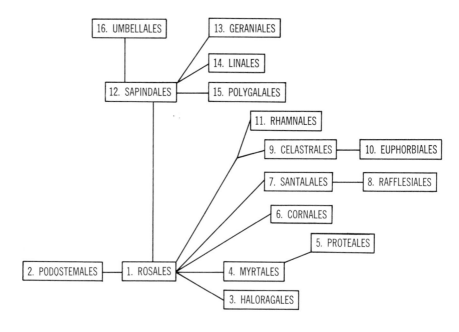

FIGURE 4.7

Probable relationships among the orders of Rosidae. Here, as in the other phyletic diagrams, the length of the line between two orders is governed by the requirements of two-dimensional representation and does not indicate the closeness of the relationship.

The Rosales are the most primitive order of their subclass. All other orders of the Rosidae appear to be derived directly or indirectly from the Rosales. If the other orders of the subclass were wiped out of existence, the Rosales could be accommodated without great difficulty as a somewhat isolated order of the Magnoliidae. A large proportion of the Rosales are apocarpous or monocarpous, as in the Magnoliidae, whereas apocarpy or monocarpy is rare in the rest of the group. On the other hand, the relationship between the Rosales and some of the more advanced orders of the subclass is so obvious, and the difficulty of drawing lines is so great, that it seems necessary to associate the Rosales with their descendants rather than with their ancestors.

### SYNOPTICAL ARRANGEMENT OF THE ORDERS OF ROSIDAE

1. Plants relatively primitive, the flowers commonly either apocarpous, or with the carpels united only toward the base or by their styles, or with more or less numerous ovules per carpel, or

with more or less numerous stamens, often with more than one of these features; vascular bundles never with internal phloem; plants never parasitic (though sometimes insectivorous) **1. Rosales**

1.  Plants more advanced in one or another respect, the flowers mostly syncarpous (or pseudomonomerous), seldom apocarpous, typically with not more than twice as many stamens as sepals or petals (stamens numerous in many Myrtales and in some others), very often with only 1 or 2 ovules per carpel; vascular bundles sometimes with internal phloem; plants autotrophic or sometimes parasitic

    2.  Vascular bundles mostly with internal (as well as external) phloem; flowers strongly perigynous to epigynous, often with numerous stamens and numerous ovules
        **4. Myrtales**

    2.  Vascular bundles without internal phloem; flowers various

        3.  Flowers pseudomonomerous, strongly perigynous
            **5. Proteales**

        3.  Flowers evidently with more than one carpel, or not strongly perigynous, or both

            4.  Plants mostly parasitic or hemiparasitic (wholly autotrophic in some Santalales)

                5.  Ovules 1–8; plants with or without chlorophyll
                    **7. Santalales**

                5.  Ovules very numerous; plants without chlorophyll
                    **8. Rafflesiales**

            4.  Plants (with few exceptions) autotrophic

                6.  Flowers of ordinary type, not markedly reduced except in a few small families such as the Garryaceae (Cornales)

                    7.  Leaves mostly compound or conspicuously lobed or cleft, less often simple and unlobed

8. Ovary superior; pollen variously binucleate or tri-nucleate

    9. Plants mostly woody        **12. Sapindales**

    9. Plants mostly herbaceous or nearly so
        **13. Geraniales**

8. Ovary inferior; pollen trinucleate    **16. Umbellales**

7. Leaves simple and entire or merely toothed, only seldom compound

  10. Ovary mostly superior

    11. Flowers regular

      12. Stamens basally connate, usually more than 5; disk usually wanting; pollen trinucleate
        **14. Linales**

      12. Stamens free from each other, seldom more than 5; disk often present

        13. Stamens alternate with the petals, or the flowers rarely diplostemonous; pollen variously binucleate or trinucleate
        **9. Celastrales**

        13. Stamens opposite the petals; pollen binucleate    **11. Rhamnales**

    11. Flowers irregular (except in Tremandraceae)
        **15. Polygalales**

  10. Ovary inferior        **6. Cornales**

6. Flowers more or less reduced and often unisexual, the perianth poorly developed or wanting

  14. Endosperm wanting; submerged aquatics
        **2. Podostemales**

  14. Endosperm present; habitat various

15. Herbs with an inferior ovary and without milky juice; flowers not forming pseudanthia
**3. Haloragales**

15. Trees, shrubs, or herbs, with superior or naked ovary and often with milky juice; flowers often grouped into pseudanthia **10. Euphorbiales**

ORDER 1. **Rosales**

The order Rosales, as here defined, consists of 17 families and nearly 20,000 species. About 13,000 of these belong to the Leguminosae, which are one of the three largest families of flowering plants. Another 5000 species belong to only three more families, the Rosaceae (3000), Crassulaceae (1400), and Saxifragaceae (700). The remaining 13 families have only about 1500 species in all, and seven of them have less than a dozen species each.

The several families which make up the order Rosales are held together by a complex pattern of overlapping similarities, but the order is morphologically diffuse and difficult to define. The plants are herbs, shrubs, or trees, with alternate to opposite or whorled, stipulate to exstipulate, simple or compound leaves, regular or irregular, hypogynous to perigynous or epigynous flowers with separate or united or no petals, numerous to few stamens, numerous to solitary carpels that are separate or more or less united, and one to many crassinucellate to tenuinucellate ovules with one or two integuments. When the pistil is compound the styles are separate or less often united. The seeds may or may not have endosperm and are borne in many different kinds of fruits.

A plant which combined all the primitive features found in various members of the Rosales would be woody, with alternate, stipulate leaves, regular, hypogynous flowers with five separate petals, numerous centripetal stamens, several separate carpels, and numerous crassinucellate, bitegmic ovules, the ovaries ripening into follicles with endspermous seeds. Such a plant would be very difficult to exclude from the subclass Magnoliidae. Aside from the absence of ethereal oil cells, it would fit well enough into the order Magnoliales, although it would not necessarily be very closely allied to any of the existing families of that order.

The evidence clearly indicates that the Rosales are derived from the Magnoliidae, from within or near the Magnoliales. The especially primitive features found in various members of the Magnoliales (e.g., uniaperturate pollen, primitively vesselless xylem, unsealed carpels, and laminar stamens) are wholly wanting from the Rosales and from the subclass Rosidae.

There is much disagreement about the limits of families within the

Rosales. The treatment here presented is conservative in grouping *Roridula* with *Byblis* in a single family, and in recognizing only one family of legumes instead of three. It is more liberal in dividing the traditional family Saxifragaceae into three families. Even as here restricted to include only the herbaceous genera, the family Saxifragaceae is still heterogeneous and might with some reason be further divided, and the family Grossulariaceae is an equally good candidate for more splitting. Some authors divide the traditional Saxifragaceae into a dozen or more families instead of only 3. These are all matters of opinion, on which botanists may reasonably differ even when they have access to the same set of facts.

There is a dichotomy of sorts between the Rosaceae, Chrysobalanaceae, Neuradaceae, and Leguminosae, on the one hand, the remaining families of the order on the other, and a further dichotomy among these latter between the woody and the herbaceous families. It would be possible to recognize two or three different orders on these bases, and some authors would further segregate the Leguminosae into an order of their own. These possible segregate orders are indicated in the key to families. Here again we are dealing with matters in which there is no clear right and wrong.

In spite of the diversity of the group, it is generally agreed that most of the families here referred to the Rosales are fairly closely related to each other. The Saxifragaceae are connected to the Crassulaceae by an intermediate genus *Penthorum*, with five separate follicular carpels, as in the Crassulaceae, but without succulence. *Penthorum* has been referred by some authors to the Crassulaceae, and by others to the Saxifragaceae. Lacking very close relatives in either family, it is often treated as a family of its own by those who prefer small and narrowly limited families.

The similarity of *Astilbe* (Saxifragaceae) to *Aruncus* (Rosaceae) in many features has been obvious to all, but the phylogenetic significance of the similarities remains to be carefully evaluated. In spite of their overall resemblance, *Aruncus* has three to four carpels, numerous stamens, and nonendospermous seeds, whereas *Astilbe* has two carpels, eight to ten stamens, and endospermous seeds. Each is morphologically at home in its own family, although *Aruncus* is a bit unusual in the Rosaceae in lacking stipules. At present it seems equally difficult to explain the similarities as a result either of convergence or of phyletic unity.

The order Rosales as here defined includes four small families (Alseuosmiaceae, Byblidaceae, Columelliaceae, and Eucryphiaceae) which have not always been referred to the rosalean cluster of families, and excludes several others. The Hamamelidaceae, Platanaceae, and Myrothamnaceae are treated as the principal members of the order Hamamelidales, in another subclass. The Krameriaceae are referred to the Polygalales, the Connaraceae to the Sapindales, and the Crypteroniaceae to the Myrtales.

The Alseuosmiaceae consist of three small genera of New Zealand and New Caledonia which have often been referred to the Caprifoliaceae. They are clearly out of place in the Caprifoliaceae, from which they differ in their alternate leaves, valvate corolla lobes, frequently hypogynous stamens, and the structure of the pollen. Several authors have noted that the plants are habitally much like the Pittosporaceae, but they differ from the Pittosporaceae in having the ovary inferior or half-inferior. They are also much like the members of the Grossulariaceae that have sometimes been segregated as the Escalloniaceae, differing most obviously in their sympetalous corolla. The Pittosporaceae, on the other hand, are more or less sympetalous. On the basis of present information, the Alseuosmiaceae seem best accommodated in the Rosales, near the Pittosporaceae and Grossulariaceae.

The Byblidaceae have often been associated with the Droseraceae, in another order, because of the glandular, insect-catching hairs. In other respects these two families are wholly different, however, and there is growing agreement that the Byblidaceae are related to the Pittosporaceae. The pittosporaceous genus *Cheiranthera*, a small group of shrubs and half-shrubs with the anthers opening by apical pores as in the Byblidaceae, is probably a collateral close relative of the Byblidaceae.

The Columelliaceae, consisting of four species of Andean shrubs or small trees, have sympetalous flowers and have usually been included in the Scrophulariales. The relatively primitive wood anatomy of the Columelliaceae is out of place in the Scrophulariales (teste William L. Stern), however, and the plants are habitally reminiscent of *Escallonia*. In the present state of knowledge the Columelliaceae seem best accommodated in the Rosales near the Pittosporaceae and Grossulariaceae.

The small South Pacific genus *Eucryphia* (five spp) constitutes a family which has often been referred to as the Theales. Both the pollen morphology and the wood anatomy suggest a relationship to the Cunoniaceae instead, and there is nothing obvious in the floral morphology to negate such a relationship. Furthermore, the stamens are reported to be centripetal, which is proper for a member of the Rosidae but would be anomalous in the Dilleniidae.

One of the interesting features about the family Rosaceae is that the carpel appears primitively to have five bundles, although the leaves have only three. The extra bundles are along the suture and supply the ovules. Similar five-bundle carpels are known in the Mimosoideae (but not the other two sub-families of Leguminosae) and a scattering of Magnoliidae. The morphologic significance of the five-bundle structure, and its bearing on the interpretation of the carpel as a megasporophyll, have yet to be elucidated.

The Rosales have been a prolific source of diversification within the angiosperms. Clearly the most primitive order in their subclass, they are directly or indirectly ancestral to some 15 other orders of the Rosidae, and possibly also to the whole subclass Asteridae. The small family Crypteroniaceae (four spp) is transitional between the Rosales and Myrtales and is here referred to the latter order. The Connaraceae, Melianthaceae, Greyiaceae, and Staphyleaceae are all more or less transitional between the Rosales and Sapindales, but are here referred to the Sapindales. The distinction between the Staphyleaceae (Sapindales) and the Cunoniaceae (Rosales) is slender and scarcely related to the more general differences between the two orders.

The Pittosporaceae are of special interest because they suggest a possible connection between the Rosidae and the Asteridae. Systematists are agreed that the Pittosporaceae belong with the mainly polypetalous rather than with the mainly sympetalous families, and most authors associate them with *Escallonia* (Grossulariaceae) or other woody segregates from the traditional Saxifragaceae. Yet most members of the Pittosporaceae have a more or less sympetalous corolla, and furthermore they resemble the Asteridae in having unitegmic, tenuinucellate ovules. The mature structure of the corolla suggests that the petals probably arise from separate primordia in the young bud, rather than from a continuous ring as in the Asteridae, but the tendency toward sympetaly is certainly there. There is no question of a direct connection between the Pittosporaceae and anything in the Asteridae, however. Among other reasons, the Pittosporaceae lack stipules, whereas the Asteridae clearly have a stipulate ancestry. The possibility of a collateral relationship, on the other hand, seems very good. We have already noted that parallelism in a series of independent characters is in itself some indication of relationship.

In view of the great morphological diversity within the Rosales, it should not be surprising that the group also lacks ecological unity. Most of the larger families of the order are also ecologically diverse. Only one of the families, the Crassulaceae, is ecologically distinctive. These are succulents with well developed leaves and without spines. Nearly cosmopolitan in distribution, they are commonest in warm-temperate or subtropical regions which have a severe or prolonged dry season but are not true deserts. Although they commonly grow in company with non-succulents of other families and orders, the Crassulaceae at least have their own way of meeting the problem of drought. In many places they are the commonest or the only leaf-succulents, but in South Africa they are much outnumbered by the Aizoaceae, in another subclass.

SYNOPTICAL ARRANGEMENT OF THE FAMILIES OF ROSALES

1. Seeds mostly with endosperm (except notably the Davidsoniaceae and a few Grossulariaceae); leaves with or without stipules

2. Plants more or less distinctly woody, not succulent (herbaceous only in some Byblidaceae, marked by the conspicuous, glandular, insect-catching hairs); carpels nearly always united into a compound pistil (some exceptions among the Cunoniaceae) (segregate order Cunoniales)

   3. Leaves mostly compound (simple in some Cunoniaceae and Eucryphiaceae), stipulate; flowers hypogynous, with axile placentation, or the carpels rarely free

      4. Leaves opposite or whorled; seeds with endosperm

         5. Carpels 5–12 (18), multiovulate; stamens numerous
            **1. Eucryphiaceae**

         5. Carpels 2–5, 1–many-ovulate; stamens as many or twice as many as the sepals, seldom numerous
            **2. Cunoniaceae**

      4. Leaves alternate; seeds without endosperm
         **3. Davidsoniaceae**

   3. Leaves simple, entire to deeply cleft, exstipulate, or seldom with vestigial stipules; flowers hypogynous to epigynous, with axile or parietal placentae

      6. Flowers hypogynous or slightly perigynous; ovules uniteg-mic and, so far as known, tenuinucellate

         7. Stamens 5; corolla polypetalous to more or less sym-petalous

            8. Leaves without insect-catching hairs; corolla usually more or less sympetalous; half-shrubs or more often shrubs or trees; anthers opening by elongate slits or less often by pores    **4. Pittosporaceae**

            8. Leaves with insect-catching hairs; corolla strictly poly-petalous; herbs or half-shrubs; anthers opening by apical pores or short slits    **5. Byblidaceae**

         7. Stamens 2; corolla sympetalous    **6. Columelliaceae**

      6. Flowers usually partly or wholly epigynous, seldom hypogy-nous; ovules diverse

9. Leaves opposite; stamens 10–numerous, seldom only 5; ovules few to many, unitegmic, tenuinucellate

**7. Hydrangeaceae**

9. Leaves alternate; stamens 4–6, sometimes accompanied by as many staminodes; ovules various

10. Petals separate, or sometimes wanting

11. Ovules mostly numerous, on axile or parietal placentae, variously unitegmic or bitegmic, crassinucellate or tenuinucellate; leaves normally developed, the plant not heatherlike

**8. Grossulariaceae**

11. Ovules 1–2 (8) in each cell of a 2–3–celled ovary, axile, unitegmic, crassinucellate; leaves mostly small, the plant often heatherlike

**9. Bruniaceae**

10. Petals connate, the corolla sympetalous

**10. Alseuosmiaceae**

2. Plants either herbaceous, or succulent, or both, without insect-catching hairs; leaves exstipulate, alternate or seldom opposite; stamens as many or twice as many as the sepals; ovules mostly bitegmic and crassinucellate (segregate order Saxifragales)

12. Plants succulent; carpels as many as the petals, free or united only at the base                    **11. Crassulaceae**

12. Plants scarcely or not at all succulent; carpels seldom the same number as the petals

13. Carpels 6, free, uniovulate; sepals 6; petals none; stamens 12; some of the leaves modified into pitchers

**12. Cephalotaceae**

13. Carpels 2–5, usually more or less united, at least below, and usually with numerous ovules on parietal placentae; flowers mostly 4–5–merous, with 1 or 2 sets of

stamens and usually with petals; leaves not modified
into pitchers                                    **13. Saxifragaceae**

1. Seeds mostly without endosperm, except in a few Rosaceae; leaves
mostly stipulate

14. Carpels 1–many, when solitary the plants with simple leaves
and only 1 or 2 ovules per carpel; stamens free, numerous
or less often only 5 or 10 (Rosales, sens. strict.)

15. Gynoecium of 1–many free or less often united carpels, never
with the carpels united only by a common style, nor with
a solitary carpel that has a gynobasic style

16. Gynoecium apocarpous, or syncarpous with 2–5 styles;
stamens usually more than 10        **14. Rosaceae**

16. Gynoecium syncarpous, with 10 styles; stamens 10
**15. Neuradaceae**

15. Gynoecium of 2–3 carpels united only by a gynobasic style,
or seemingly of a single carpel with a gynobasic style
**16. Chrysobalanaceae**

14. Carpel nearly always solitary, and usually with more than 2
ovules; leaves usually compound; stamens (5) 10–many,
often connate by the filaments (segregate order Fabales)
**17. Leguminosae**

## SELECTED REFERENCES

Bange, G. G. J., A new family of dicotyledons: Davidsoniaceae. Blumea 7: 293–296. 1952.

Bausch, J., A revision of the Eucryphiaceae. Kew Bull. 1938: 317–349.

Birdsong, B. A., R. Alston, and B. L. Turner, Distribution of cavanine in the family Leguminosae as related to phyletic groupings. Can. Jour. Bot. 38: 499–505. 1960.

Gelius, L., Studien zur Entwicklungsgeschichte an Blüten der Saxifragales sensu lato mit besonderer Berücksichtigung des Androeceums. Bot. Jahrb. 87:253–303. 1967.

Ingle, H. D., and H. E. Dadswell, The anatomy of the timbers of the South-West Pacific area. IV. Cunoniaceae, Davidsoniaceae, and Eucryphiaceae. Austral. Jour. Bot. 4: 125–151. 1956.

Leinfellner, W., Uber die falsche Sympetalie bei Lonchostoma. Österr. Bot. Zeits. 111: 345–353. 1964.

Mauritzon, J., Contributions to the embryology of the orders Rosales and Myrtales. Lunds Univ. Årsskr. N. F. II. 35 (2): 1–121. 1939.

Morf, E., Vergleichend-morphologische Untersuchungen am Gynoeceum der Saxifragaceen. Ber. Schweiz. Bot. Ges. 60: 516–590. 1950.

Prance, G. T., A synopsis of Chrysobalanaceae. Oxford University, in press 1968.

Saxena, N. P., Studies in the family Saxifragaceae. I. A contribution to the morphology and embryology of Saxifraga diversifolia Wall. II. Development of the ovule of Parnassia nubicola Wall. Proc. Indian Acad. Sci.: B60:38-51; 196-202. 1964.

Schaeppi, H., and F. Steindl, Vergleichend-morphologische Untersuchungen am Gynoecium der Rosoideen. Ber. Schweiz. Bot. Ges. 60: 693–699. 1950.

Sterling, C., Developmental anatomy of the fruit of Prunus domestica L. Bull. Torrey Club 80: 457–477. 1953.

————, Comparative morphology of the carpel in the Rosaceae. VIII and IX. Spiraeoideae. Am. Jour. Bot. 53: 521–530; 951–960. 1966.

ORDER 2. **Podostemales**

The order Podostemales consists of the single family Podostemaceae, with about 200 species. They are submerged aquatics with small, perfect flowers, a much reduced, haplochlamydeous perianth, a syncarpous, superior ovary, and numerous small seeds without endosperm.

The Hydrostachyaceae, another family of submerged aquatics with reduced flowers, have often been included in the Podostemales. Recent studies, however, indicate that they are probably better referred to the Scrophulariales.

It is fairly generally agreed that the Podostemales are related to the Crassulaceae and/or the Saxifragaceae, in the Rosales. The embryological features, in particular, have been adduced to show a connection to the Crassulaceae. The Crassulaceae might seem an odd starting point for a group of aquatics, but Tillaea aquatica (Crassulaceae) is semiaquatic, Tillaea is not to be regarded as on the direct line of evolution to the Podostemaceae, however; it merely indicates a potentiality within the Crassulaceae to adapt to aquatic habitats.

### SELECTED REFERENCES

Jäger-Zürn, I., Embryologische Untersuchungen an vier Podostemaceen. Österr. Bot. Zeits. 114: 20–45. 1967.

Mauritzon, J., Contributions to the embryology of the orders Rosales and Myrtales. Lunds Univ. Årsskr. Avd. 2. 35 (2): 1–120. 1939.

ORDER 3. **Haloragales**

The Haloragales as here defined consist of four families, with less than

175 species. The Haloragaceae are estimated at 125 species, the Gunner-aceae at 35, the Theligonaceae at three, and the Hippuridaceae at only one. It is generally agreed that the Haloragaceae, Gunneraceae, and Hippuri-daceae are related to each other. Some authors even include them all in one family, although they are distinguished by abundant characters.

The Theligonaceae are more doubtful. They have sometimes been re-ferred to the Caryophyllales, where they have no close relatives and are anomalous in having unitegmic ovules and in lacking both perisperm and betacyanins. They share a number of features with the other families of Haloragales, but they also differ in several features, as shown in the key.

The Haloragales are herbs with perfect or often unisexual, more or less reduced flowers. The perianth is minute or vestigial and was probably originally tetramerous. The ovary is inferior, with only one ovule in each locule, and the seeds have a more or less abundant endosperm. Many of the species are aquatic. Entomophily has been largely or wholly aban-doned, and the pollen is commonly distributed by wind or water.

Most authors have considered the Haloragales to be reduced derivatives of the Myrtales, and I agree that the two orders are probably related. *Trapa*, in the Myrtales, is especially suggestive of such a relationship, being an aquatic with a single pendulous ovule in each locule. *Trapa* is not to be taken as directly ancestral to the Haloragales, however. Instead it merely indicates that the possibility for this sort of evolutionary development exists in the Myrtales. The presence of well developed endosperm and the absence of internal phloem in the Haloragales militate against a direct ancestry of the Haloragales in the Myrtales. Probably the two orders have a common origin in the Rosales.

### SYNOPTICAL ARRANGEMENT OF THE FAMILIES OF HALORAGALES

1. Ovule apical; embryo straight; stamens 1–8

   2. Styles 2–4; ovule bitegmic; stamens (1) 2–8

      3. Ovary 2–4-locular; terrestrial or aquatic herbs with small, al-ternate or more often opposite or whorled, exstipulate leaves and small inflorescences; fruit not drupaceous; petals present or absent **1. Haloragaceae**

      3. Ovary unilocular; terrestrial herbs with large, alternate, stipu-late leaves and large inflorescences; fruit drupaceous; pet-als none **2. Gunneraceae**

   2. Style 1; ovule unitegmic; stamen 1; emergent aquatics; leaves whorled, exstipulate **3. Hippuridaceae**

1. Ovule basal, unitegmic; embryo strongly curved; stamens 6–28; terrestrial herbs with stipulate leaves that are opposite below and alternate above; ovary unilocular, with a single style
**4. Theligonaceae**

SELECTED  REFERENCE

Kapil, R. N., and P. R. Mohana Rao, Embryology and systematic position of *Theligonum* Linn. Proc. Nat. Inst. Sci. India B32: 218–232. 1966.

ORDER 4. **Myrtales**

The order Myrtales as here defined consists of 13 families and more than 9000 species. About 7000 of the species belong to only two of the families, the Melastomataceae (4000) and the Myrtaceae (3000). The Onagraceae and Thymelaeaceae are estimated at about 650 species each, and the Lythraceae and Combretaceae at about 500 each. The remaining seven families have scarcely 50 species amongst them. The Heteropyxidaceae, sometimes treated as a separate family because of their alternate, glandular-punctate leaves and superior ovary, are here referred to the Myrtaceae, following Stern and Brizicky (cited below).

The most characteristic feature of the Myrtales is the presence of internal phloem, which is otherwise very rare in the subclass Rosidae, although it is fairly common in the Asteridae. Occasional genera in some families of the order lack internal phloem, but no family except the mono-typic Dialypetalanthaceae is wholly (or even chiefly) without it. It may be noted that the mutual affinity amongst most of the families of the order had been perceived long before the taxonomic significance of internal phloem became apparent.

Members of the Myrtales usually have opposite, simple, commonly entire leaves, and perigynous to epigynous flowers with compound pistil and axile or sometimes apical, rarely parietal placentation. The seeds have little or no endosperm. The stamens, when numerous, develop in centripetal sequence. The majority of the species have tetramerous flowers, which are otherwise rather uncommon in the subclass and indeed in the angiosperms as a whole. Two smaller orders (Proteales and Haloragales) which are related to the Myrtales also have tetramerous flowers, as do many of the Capparales in the subclass Dilleniidae.

The limits of the order have not been fully agreed on in the past, but among the larger families here included only the Thymelaeaceae have occasioned much dispute. In my opinion the Thymelaeaceae are wholly at home in the Myrtales, especially inasmuch as the more primitive genera of the family have an obviously compound pistil. Several other families (notably the Lecythidaceae, Rhizophoraceae, and Geissolomataceae) which

have sometimes been referred to the Myrtales are here excluded and referred to diverse other orders. They all lack internal phloem and are out
of harmony with the Myrtales in other respects as well. The Lecythidaceae
in particular differ in so many respects (alternate leaves, centrifugal stamens, no internal phloem, and a series of embryological features) from
characteristic members of the Myrtales that their exclusion seems to be
required. They are here treated as a separate order in the subclass
Dilleniidae.

The Dialypetalanthaceae (one sp) are referred by most authors to the
Myrtales, but they have also been thought to be related to the Rubiaceae
or Loganiaceae. They are unusual in the order in lacking internal phloem,
and the pollen suggests that of the Rhizophoraceae rather than any truly
myrtalean family. In other respects they seem at home in the Myrtales,
and the absence of internal phloem should not be taken too seriously, inasmuch as a few other genera which undoubtedly belong to the order also
lack it. The numerous stamens and polypetalous corolla of *Dialypetalanthus* would be highly anomalous in or near the Rubiaceae or Loganiaceae,
and the capsular, many-seeded fruits would be equally out of place in the
Cornales, to which the Rhizophoraceae are here referred. Admitting as
always the possibility of error, I think the family is best included in the
Myrtales.

The ancestry of the Myrtales pretty clearly lies in the Rosales. Indeed
the only character that is widespread in the Myrtales which is not also
well known in the Rosales is internal phloem. It may be significant that
the Crypteroniaceae, here included in the Myrtales, have sometimes been
referred to the Rosales. The Crypteroniaceae, with haplostemonous,
apetalous flowers, do not appear to be primitive among the Myrtales,
however.

We may reasonably visualize the ancestral prototype of the order as a
woody plant with simple, opposite, minutely stipulate leaves, perigynous
flowers, well developed petals, numerous, centripetal stamens, a compound
pistil composed of several or many multiovulate carpels with a single
common style, and a many-seeded, capsular fruit. Among the existing
families, only the Sonneratiaceae meet these specifications.

Among the Rosales, only the Rosaceae have the characters from which
those of the Myrtales might reasonably have evolved. The relationship is
certainly not close, however. The only members of the Rosaceae which
have a compound pistil (the Maloideae) still have separate styles but have
already become epigynous. A separate syncarpous line leading to the
Myrtales must be envisaged.

The Myrtales are ecologically diverse. The two largest families and
several of the smaller ones are chiefly tropical, but this is nothing unusual
among angiosperms in general. The Onagraceae, on the other hand, are

best developed in temperate regions. Neither the order as a whole nor any of the larger families is confined to any particular habitat or meets its ecological needs in any unusual way. The small family Trapaceae consists of free-floating annual aquatics, but here we are dealing with a single genus consisting of only three species.

The adaptive significance of the characters which mark the Myrtales as a group is obscure. Some botanists have thought to see survival value in such characters as epigyny and loss of endosperm, but tetramerous flowers and simple, opposite leaves pose greater difficulties, and internal phloem still escapes the most valiant attempt at a Darwinian interpretation.

### SYNOPTICAL ARRANGEMENT OF THE FAMILIES OF MYRTALES

1. Ovary superior

    2. Plants woody or less often herbaceous, mostly terrestrial, when aquatic not free-floating

        3. Ovules 2–many per carpel; ovary 2–several-locular; petals present or absent; fruit capsular

            4. Stamens numerous (seldom only 12); carpels 4–20; ovules numerous                                          **1. Sonneratiaceae**

            4. Stamens 4–8 (12 in some Lythraceae)

                5. Flowers perfect; petals present or absent

                    6. Filaments elongate; carpels 2–3, seldom more numerous, each with 2–many ovules; mostly diplostemonous                                    **2. Lythraceae**

                    6. Filaments very short; carpels 4, each with 2–4 ovules; haplostemonous                                **3. Penaeceae**

                5. Plants polygamo-dioecious; petals none; carpels 2, multiovulate                                        **4. Crypteroniaceae**

        3. Ovule solitary in a pseudomonomerous pistil, or the ovules solitary in each of 2 or more locules; petals scale-like or none; fruit seldom capsular; mostly diplostemonous                                **5. Thymelaeaceae**

2. Plants free-floating annual aquatics; ovary bilocular, with a single pendulous ovule in each locule, only one locule maturing; haplostemonous    **6. Trapaceae**

1. Ovary in most species inferior

7. Placentation axile, seldom parietal; fruit seldom 1-seeded and indehiscent

8. Stamens numerous

9. Carpels 2–3, seldom more; leaves glandular-punctate (inconspicuously so in Dialypetalanthaceae)

10. Fruit a many-seeded capsule; anthers opening by terminal pores; no internal phloem
**7. Dialypetalanthaceae**

10. Fruit few-seeded (even when the ovary has many ovules), a berry, drupe, or capsule; anthers opening by slits or less often by terminal pores; mostly with internal phloem    **8. Myrtaceae**

9. Carpels more or less numerous, commonly about 9; leaves not glandular-punctate; fruit a many-seeded berry
**9. Punicaceae**

8. Stamens up to twice as many as the petals (or sepals)

11. Anthers opening by longitudinal slits; connective not much modified; leaves mostly without subparallel longitudinal ribs

12. Embryo sac 4-nucleate; endosperm diploid; most species herbaceous; ovules usually many; fruit usually capsular, sometimes a berry or drupe
**10. Onagraceae**

12. Embryo sac 8-nucleate; endosperm triploid; plants woody; ovules 2–3 per carpel; fruit drupaceous
**11. Oliniaceae**

11. Anthers mostly opening by terminal pores; connective

variously modified; leaves mostly with prominent subparallel longitudinal ribs          **12. Melastomataceae**

7. Placentation apical in a 1–celled ovary; ovues 2 (–6); fruit 1-
seeded, indehiscent          **13. Combretaceae**

## SELECTED REFERENCES

Exell, A. W., Space problems arising from the conflict between two evolutionary tendencies in the Combretaceae. Bull. Soc. Bot. Belg. 95: 41–49. 1962.

Graham, S., The genera of Lythraceae in the southeastern United States. Jour. Arn. Arb. 45: 235–250. 1964.

Heinig, K., Studies in the floral morphology of the Thymelaeaceae. Am. Jour. Bot. 38: 113–132. 1951.

Mauritzon, J., Contributions to the embryology of the orders Rosales and Myrtales. Lunds Univ. Årsskr. N. F. II. 35 (2): 1–121. 1939.

Miki, S., Evolution of *Trapa* from ancestral *Lythrum* through *Hemitrapa*. Proc. Jap. Acad. 35: 289–294. 1959.

Nevling, L. I., Jr., A revision of the genus *Daphnopsis*. Ann. Mo. Bot. Gard. 46: 257–358. 1959.

————, The Thymelaeaceae in the southeastern United States. Jour. Arn. Arb. 43: 428–434. 1962.

Ram, M., Floral morphology and embryology of *Trapa bispinosa* Roxb. with a discussion of the systematic position of the genus. Phytomorph. 6: 312–323. 1956.

Stern, W. L., and G. K. Brizicky, The comparative anatomy and taxonomy of *Heteropyxis*. Bull. Torrey Club 85: 111–123. 1958.

Weberling, F., Ein Beitrag zur systematischen Stellung der Geissolomataceae, Penaeaceae and Oliniaceae sowie der Gattung *Heteropyxis* (Myrtaceae) Bot. Jahrb. 82: 119–128. 1963.

Wilson, K., The genera of Myrtaceae in the southeastern United States. Jour. Arn. Arb. 41: 270–278. 1960.

Order 5. **Proteales**

The order Proteales as here defined includes only two families, the Proteaceae, with about 1400 species, and the Elaeagnaceae, with perhaps 50.
Each of the two families is generally regarded as taxonomically isolated
and of uncertain affinities. My 1957 proposal to associate them has been
followed in 1966 by Takhtajan, who considers each to constitute a monotypic order, the two orders collectively forming a superorder. The difference
between his treatment and mine lies only in the taxonomic rank at which
the agreed relationships are recognized.

The Proteaceae and Elaeagnaceae have an impressive number of characters in common. They both have strongly perigynous flowers in which the petals have been reduced or lost and the number of carpels has been reduced to one. The calyx is tetramerous, valvate, and often corolloid in appearance. The seeds have little or no endosperm. The stem has tanniferous secretory cells, and the leaves are exstipulate. There is also a considerable list of differences, as shown in the following key. Pollen morphology and wood anatomy throw little light on the relationships of either family.

The similarities between the Proteaceae and Elaeagnaceae do not prove a very close relationship, but at least the two families must have undergone a series of parallel changes from more or less similar ancestors. There is no indication of a relationship for either family that would be incompatible with a relationship between them. At the present state of knowledge I think it more useful to couple the two in one order than to treat either or both of them as separate orders.

The most likely origin of the Proteales is near the Thymelaeaceae, in the Myrtales. A relationship between the Thymelaeaceae and Elaeagnaceae has often been suggested. The Thymelaeaceae show a clear progression from a syncarpous to a pseudomonomerous gynoecium. Flowers of pseudomonomerous members of the Thymelaeaceae are much like those of the Proteaceae and Elaeagnaceae, having a prominent hypanthium (calyx tube), reduced or no petals, and an often corolloid, frequently tetramerous calyx. Although the gynoecium in both the Proteaceae and Elaeagnaceae appears to be strictly monocarpellate, the possibility that it is actually pseudomonomerous is perfectly reasonable at the present level of information.

Origin of the Proteales from the Myrtales would imply that the internal phloem characteristic of the latter order has been lost. This poses no great difficulty, inasmuch as a similar loss (or original absence?) must be admitted among various members of the Myrtales. Several of the genera which are customarily referred to the Thymelaeaceae also lack internal phloem.

If the gynoecium of the Proteales were regarded as truly apocarpous, the Myrtales would be excluded as possible ancestors. The most likely origin of the Proteales would then be in the Rosales. No family of the Rosales appears at all close to the Proteales, however, and the prospect of a rosalean affinity is much less appealing than a myrtalean one. The Myrtales themselves are, as we have noted, probably derived from the Rosales, so an eventual ancestry of the Proteales in the Rosales is indicated in either case.

## SYNOPTICAL ARRANGEMENT OF THE FAMILIES OF PROTEALES

1. Plants with a pubescence of peltate scales; flowers not aggregated into Compositae-like bracteate heads; fruit a pseudodrupe,

the dry achene surrounded by the persistent fleshy hypanthium
or calyx-tube; nodes unilacunar              **1. Elaeagnaceae**

1. Plants without peltate scales; flowers often aggregated into Compositae-like bracteate heads; fruit diverse, but not as in the
Elaeagnaceae; nodes trilacunar              **2. Proteaceae**

## SELECTED REFERENCES

Haber, J. M., The comparative anatomy and morphology of the flowers and
inflorescences of the Proteaceae. I. Some Australian taxa. Phytomorph. 9:
325–358. 1959. II. Some American taxa. Phytomorph. 11: 1–16. 1961.

Graham, S., The Elaeagnaceae in the southeastern United States. Jour. Arn.
Arb. 45: 274–278. 1964.

Johnson, J. A. S., and B. G. Briggs, Evolution in the Proteaceae. Austral. Jour.
Bot. 11: 21–61. 1963.

Venkata Rao, C., Morphology of the nectary in Proteaceae. New Phytol. 66:
99–107. 1967.

ORDER 6. **Cornales**

The order Cornales as here defined consists of six families and about
250 species. More than 200 of the species belong to only two of the
families, the Rhizophoraceae (120) and the Cornaceae (90). The
Alangiaceae are estimated at 18 species, the Garryaceae at 15, and Nyssaceae
at about eight, and the Davidiaceae at one.

The close relationship among the Alangiaceae, Cornaceae, Davidiaceae,
and Nyssaceae has been evident to all, and indeed some botanists have
included them all in the single family Cornaceae. In recent years the
Garryaceae have also been generally conceded to be of this affinity. These
several families consist of simple-leaved plants with epigynous flowers, a
single unitegmic ovule in each locule, well developed endosperm, and indehiscent, fleshy fruits. The petals are imbricate or valvate, or sometimes
wanting. The stamens are in one to four cycles. Almost all of the species
are woody, and none of them is known to have internal phloem. A series of
extensive serological tests by Fairbrothers and his associates support the
concept, developed on classical morphological and anatomical grounds, that
these several families are related among themselves.

The position of the Rhizophoraceae is more controversial. Like the
other families of the order, they are woody plants with simple leaves,
epigynous flowers, unitegmic ovules, a well developed endosperm (some
genera excepted), and fleshy fruits, but without internal phloem. They
differ from the other families in having two (seldom many) ovules in each
locule (though usually only one matures), and in having the petals folded

or convolute in bud. They have often been referred to the Myrtales, where they are anomalous in lacking internal phloem and in commonly having a copious endosperm. The pollen is compatible with a myrtalean relationship, being more or less similar to that of *Dialypetalanthus* and some of the Lythraceae, but both the Myrtales and the Cornales include such a wide range of pollen types that the pollen morphology is hardly definitive. Studies of the floral anatomy by Eyde are still (1967) in a preliminary stage, and serological tests have not yet been undertaken.

At the present state of our knowledge, I think it most useful to include the Rhizophoraceae in the Cornales as a near-basal side-branch not far distant from the Myrtales. If one prefers small orders and does not object to monotypes, the establishment of a separate order might perhaps be defended.

The Cornales are in my opinion pretty clearly of rosalean origin, parallel to the Myrtales. The possibility of deriving them directly from the Myrtales is not attractive. One might conceivably maintain that the absence of internal phloem from the Cornales represents a secondary loss, but he would also have to maintain that the presence of well developed endosperm represents a reversion from the virtual absence of endosperm in the Myrtales. These same considerations militate against the inclusion of the Rhizophoraceae in the Myrtales, especially in view of the fact that some members of the Myrtales are more primitive than the Rhizophoraceae in other characters (numerous stamens, superior ovary).

It may be significant that the small (six spp) South Pacific genus *Corokia* has by different authors been referred to the Cornaceae or to the Saxifragaceae sens. lat. (including Grossulariaceae). On classical morphological characters it is not out of place in the Cornaceae, and within the subclass Rosidae it runs easily to the order Cornales and the family Cornaceae in the keys here presented. On the other hand, more detailed anatomical and morphological studies by Eyde (cited below) reveal some differences from the Cornaceae and a series of similarities to *Argophyllum,* a genus of presumed escallonioid affinity that is here referred (along with the rest of the Escalloniaceae) to the Grossulariaceae. The wood anatomy and pollen morphology are compatible with either relationship. *Corokia* shows a rather small serological correspondence with some members of the Cornales, and has not been tested against other groups. Its plurilocular, drupaceous fruits are without parallel in the Grossulariaceae but are similar to, if perhaps more primitive than, those of some Cornaceae. *Argophyllum,* on the other hand, has a half-inferior ovary, many ovules per locule, and capsular fruits. The position of *Corokia* need not be established in a survey such as this, but its status as a possible nonmissing link merits consideration.

The adaptive significance of most of the characters which mark the order Cornales is obscure. Simple leaves, epigynous flowers, one or few sets of stamens, uniovulate or biovulate locules and endospermous seeds are all common enough among angiosperms in general, but their selective significance is still speculative. Certainly the particular combination of characters presented by the Cornales has no obvious advantage distinct from the possible advantage of each character separately. Only the fleshy fruits, in most species adapted to seed dispersal by animals, are clearly of selective importance.

The Cornales show no obvious ecologic unity, beyond the fleshy fruits and almost uniformly woody habit. Even the single genus *Cornus*, if taken in the broad sense, is ecologically diverse. *Cornus florida*, *C. canadensis*, and *C. stolonifera*, for example, are very different in appearance and ecologic role. *Cornus* as a whole occurs chiefly in temperate or boreal regions, but some other genera of the family, such as *Mastixia*, are wholly tropical. The Garryaceae are amentiferous and wind-pollinated. *Rhizophora*, but not other genera of its family, has developed the mangrove habit. This habit, although distinctive, is not limited to any one taxonomic group. Some genera of mangroves, such as *Avicennia* in the Verbenaceae, are taxonomically far removed from the Cornales.

### SYNOPTICAL ARRANGEMENT OF THE FAMILIES OF CORNALES

1. Flowers perfect or unisexual, with more or less well developed sepals or petals or both, not borne in catkins

    2. Ovules 2 (seldom more) per locule; petals folded or convolute in bud; leaves opposite, stipulate; stamens more numerous than the petals; fruit a berry      **1. Rhizophoraceae**

    2. Ovules solitary in each locule (or some locules empty); petals valvate or imbricate; leaves opposite or alternate, exstipulate; fruit mostly drupaceous

        3. Stamens more numerous than the petals (or sepals) except in some Alangiaceae; leaves strictly alternate; embryo large

            4. Ovary 6–9-locular      **2. Davidiaeeae**

            4. Ovary 1–2-locular

                5. Petals imbricate; flowers perfect or often unisexual; ovary unilocular      **3. Nyssaceae**

                5. Petals valvate; flowers perfect; ovary 1–2-locular      **4. Alangiaceae**

3. Stamens as many as the petals; leaves opposite, seldom alternate; embryo small                                    **5. Cornaceae**

1. Flowers unisexual (plants dioecious), the pistillate ones with vestigial or no perianth, the staminate ones with a perianth of 4 sepals, both types borne in catkins                    **6. Garryaceae**

## SELECTED REFERENCES

Adams, J. E., Studies in the comparative anatomy of the Cornaceae. Jour. Elisha Mitchell Soc. 65: 218–244. 1949.

Chopra, R. N., and H. Kaur, Some aspects of the embryology of Cornus. Phytomorph. 15: 353–399. 1965.

Eyde, R. H., Morphological and paleobotanical studies of the Nyssaceae. I. A survey of the modern species and their fruits. Jour. Arn. Arb. 44: 1–59. 1963.

————, Inferior ovary and generic affinities of Garrya. Am. Jour. Bot. 51: 1083–1092. 1964.

————, Systematic anatomy of the flower and fruit of Corokia. Am. Jour. Bot. 53: 833–847. 1966.

————, The Nyssaceae in the southeastern United States. Jour. Arn. Arb. 47: 117–125. 1966.

Eyde, R. H., and E. S. Barghoorn, Morphological and paleobotanical studies of the Nyssaceae. II. The fossil record. Jour. Arn. Arb. 44: 328–376. 1962.

Fairbrothers, D. E., and M. E. Johnson, "Comparative Serological Studies Within the Families Cornaceae and Nyssaceae" in Ch. Leone, ed., Taxonomic Biochemistry and Serology (New York: Ronald Press, 1964), pp. 305–318.

Ferguson, I. K., The Cornaceae in the southeastern United States. Jour. Arn. Arb. 47: 106–116. 1966.

Li, H. L., Davidia as the type of a new family Davidiaceae. Lloydia 17: 329–331. 1955.

Li, H. L., and C. Chao, Comparative anatomy of the woods of the Cornaceae and allies. Quart. Jour. Taiwan Mus. 7: 119–136. 1954.

Moseley, M. F., and R. M. Beeks, Studies of the Garryaceae. I. The comparative morphology and phylogeny. Phytomorph. 5: 314–346. 1955.

Titman, P. W., Studies in the woody anatomy of the family Nyssaceae. Jour. Elisha Mitchell Soc. 65: 245–261. 1949.

## Order 7. **Santalales**

The order Santalales as here defined consists of ten families and about 2200 species. More than 2000 of these belong to only three families, the

Loranthaceae (1400), Santalaceae (400), and Olacaceae (230), and another hundred belong to the Balanophoraceae. The Opiliaceae have about 60 species, the Grubbiaceae only five, the Misodendraceae two, and the Medusandraceae, Dipentodontaceae, and Cynomoriaceae are monotypic.

It is widely agreed that the Olacaceae, Opiliaceae, Santalaceae, Misodendraceae, and Loranthaceae are closely related. The Balanophoraceae are also pretty clearly of this affinity, as shown inter alia by the morphological and embryological studies of Fagerlind (cited below). The relationships of the Grubbiaceae and the three monotypic families are still debatable.

Cynomoria is habitally very much like the Balanophoraceae and has often been included in that family. It is more primitive than the Balanophoraceae in having a well developed ovular integument, but more advanced in its inferior ovary. Its pollen is unlike that of anything in the Santalales. However, it has no apparent relatives in any other order, and except for the pollen morphology it introduces no significant new characters to the Santalales. Adoxa and Salix, which have been mentioned as having pollen similar to that of Cynomoria, are certainly far removed taxonomically from Cynomoria as well as from each other.

The Grubbiaceae, with only the small genus Grubbia, are autotrophic South African shrubs of somewhat ericoid habit. They also resemble the Ericales in some embryological features, although they lack the tetradinous pollen and appendiculate, poricidal anthers of the more typical members of that order. The gynoecium is so typically santalalean that the family is retained in the Santalales in spite of its autotrophism. The ovary is inferior and at first more or less bilocular, but by the time of anthesis it is unilocular with a free-central placenta that bears two pendulous ovules at the tip.

The Medusandraceae and Dipentodontaceae are here included in the Santalales only with some hesitation. Recent authors have mostly thought their relationships to be here. The free-central placenta with a few pendulous, terminal ovules is very much like that of some other santalalean families, and the pollen of Medusandra, at least, is similar to that of the Olacaceae. The inclusion of Medusandra and Dipentodon in the Santalales introduces two new characters into the order — stipules and dehiscent fruits. The fruits are, however, one-seeded like those of the other families of Santalales, and if they merely failed to dehisce they would be perfectly normal for the order.

Alternatively, one might establish a separate order for the Medusandraceae and Dipentodontaceae. The name Medusandrales has already been proposed for Medusandra alone and is available also as an ordinal name for the two families together. Such an order would presumably stand alongside the Santalales in the system, near also to the Celastrales and Rhamnales. One might consider including Medusandra and Dipentodon in either the Celastrales or the Rhamnales, but they would be discordant there, and furthermore on the basis of the staminal position it would presumably be necessary to refer Dipentodon to the Celastrales and Medusandra to the Rhamnales.

The Santalales are a group characterized by progressive adaptation to parasitism. This is accompanied by simplification of the ovules, culminating in the condition in the Loranthaceae, in which the ovules are scarcely differentiated from the massive placenta. The Grubbiaceae, Medusandraceae, and Dipentodontaceae (seven species in all) are the only wholly autotrophic families in the order. The Olacaceae contain many autotrophic species as well as many hemiparasites with green leaves and haustorial connections to the roots of other plants. The Opiliaceae probably also have both types. The remaining five families, including nearly $\frac{9}{10}$ of the species in the order, contain only parasites and hemiparasites, some with and some without chlorophyll, some rooted in the ground and some borne on the stem of the host.

It is not certain how much of the vegetative reduction in the Santalales has positive survival value in correlation with the parasitic habit, and how much has merely been permitted by the relaxation of selection, resulting in the accumulation of loss-mutations. It may be suggested that the ovular reduction is a mere side-effect of the vegetative reduction, being controlled by some of the same genes.

Recent authors have mostly thought the Santalales to be related to the Celastrales and Rhamnales. The Proteales have also been suggested as possible allies, but the relationship there cannot be very close. Any possible common ancestor of the Santalales and Celastrales would have to have several cycles to stamens, as in some of the Olacaceae, and a fully plurilocular ovary with more or less numerous ovules in each cell, as in some of the Celastraceae. Such a plant would not fit well into either order, but would belong instead to the rosalean plexus from which the Santalales, Celastrales, Rhamnales, and Sapindales appear to have been derived. The Rhamnales are likewise no more than collateral relatives of the Santalales.

In a paper now awaiting publication, Job Kuijt restricts the order Santalales to the families here treated as Olacaceae, Opiliaceae, Santalaceae, Loranthaceae, and Misodendraceae. He excludes the Grubbiaceae, Balanophoraceae, and by implication the Cynomoriaceae, and does not consider the Medusandraceae and Dipentodontaceae. No attempt is here made to evaluate his argument and conclusions, but it may be noted that the families he excludes contain barely over a hundred species in all, hardly five percent of the total number in the order.

SYNOPTICAL ARRANGEMENT OF THE FAMILIES OF SANTALALES

1. Fruit capsular; ovules 6–8; leaves with small stipules; plants autotrophic

   2. Fertile stamens opposite the petals, alternating with well developed staminodes; petals and sepals well differentiated; flowers in slender, catkin-like racemes  **1. Medusandraceae**

   2. Fertile stamens alternate with the petals, alternating with a set of nectary glands; petals and sepals very much alike; flowers in globose umbels  **2. Dipentodontaceae**

1. Fruit indehiscent; ovules 1–5; leaves without stipules; plants autotrophic or more often partly or wholly parasitic

   3. Plants with chlorophyll; flowers perfect or sometimes unisexual; fruit dry or fleshy

      4. Ovules mostly well differentiated; leaves alternate except in the Grubbiaceae and most Santalaceae; flowers mostly small and not very showy; plants rooted in the ground

         5. Petals present; stamens, when in a single cycle, opposite the petals; ovary superior to occasionally inferior

            6. Ovary 2–5-celled below, with 1 ovule per locule; integuments 0-2  **3. Olacaceae**

            6. Ovary 1-celled, with 1 apical or basal ovule; integument none  **4. Opiliaceae**

         5. Petals absent; stamens, when in a single cycle, opposite the sepals; ovary inferior

            7. Plants fully autotrophic; stamens 8, in 2 cycles; ovules unitegmic  **5. Grubbiaceae**

            7. Plants partly or wholly parasitic; stamens 3–6, in a single cycle; ovules without integument  **6. Santalaceae**

      4. Ovules scarcely differentiated from the large, central placenta of the 1-celled ovary, without integument; leaves oppo-

site or whorled; perianth often large and showy; plants
rarely rooted in the ground **7. Loranthaceae**

3. Plants without chlorophyll; flowers mostly unisexual; fruit dry,
nutlike

2. Plants aerial, attached to the stem of the host; fruit 3-winged;
ovules without integument **8. Misodendraceae**

8. Plants terrestrial, attached to the roots of the host; fruit not
winged

9. Ovary superior; ovules 1–5, without integument; stamens
usually more than one **9. Balanophoraceae**

9. Ovary inferior; ovule solitary, unitegmic; stamen 1
**10. Cynomoriaceae**

## SELECTED REFERENCES

Agarwal, S., Morphological and embryological studies in the family Olacaceae.
I. *Olax* L. Phytomorph. 13: 185–196. 1963.

Bhatnagar, S. P., Morphological and embryological studies in the family Santalaceae. IV. *Mida salicifolia* A. Cunn. Phytomorph. 10: 198–207. 1960.

Bhatnagar, S. P., and S. Agarwal, Morphological and embryological studies in the family Santalaceae. VI. *Thesium.* Phytomorph. 11: 273–282. 1961.

Dixit, S. N., Morphological and embryological studies in the family Loranthaceae. VIII. *Tolypanthus* Bl. Phytomorph. 11: 335–345. 1961.

Fagerlind, F., Blüte und Blütenstand der Gattung *Balanophora.* Bot. Notis. 1945: 330–350.

⸻, Gynoceummorphologische und embryologische Studien in der Familie Olacaceae. Bot. Notis. 1947: 207–230.

⸻, Die systematische Stellung der Familie Grubbiaceae. Svensk. Bot. Tidskr. 41: 315–320. 1947.

⸻, Beiträge zur Kenntnis der Gynäceummorphologie und Phylogenie der Santalales-Familien. Svensk. Bot. Tidskr. 42: 195-229. 1948.

⸻, Development and structure of the flower and gametophytes in the genus *Exocarpus.* Svensk. Bot. Tidskr. 53: 257–282. 1959.

Joshi, P. C., Morphological and embryological studies in the family Santalaceae. V. *Osyris wightiana* Wall. Phytomorph. 10: 239–248. 1960.

Kuijt, J., Mutual affinities of Santalalean families. Brittonia, in press, 1968.

Metcalfe, C. R., *Medusandra richardsiana* Brenan. Anatomy of the leaf, stem and wood. Kew Bull. 1952: 237–244.

Narayana, R., Morphological and embryological studies in the family Lorantha-ceae. III. *Nuytsia floribunda* (Labill.) R. Br. Phytomorph. 8: 306–323. 1958.

Piehl, M. A., The natural history and taxonomy of *Comandra* (Santalaceae). Mem. Torrey Club 22: 1–97. 1965.

Ram, M., Morphological and embryological studies in the family Santalaceae. I. *Comandra umbellata* (L.) Nutt. Phytomorph. 7: 24–35. 1957. II. *Exo-carpus*, with a discussion on its systematic position. Phytomorph. 9: 4–19. 1959. III. *Leptomeria*. Phytomorph. 9: 20–33. 1959.

Reed, C., The comparative morphology of the Olacaceae, Opiliaceae and Oc-toknemaceae. Mem. Soc. Brot. 10: 29–79. 1955.

Schaeppi, H., and F. Steindl, Blütenmorphologische und embryologische Unter-suchungen an Loranthoideen. Viert. Naturf. Ges. Zürich 87: 301–372. 1942.

——————, Blütenmorphologische und embryologische Untersuchungen an einigen Viscoideen. Viert. Naturf. Ges. Zürich 90, Beih. 1: 1–46. 1945.

Smith, F. C., and E. C. Smith, Floral anatomy of the Santalaceae and some related forms. Oreg. State Monogr. Stud. Bot. 5: 1–93. 1942.

Stauffer, H. U., Beiträge zum Blütendiagramm der Santalales. Verh. Schweiz. Naturf. Ges. 141: 123–125. 1961.

Swamy, B. G. L., The comparative morphology of the Santalaceae: node, secondary xylem and pollen. Am. Jour. Bot. 36: 661–673. 1949.

## Order 8. Rafflesiales

The order Rafflesiales consists of three families and fewer than a hundred species. The Rafflesiaceae are estimated at about 50 species, the Hydnora-ceae at about 20, and the Mitrastemonaceae at only two. They are all tropical or subtropical.

The Rafflesiales are highly specialized, nongreen, rootless parasites which grow from the roots of the host. They have few or solitary, rather large to very large flowers, with a single set of tepals that are commonly united into a conspicuous, corolloid calyx, somewhat in the manner of the Aristolochiaceae. *Rafflesia* is famous for its gigantic flowers, which in *R. arnoldii* are about 1 m. across. No other family of plants has individual flowers even approaching this size. The pollen of the Rafflesiales is some-what similar to that of the Balanophoraceae, in the Santalales.

The relationships of the Rafflesiales are disputed and doubtful. Tradi-tionally they have been associated with the Aristolochiaceae, presumably because of the similarity in the perianth. In other respects, however, the two groups are very different. The Rafflesiales are so highly specialized that they must have a long history of parasitism, of which there is not a whisper in the Aristolochiaceae. The pollen of the Aristolochiaceae belongs to the uniaperturate series, whereas that of the Rafflesiales belongs to the triaperturate series. The ethereal oil cells of the Aristolochiaceae find no parallel in the Rafflesiales. Syncarpous members of the Aristolochiaceae have axile placentation, whereas in the Rafflesiales the placentation is parietal or sometimes apical. In my opinion the Rafflesiales are singularly misplaced in the Aristolochiales.

The most likely relatives of the Rafflesiales are the Santalales. Aside from the size of the flowers, the most notable difference between the two orders lies in the gynoecium. The Santalales have only a few ovules, these mostly on a basal or free-central placenta, or sometimes apical or axile in a plurilocular ovary; the fruit has only one seed. The Rafflesiales, on the other hand, have numerous ovules on parietal or sometimes apical placentae, and the fruit has many seeds. In order to see any relationship between the two groups, short of a possible common ancestry in multiovulate, autotrophic Rosales, one must maintain that the multiovulate condition in the Rafflesiales is secondary rather than primitive. This may not be so unlikely as it at first sounds, inasmuch as some of the Rafflesiaceae have apical rather than parietal placentation, and the presence of as many as 20 parietal placentae in some members of the family surely reflects a secondary increase rather than an originally high number. Even aside from the structure of the gynoecium, however, the Rafflesiales cannot well be derived from any of the more advanced members of the Santalales. Some of the Rafflesiaceae have bitegmic ovules, and nothing in the Santalales more advanced than the Olacaceae has more than one integument — indeed most of them have no integument at all.

If any other families or orders merit serious consideration as possible allies of the Rafflesiales, the fact is not immediately evident. All of the other parasitic groups are clearly of very different affinity, and none of the autotrophic families with numerous ovules on parietal placentae presents itself to mind as a likely ancestor.

### SYNOPTICAL ARRANGEMENT OF THE FAMILIES OF RAFFLESIALES

1. Ovary superior; scale-leaves opposite; flowers perfect
   **1. Mitrastemonaceae**

1. Ovary inferior or half-inferior; scale-leaves alternate or none

   2. Flowers mostly perfect; leaves none          **2. Hydnoraceae**

   2. Flowers unisexual; scale-leaves alternate          **3. Rafflesiaceae**

### SELECTED REFERENCE

Olah, L. von, Cytological and morphological investigation in *Rafflesia arnoldii* R. Br. Bull. Torrey Club 87: 406–416. 1960.

ORDER 9. **Celastrales**

The Celastrales as here defined consist of ten families and about 2300 species. The vast majority of them belong to only five families, the Celastraceae (850), Aquifoliaceae (450), Icacinaceae (400), Hippocrateaceae (300), and Dichapetalaceae (250). The remaining five families have less than 50 species amongst them.

The Celastraceae, Hippocrateaceae, Aquifoliaceae, and Icacinaceae form the core of the order and are undoubtedly closely allied. The Stackhousiaceae (22 spp) are anomalous in being herbaceous, but in other respects, including the anatomy and the pollen morphology, they fit comfortably enough into the order.

Several families which have sometimes been included in the Celastrales are here excluded and referred to other orders. The Buxaceae, Pandaceae, and Aextoxicaceae are referred to the Euphorbiales, and the Staphyleaceae to the Sapindales. These two orders are closely allied to the Celastrales, and the shifting of families from one to the other does not reflect any great difference in concepts of relationship. The Corynocarpaceae are here referred to the Ranunculales, the Pentaphylacaceae to the Theaceae, the Scyphostegiaceae to the Violales, and the Cyrillaceae to the Ericales. These would all be isolated and anomalous groups if placed in the Celastrales.

Among the families here retained in the Celastrales, only the Geissolomataceae (one sp), Cardiopteridaceae (three), and Dichapetalaceae are controversial. *Geissoloma* has often been referred to the Myrtales, but it is scarcely perigynous and lacks internal phloem. Its pollen is much like that of the Celastraceae.

The Cardiopteridaceae have sometimes been associated with the Olacaceae, in the Santalales, mainly because of some palynological similarities and because of the well developed latex system which resembles that of some of the Olacaceae. The Hippocrateaceae and some of the Celastraceae also have a latex system, however, and the Cardiopteridaceae do not seem out of place in the Celastrales. The Santalales are characterized by progressive specialization as parasites, beginning with the Olacaceae, some of which are hemiparasites and some of which are autotrophic. There is no indication that *Cardiopteris* is parasitic. Even though a collateral relationship between the Cardiopteridaceae and the Olacaceae seems likely, I think it more useful to retain the Cardiopteridaceae in the Celastrales, where they also have collateral relatives such as the Hippocrateaceae and Icacinaceae.

The Dichapetalaceae are reminiscent in some respects of the Euphorbiaceae and have sometimes been associated with that family, but they have perfect flowers which are not nearly so much reduced as those of the

Euphorbiaceae and other members of the Euphorbiales. On balance, the Dichapetalaceae seem much better placed in the Celastrales, even though they have no very close relatives there. They would be at least equally isolated and much more anomalous in the Euphorbiales.

The Celastrales are woody plants, or seldom herbs, with simple leaves, perfect or occasionally unisexual, hypogynous to perigynous, usually polypetalous, mostly haplostemonous flowers, the stamens alternating with the petals. They often have an intra- or extra-staminal nectariferous disk which represents a reduced whorl of stamens. There are usually only one or two ovules in a locule, although some genera of the Celastraceae and Hippocrateaceae have more.

The Celastrales are related to the Rosales, Sapindales, Rhamnales, Euphorbiales, and probably the Santalales. The Euphorbiales are probably derived from the Celastrales by floral reduction. The Rhamnales are a parallel group derived from a diplostemonous ancestor, with the antipetalous instead of the antisepalous stamens retained. The Sapindales are also parallel in some respects to the Celastrales, and the only solid difference between them is that the Celastrales consistently have simple leaves, whereas most of the Sapindales have compound (or at least conspicuously lobed) leaves. The multiovulate locules of some of the Celastraceae and Hippocrateaceae rule out as possible ancestors all of the Sapindales except some of the primitive, likewise multiovulate families, such as the Staphyleaceae. It may be significant that the pollen of the Staphyleaceae is much like that of the Celastraceae. Unlike the simple-leaved Polygalales, the Celastrales show no trace of a compound-leaved ancestry. It is possible that the Celastrales are derived from the more primitive members of the Sapindales, but it seems more prudent to consider the two orders as probably derived from a common source in the Rosales. If the Celastrales are derived from the Rosales, one need not be surprised at the existence of such a genus as *Geissoloma*, which suggests both the Celastrales and the Myrtales, because the Myrtales also have a rosalean ancestry.

There is nothing ecologically unusual about the Celastrales. If the characters which mark them as a group have anything to do with fitting them to a particular habitat or way of life, it is not obvious. Neither do the larger families of the order appear to be ecologically characterized.

## SYNOPTICAL ARRANGEMENT OF THE FAMILIES OF CELASTRALES

1. Flowers diplostemonous; petals none; disk none; habit ericoid

**I. Geissolomataceae**

1. Flowers haplostemonous (except a few Celastraceae); petals generally well developed; disk often present; habit not ericoid

2. Ovules erect and basal, or several and superposed in 2 rows; disk present except in Salvadoraceae; integuments 2 except in many Salvadoraceae; leaves opposite or less often alternate

   3. Flowers with a disk; carpels more than 2 except in some Celastraceae; flowers pentamerous except in some Celastraceae

      4. Woody plants; flowers hypogynous; leaves opposite or alternate

         5. Disk extrastaminal; stamens mostly 3 (2–5); leaves mostly opposite; seeds without endosperm; latex system well developed     **2. Hippocrateaceae**

         5. Disk intrastaminal, or the stamens seated on the disk; stamens 4–5 (seldom 6–10); seeds mostly with endosperm; latex system seldom well developed

            6. Disk of ordinary type, merely surrounding the ovary at the base; leaves opposite or sometimes alternate; locules 1–5     **3. Celastraceae**

            6. Disk much enlarged, enclosing the ovary; leaves alternate; locules 5     **4. Siphonodontaceae**

      4. Herbs; flowers distinctly perigynous, the disk lining the hypanthium; leaves alternate; seeds with endosperm     **5. Stackhousiaceae**

   3. Flowers without a disk, but the stamens sometimes alternating with nectariferous glands; carpels 2; flowers tetramerous; seeds without endosperm     **6. Salvadoraceae**

2. Ovules pendulous, apical or nearly so, 1–2 per locule, unitegmic (except in Dichapetalaceae); leaves alternate

   7. Stipules wanting or vestigial; seeds with endosperm except in some Icacinaceae; disk wanting

      8. Corolla mostly polypetalous; fruit drupaceous; woody plants without milky juice

9.  Locules 4–6 or more; petals imbricate; pollen binucleate
    **7. Aquifoliaceae**

9.  Locules 3, or more often 1 by abortion; petals valvate;
    pollen trinucleate **8. Icacinaceae**

8.  Corolla sympetalous; fruit samaroid; climbing herbs with
    milky juice **9. Cardiopteridaceae**

7.  Stipules present; endosperm wanting; disk present and in-
    trastaminal, or represented by nectariferous glands alter-
    nating with the stamens **10. Dichapetalaceae**

## SELECTED REFERENCES

Bailey, I. W., and R. A. Howard, The comparative morphology of the Icacinaceae. Jour. Arn. Arb. 22: 125–132; 171–187; 432–442; 556–568. 1941.

Berkeley, E., Morphological studies in the Celastraceae. Jour. Elisha Mitchell Soc. 69: 185–206. 1953.

Brizicky, G. K., The genera of Celastrales in the southeastern United States. Jour. Arn. Arb. 45: 206–234. 1964.

Dahl, A. O., The comparative morphology of the Icacinaceae. VI. The pollen. Jour. Arn. Arb. 33: 252–295. 1952.

Fagerlind, F., Bau des Gynöceums, der Samenanlage und des Embryosackes bei einigen Repräsentaten der Familie Icacinaceae. Svensk. Bot. Tidsk. 39: 346–364. 1945.

Herr, J. M., Jr., The development of the ovule and megagametophyte in the genus *Ilex*. Jour. Elisha Mitchell Soc. 75: 107–128. 1959.

———, Endosperm development and associated ovule modifications in the genus *Ilex*. Jour. Elisha Mitchell Soc. 77: 26–32. 1961.

Mauritzon, J., Zur Embryologie und systematischen Abgrenzung der Reihen Terebinthales und Celastrales. Bot. Notis. 1936: 161–212.

Merrill, E. D., and F. L. Freeman, The old world species of the celastraceous genus *Microtropis* Wallich. Proc. Am. Acad. 73: 271–310. 1940. (For reduction of the Chingithamnaceae)

Narang, N., The life-history of *Stackhousia linariaefolia* A. Cunn. with a discussion on its systematic position. Phytomorph. 3: 485–493. 1953.

Stant, M. Y., Notes on the systematic anatomy of *Stackhousia*. Kew Bull. 1951: 309–318.

ORDER 10. **Euphorbiales**

The order Euphorbiales as here defined consists of five families. One of these, the Euphorbiaceae, is estimated to have about 7500 species. The other four, the Buxaceae (60), Daphniphyllaceae (35), Pandaceae (35), and Aextoxicaceae (one), are minuscule by comparison. The Buxaceae, Daphniphyllaceae, and Aextoxicaceae have often been included in the Euphorbiaceae, but they seem amply distinct.

Forman (cited below) has convincingly shown the relationship of *Panda* (hitherto regarded as forming a monotypic family) to *Galearia* and *Microdesmis*, genera which have traditionally been included in the Euphorbiaceae. I agree with his conclusion that it is better to hold these three genera as a distinct family Pandaceae than to include them all in the Euphorbiaceae, where *Panda*, especially, would be morphologically aberrant.

All members of the Euphorbiales, with very minor exceptions in the Euphorbiaceae, have unisexual flowers. The flowers of most genera lack a corolla, and they often lack a calyx as well. They all have one or two pendulous ovules in each locule (or one locule may be empty). The ovules have two integuments, and the seeds have a well developed endosperm. The pollen in all four of the smaller families is similar to that of at least some genera of the Euphorbiaceae. The Euphorbiaceae and Buxaceae typically have three carpels, and the seeds of these two families usually have a caruncle.

The Euphorbiales may be regarded as a group of few-ovulate, mostly simple-leaved Rosidae in which the flowers have become unisexual and then undergone further reduction accompanied by aggregation. The trend toward reduction and aggregation culminates in the pseudanthia of *Euphorbia*, which have a pistillate flower consisting of a naked pistil, surrounded by several staminate flowers each of which consists of a single stamen. The illusion that this cluster constitutes a single flower is fostered also by the cupulate, hypanthium-like involucre which often bears showy, petal-like appendages.

In spite of the similarities which we have noted, the relationship of every family in the order is in dispute. A full scale consideration of the ideas and arguments would take many pages. I regard all of them as collateral relatives with a common origin in or near the Celastrales. Each of them has a combination of primitive and advanced characters which makes it unlikely to be directly ancestral to any of the others.

The Dichapetalaceae, another collateral relative of the families of the Euphorbiales, are here retained in the Celastrales. They have perfect, dichlamydeous flowers and nonendospermous seeds, and their inclusion in the Euphorbiales would vitiate the conceptual unity of the order Their inclusion in the Celastrales, on the other hand, poses no serious problems.

The important paper by Webster cited in the bibliography appeared after this manuscript was completed, and the consequences of his comments have not been fully explored here. It may be noted that he maintains the Aextoxicaceae, Buxaceae, Daphniphyllaceae, and Pandaceae as distinct from the Euphorbiaceae, and that of these he thinks only the Pandaceae are closely allied to the Euphorbiaceae. He also thinks that *Picrodendron*, here doubtfully referred to the Juglandaceae as a distinct family, may be truly at home in the Euphorbiaceae. His tentative conclusion that the Euphorbiaceae can properly be retained in the Englerian order Geraniales, and that the admitted resemblances to Malvalean families are probably secondary, is in general accord with my placement of the Euphorbiales in the subclass Rosidae rather than in the Dilleniidae.

The Euphorbiales as a whole have no ecological unity. Even the single genus *Euphorbia* (sens. lat.) embraces a wide variety of growth forms and occurs in highly diverse habitats. Many African species of *Euphorbia* are stem-succulents which occupy the niche generally filled by cacti in America, but these are ecologically far removed from the prostrate, weedy species of the section Chamaesyce, or the broad-leaved, mesophytic species of the section Poinsettia. *Hevea brasiliensis*, a well known tree of the Amazonian forest, represents still another habit and habitat for the family, and the list of diverse ecological types could be extended almost indefinitely. The Euphorbiaceae are also highly diverse in anatomy and pollen morphology, yet the family resists all attempts to fragment it. The other families of the order, being smaller, are individually less diverse, but they are in no way ecologically distinctive.

The selective significance of the pseudanthia of *Euphorbia* is debatable. At first thought it seems logical to suppose that the organization of separate, inconspicuous, reduced flowers into a pseudanthium reflects a return to ordinary entomophily, driven by selective pressure. The next question is why the original floral reduction occurred. A comprehensive theory of how selection could drive the inflorescence through a full cycle from normal flowers to separate, reduced, unisexual flowers to pseudanthia has not yet been presented.

## SYNOPTICAL ARRANGEMENT OF THE FAMILIES OF EUPHORBIALES

1.  Fruit mostly dehiscent and with more than 1 seed, seldom drupaceous; seeds usually with a caruncle

    2.  Ovules apotropous, the raphe dorsal; fruit a loculicidal capsule, or sometimes drupaceous; small trees and shrubs, with neither stipules nor a milky juice; disk wanting    **1. Buxaceae**

    2.  Ovules epitropous, the raphe ventral; fruit mostly schizocarpic, each carpel eventually dehiscing ventrally to release its solitary seed; plants of very diverse habit, often with a milky juice, and usually with stipules; disk often present
        **2. Euphorbiaceae**

1. Fruit indehiscent, drupaceous, often 1-seeded; seeds mostly without a caruncle

3. Ovules 2 in each locule, or 1 locule empty; leaves without stipules

4. Ovules epitropous, present in each of the 2 (–4) locules; petals none; stamens 6–12; disk wanting; embryo minute
   **3. Daphniphyllaceae**

4. Ovules apotropous, borne in only 1 of the 2 locules; petals 5; stamens 5, alternating with the disk-glands; embryo not especially small
   **4. Aextoxicaceae**

3. Ovules solitary in each of the 2–5 locules; leaves with stipules; petals 5; stamens 5, 10, or 15; disk none
   **5. Pandaceae**

## SELECTED REFERENCES

Forman, L. L., The reinstatement of Galearia Zoll & Mor. and Microdesmis Hook.f. in the Pandaceae. Kew Bull. 20: 309–321. 1966.

Heimsch, C., Comparative anatomy of the secondary xylem in the "Gruinales" and "Terebinthales" of Wettstein with reference to taxonomic grouping. Lilloa 8: 83–198. 1942.

Nair, N., and V. Abraham, Floral morphology of a few species of Euphorbiaceae. Proc. Indian Acad. Sci. B 56: 1–12. 1962.

Singh, R. P., Structure and development of seeds in Euphorbiaceae: Ricinus communis L. Phytomorph. 4: 118–123. 1954.

Webster, G. L., The genera of Euphorbiaceae in the southeastern United States. Jour. Arn. Arb. 48: 303–361. 1967.

ORDER 11. **Rhamnales**

The order Rhamnales consists of three families, the Rhamnaceae, with perhaps 900 species, the Vitaceae, with about 700, and the Leeaceae, with about 70. Botanists are agreed that the three families collectively form a natural group. The Leeaceae have often been included in the Vitaceae, but the differences are as substantial as those which separate the Vitaceae from the Rhamnaceae. There are no connecting forms between any two of the three families.

The Rhamnales are Rosidae with a single set of stamens, these opposite the petals, alternate with the sepals. There is usually a well developed intrastaminal disk. The ovary has two to several locules, with one or two erect, bitegmic ovules from the base in each locule, and the seeds have a well developed endosperm.

The three families of the Rhamnales are siblings from a common ancestor. Each is in its own way too advanced to be ancestral to any of the others. Their status as a group parallel to the Celastrales has been mentioned under that order.

The origin of the Rhamnales is to be sought in the Rosales, in a common complex with the ancestors of the Celastrales and Sapindales. There may well have been a displostemonous common ancestor of the Rhamnales and Celastrales (a very few of the Celastrales are in fact displostemonous). Differentiation of the two orders from this common ancestor involved the loss of the antisepalous stamens by the Rhamnales, and loss of the antipetalous stamens by the Celastrales.

The Vitaceae are mostly climbers, often with tendrils and a sympodial stem. Some of the Rhamnaceae are also climbers, but they seldom have tendrils. Otherwise the members of the order are trees or shrubs, rarely herbs. Aside from the climbing habit of some members, there is nothing ecologically distinctive about the order. Tendriliferous climbers also occur in groups of very different affinity, such as the Passifloraceae in the subclass Dilleniidae.

### SYNOPTICAL ARRANGEMENT OF THE FAMILIES OF RHAMNALES

1.  Ovule solitary in each locule; carpels mostly 3–8, seldom only 2; shrubs or small trees, sometimes climbing, but seldom with tendrils

    2.  Stamens free; leaves simple, entire or merely toothed, usually stipulate; embryo large; flowers hypogynous to perigynous or epignous; fruits diverse                **1. Rhamnaceae**

    2.  Stamens with the filaments united into a tube; leaves simple to compound, exstipulate; embryo small; flowers hypogynous; fruit a berry                **2. Leeaceae**

1.  Ovules 2 in each locule; carpels 2; mostly tendril-bearing climbers; leaves stipulate, usually compound or lobed; fruit a berry.
                **3. Vitaceae**

### SELECTED REFERENCES

Brizicky, G. K., The genera of Rhamnaceae in the southeastern United States. Jour. Arn. Arb. 45: 439–463. 1964.

————, The genera of Vitaceae in the southeastern United States. Jour. Arn. Arb. 46: 48–67. 1965.

Nair, N. C., and K. V. Mani, Organography and floral anatomy of some species of Vitaceae. Phytomorph. 10: 138–144. 1960.

Nair, N. C., and V. S. Sarma, Organography and floral anatomy of some members of the Rhamnaceae. Jour. Indian Bot. Soc. 40: 47–55. 1961.

Prichard, E. D., Morphological studies in Rhamnaceae. Jour. Elisha Mitchell Soc. 71: 82–106. 1955.

## ORDER 12. **Sapindales**

Most botanists have agreed that all or nearly all of the families here assigned to the orders Sapindales, Geraniales, Linales, and Polygalales are related among themselves, but there has been no agreement as to how these families should be parceled out into orders. The system here proposed is perhaps best compared to that of Takhtajan, which differs chiefly in that an order Rutales is segregated from the Sapindales, and the order Linales is submerged in the Geraniales. In addition, the Zygophyllaceae, here referred to the Sapindales, are referred by Takhtajan to the Geraniales; and the Malpighiaceae, here referred to the Polygalales, are referred by him to the Geraniales.

The present arrangement visualizes the Sapindales as the most primitive order of the group, from which the others are separately derived. The Sapindales are, with few exceptions, woody; most of them have compound leaves, and at least the more primitive families among them have regular flowers with two or more sets of stamens. The Geraniales are here regarded as an herbaceous offshoot of the Sapindales. The leaves of the Geraniales range from compound to simple but deeply cleft, to simple and entire, and the more advanced families have highly irregular flowers. The Linales and Polygalales are visualized as simple-leaved offshoots of the Sapindales which are at first woody, but whose more advanced members are herbaceous. The Linales have regular flowers, and the Polygalales have irregular flowers. The leaf blade of the Polygalales, at least, probably represents the terminal leaflet of a compound leaf. The most primitive family of the Polygalales, the Malpighiaceae, often has a jointed petiole.

Much has been made of the orientation of the ovules in this group of orders. They are said to be apotropous (or sapindaceous) if they are so oriented that the raphe is ventral (between the partition and the body of the ovule) in a basal, erect ovule, or dorsal in an apical, pendulous ovule. They are said to be epitropous (or geraniaceous) if they are so oriented that the raphe is dorsal in a basal, erect ovule, or ventral in an apical, pendulous ovule. The Anacardiaceae and Burseraceae have been put in different orders on this account, but on the basis of other characters they appear to be closely allied. Both apotropous and epitropous ovules occur in the Cornaceae, where the character is considered to be of only sub-familial importance. As here defined, the Sapindales include both apotropous and epitropous families, the Geraniales are epitropous except for the Balsaminaceae, and the Linales and Polygalales are wholly epitropous. It would appear that in this group the apotropous orientation is primitive, and the epitropous orientation secondary. The more primitive members of the Sapindales, and at least the core families of the related order Celastrales, are apotropous.

The order Sapindales as here defined consists of 17 families and perhaps 6700 species. About 4500 of these belong to only three families of nearly equal size, the Rutaceae (1600), Sapindaceae (1500), and Meliaceae (1400). Another 2100 species belong to only six families, the Anacardiaceae (600), Burseraceae (600), Connaraceae (350), Zygophyllaceae (250), Simaroubaceae (170), and Aceraceae (150). The remaining eight families have only a little more than 120 species amongst them. The monotypic Chinese genus *Bretschneideria*, often treated as a separate family, is here regarded as a peripheral member of the Sapindaceae.

The mutual interrelationship of ten of the families (including the five largest ones) here referred to the Sapindales seems well established. These are the Aceraceae, Anacardiaceae, Burseraceae, Cneoraceae, Hippocastanaceae, Julianaceae, Meliaceae, Rutaceae, Sapindaceae, and Simaroubaceae. In addition to their similarity in classical morphological features, they have much in common anatomically, as shown by the studies of Heimsch (cited below).

The remaining seven families are unusual or peripheral or doubtful in the order. These are the Akaniaceae, Connaraceae, Greyiaceae, Melianthaceae, Staphyleaceae, Stylobasiaceae, and Zygophyllaceae. Although the list of families seems long, it embraces only about ten percent of the species of the order.

The Staphyleaceae consist of about 50 species, widespread in the northern hemisphere and extending into tropical South America and southern Asia. They have relatively primitive wood and are not closely allied to any other family of the order. They are also primitive and somewhat unusual in the order in having well developed stipules, usually several ovules in each locule, and a well developed endosperm. They cannot be regarded as directly ancestral to the rest of the order, however, because they have only a single set of stamens, whereas some other members of the order have two or more sets.

The Staphyleaceae have sometimes been referred to the Celastrales, and indeed their pollen is somewhat like that of the Celastraceae, but their compound leaves make them anomalous in that order. Their endospermous seeds and pluriovulate locules would make them relatively primitive in the Celastrales, if they were referred to that order. Inasmuch as both the

Celastrales and the Sapindales are here regarded as being derived from the Rosales, and inasmuch as we have promulgated a relaxed view of the concept of monophylesis, it is perhaps of no great moment to which of these two derived, related orders the Staphyleaceae are referred. Conceptually it is easier to keep them with the compound-leaved group, i.e., the Sapindales.

The Melianthaceae consist of about 35 species found in tropical and southern Africa. They combine a specialized wood structure with a relatively primitive gross morphology, having stipulate leaves, endospermous seeds, and sometimes pluriovulate locules. They have the somewhat irregular flowers and unilateral extrastaminal disk of the Sapindaceae, and their pollen is much like that found in such sapindalean families as the Rutaceae and Anacardiaceae. It is generally agreed that the Melianthaceae belong with the Sapindales, even though they are peripheral to the group.

The Greyiaceae, consisting of only three species of the South African genus Greyia, have sometimes been included in the Melianthaceae. They differ from the Melianthaceae, however, as much as other recognized families of the order differ from each other, and it may be significant that the wood is also distinctive. Their pollen, however, is very like that of the Melianthaceae. Although most authors have been content to ally Greyia with the Melianthaceae, Hutchinson has removed the Greyiaceae to a position near the Escalloniaceae. The Escalloniaceae, one of the segregates from the traditional Saxifragaceae, are here submerged in the Grossulariaceae, in the order Rosales. Anatomically Greyia fits no better with the Grossulariaceae and their allies than with the core families of the Sapindales. Although I concur in the general opinion that the Greyiaceae are best associated with the Melianthaceae in the order Sapindales, they would not be badly out of place in the Rosales. If the rest of the Sapindales were wiped out of existence, the Greyiaceae could be accommodated in the Rosales without difficulty.

The Connaraceae, a pantropical family of about 350 species, provide another connecting link between the Rosales and Sapindales. They have by different authors been referred to both of these orders. Embryologically they are much like the Cunoniaceae. They are unusual — almost unique — in the Sapindales in being strictly apocarpous. Their wood structure is fairly advanced, being comparable to that of the Sapindaceae rather than to the rosalean families, and at least some of them have the secretory cells so often found in the Sapindales. Their seeds are arillate, like those of many Sapindaceae. Their position in either order could be defended, but on balance I prefer to keep them with the Sapindales.

The Akaniaceae, consisting of a single species from Australia, are unique in the Sapindales in their very wide wood rays, and they are somewhat unusual in lacking a floral disk and having endosperm. Their pollen is like that of some of the Sapindaceae, and they commonly have eight stamens and a pentamerous perianth, like the Sapindaceae. Although it has sometimes been included in the Sapindaceae, *Akania* seems amply to merit status as a separate family. On the other hand, there is nothing to indicate that it would be better placed in any other order.

The Zygophyllaceae, with about 250 species, are cosmopolitan but best developed in warm regions. Most of them are shrubs or half-shrubs, but a few are trees or herbs. Many are halophytic or xerophytic. The Zygophyllaceae occupy an isolated position in the Sapindales. Their wood is more like that of the Linales and Polygalales than like that of the core families of the Sapindales. This is perhaps not fatal to their inclusion in the Sapindales, inasmuch as the Linales and Polygalales are here regarded as derived from the Sapindales, and what has happened twice within the complex might well have happened three times. The pollen of the Zygophyllaceae is heterogeneous, some of it much like that of one or another family of the Sapindales. Although the Zygophyllaceae apparently have no very close allies, their affinities clearly lie with the Sapindales-Geraniales-Linales-Polygalales complex. Within this complex, they are best retained in the primitive and already somewhat heterogeneous order Sapindales. They would be misplaced in any of the other orders (each of which is more homogeneous), and they are not sufficiently distinctive to stand as an order of their own.

The Stylobasiaceae as here defined consist of two monotypic genera which have not usually been associated. *Stylobasium* has usually been included in the Chrysobalanaceae, from which it has recently been excluded by Prance. *Suriana* has usually been included in the Simaroubaceae, from which it has recently been excluded by Gutzwiller. Chemical studies by Nooteboom further emphasize the distinction of *Suriana* from the Simaroubaceae; his studies did not include *Stylobasium*. The two genera are habitally similar and share a number of technical features in common, including the apocarpous flowers with basal style, small, indehiscent fruit, and exarillate seeds. Both have simple, rather small leaves, an unusual feature in the Sapindales. A point by point comparison of the micromorphology, which has yet to be attempted, should provide critical evidence as to whether the two genera should properly be associated. Even if the relationship is confirmed, the possibility remains that they, as well as the Connaraceae, might better be considered advanced Rosales than primitive Sapindales.

The adaptive significance of the characters which mark the Sapindales and the included families is obscure. The secretory cells of many of the families, the bitter bark of the Simaroubaceae, and the specialized resin ducts of the Anacardiaceae, Burseraceae, and Julianaceae are still of uncertain selective importance. It is easy to assume that these features must confer some selective advantage, but neither the fact nor the nature of the advantage is obvious in the field. Some of the Anacardiaceae are notoriously allergenic to man, and some of the alkaloids found in the bark of species of Simaroubaceae are useful in medicine, but neither of these features appears to be of much importance to the plants.

### SYNOPTICAL ARRANGEMENT OF THE FAMILIES OF SAPINDALES

1. Ovules usually more than 2 per locule or carpel; endosperm well developed; ovules apotropous; flowers usually perfect

    2. Leaves stipulate, mostly pinnately compound; flowers haplostemonous, with 2–4 carpels and axile placentation

        3. Leaves mostly opposite; disk annular, intrastaminal, or the stamens seated on the disk; flowers regular; carpels 2–3, seldom 4; pollen binucleate        **1. Staphyleaceae**

        3. Leaves alternate; disk unilateral, extrastaminal; flowers irregular; carpels mostly 4; pollen trinucleate **2. Melianthaceae**

    2. Leaves exstipulate, simple, alternate; flowers diplostemonous, with mostly 5 carpels, the placentation parietal; disk annular, extra-staminal        **3. Greyiaceae**

1. Ovules seldom more than 2 per carpel or locule; endosperm in most families wanting or scanty; ovules variously apotropous or epitropous; flowers perfect or unisexual

    4. Flowers strictly apocarpous

        5. Style terminal; fruit dehiscent; seeds arillate; leaves mostly compound        **4. Connaraceae**

        5. Style basal; fruit indehiscent; seeds exarillate; leaves simple
            **5. Stylobasiaceae**

    4. Flowers nearly always more or less strongly syncarpous (some exceptions in Simaroubaceae)

        6. Disk mostly extrastaminal (and often unilateral) or wanting (instrastaminal in some Aceraceae); ovules apotropous except in Akaniaceae; flowers very often irregular

7.  Leaves mostly alternate; stamens typically 8 (5–many)

    8.  Seeds with endosperm and without an aril; wood rays very wide; disk wanting; ovules epitropous, 2 in each locule    **6. Akaniaceae**

    8.  Seeds without endosperm and commonly with an aril; wood rays narrow; disk present, often unilateral; ovules apotropous, 1 or less often 2 in each locule    **7. Sapindaceae**

7.  Leaves opposite

    9.  Leaves palmately compound; fruit a tricarpellate, usually 1-seeded capsule; flowers irregular; stamens 5–8    **8. Hippocastanaceae**

    9.  Leaves simple and palmately lobed, or pinnately compound; fruit a bicarpellate, 2-seeded samara; flowers regular, stamens 4–10, most often 8    **9. Aceraceae**

6.  Disk intrastaminal, annular or often developed into a gynophore (wanting in the Julianaceae); ovules epitropous except in the Anacardiaceae and Julianaceae

    10.  Bark and wood with prominent, specialized resin ducts

        11.  Ovary 2–5-celled, with 2 (seldom only 1) epitropous ovules per locule    **10. Burseraceae**

        11.  Ovary 1-celled (by abortion), with a single apotropous ovule

            12.  Flowers normally developed, perfect or sometimes unisexual, with an evident perianth; intrastaminal disk present    **11. Anacardiaceae**

            12.  Flowers reduced, unisexual, the pistillate ones without a perianth; disk none    **12. Julianaceae**

    10.  Bark and wood without prominent resin ducts

        13.  Leaves mostly exstipulate, the stipules, if any, small and soon deciduous; leaves mostly alternate

14. Stamens mostly free

    15. Leaves not glandular-punctate

        16. Seeds without endosperm; leaves mostly compound, seldom simple; bark bitter; styles often separate; flowers often 5-merous; pollen binucleate    **13. Simaroubaceae**

        16. Seeds with endosperm; leaves simple; bark not bitter (?); styles united; flowers 3-merous or seldom 4-merous; pollen trinucleate
                **14. Cneoraceae**

    15. Leaves mostly glandular-punctate    **15. Rutaceae**

14. Stamens mostly connate by their filaments
                **16. Meliaceae**

13. Leaves with well developed, persistent stipules, mostly opposite; seeds often with endosperm
                **17. Zygophyllaceae**

### SELECTED REFERENCES

Brizicky, G. K., The genera of Rutaceae in the southeastern United States. Jour. Arn. Arb. 43: 1–22. 1962.

————, The genera of Simaroubaceae and Burseraceae in the southeastern United States. Jour. Arn. Arb. 43: 173–186. 1962.

————, The genera of Anacardiaceae in the southeastern United States. Jour. Arn. Arb. 43: 359–375. 1962.

————, The genera of Sapindales in the southeastern United States. Jour. Arn. Arb. 44: 462–501. 1963.

Copeland, H. F., The reproductive structures of *Schinus molle* (Anacardiaceae). Madrono 15: 14–25. 1959.

Forman, L. L., A new genus from Thailand. Kew Bull. 1953: 555–564. 1954. (For inclusion of Podoaceae in Anacardiaceae)

Gut, B. J., Beiträge zur Morphologie des Gynoeceums and der Blütenachse einiger Rutaceen. Bot. Jahrb. 85: 151–247. 1966.

Gutzwiller, M., Die phylogenetische Stellung von *Suriana maritima* L. Bot. Jahrb. 81: 1–49. 1961.

Hall, B. A., The floral anatomy of the genus Acer. Am. Jour. Bot. 38: 793–799. 1951.

————, The floral anatomy of *Dipteronia*. Am. Jour. Bot. 48: 918–924. 1961.

Heimsch, C., The comparative anatomy of the secondary xylem in the "Gruinales" and "Terebinthales" of Wettstein with special reference to taxonomic grouping. Lilloa 8: 83–198. 1942.

Nair, N. C., Studies in the Meliaceae. V. Morphology and anatomy of the flower of the tribes Melieae, Trichileae and Swietenieae. Jour. Indian Bot. Soc. 41: 226–242. 1962. VI. Morphology and anatomy of the flower of the tripe Cedrelieae and discussion on the floral anatomy of the family. Jour. Indian Bot. Soc. 42: 177–189. 1963.

Nair, N. C., and R. K. Joshi, Floral anatomy of some members of the Simaroubaceae. Bot. Gaz. 120: 88–99. 1958.

Narayana, L. L., Floral anatomy of Meliaceae. Jour. Indian Bot. Soc. 37: 365–374. 1958. 38: 288–295. 1959.

Narayana, L. L., and M. Sayeeduddin, Floral anatomy of Simarubaceae. Jour. Indian Bot. Soc. 37: 517–522. 1958.

Nooteboom, H. P., Flavonols, leuco-anthocyanins, cinnamic acids, and alkaloids in dried leaves of some Asiatic and Malesian Simaroubaceae. Blumea 14: 253–356. 1966.

Prance, G. T., The systematic position of *Stylobasium* Desf. Bull. Jard. Bot. Brux 35: 435–448. 1965.

Stern, W. L., The comparative anatomy of the xylem and the phylogeny of the Julianaceae. Am. Jour. Bot. 39: 220–229. 1952.

Stern, W. L., and G. K. Brizicky, The morphology and relationships of *Diomma*, gen. inc. sed. Mem. N. Y. Bot. Gard. 10 (2): 38–57. 1960.

Weberling, F., and P. W. Leenhouts, Systematisch-morphologische Studien an Terebinthales-Familien. Abh. Akad. Wiss. Lit. Mainz, Math.-Naturw. 1965: 499–584.

## ORDER 13. Geraniales

The order Geraniales as here defined consists of five families and about 2200 species. Of these, the Oxalidaceae are estimated at 950 species, the Geraniaceae at 750, the Balsaminaceae at 450, the Tropaeolaceae at 80, and the Limnanthaceae at only eight species. The order has sometimes been defined much more broadly, to include some families here referred to the Linales, Polygalales, Sapindales, and Euphorbiales. This matter is discussed under the order Sapindales. The definition here adopted is identical to that of Hutchinson.

The mutual relationship of the Oxalidaceae, Geraniaceae, and Tropaeolaceae has been evident to most botanists. The Oxalidaceae and Geraniaceae are indeed not well separated from each other when all of the genera are considered, although the typical forms are different enough. The Tropaeolceae are more distinctive, but their relationships have not occasioned much dispute. The characteristic spur on one of the sepals in the Tropaeolaceae is apparently homologous with a similar spur found in some of the Geraniaceae (*Pelargonium*), the chief difference being that in the Geraniaceae the spur is adnate to the pedical and is therefore easily overlooked. Cytological studies also support the concept that the Oxalidaceae, Geraniaceae, and Tropaeolaceae are closely related.

The Limnanthaceae and Balsaminaceae are more controversial. Both are morphologically and cytologically somewhat removed from the core families of the order (and from each other), without having any obvious affinities elsewhere. The Limnanthaceae have a unique type of pollen, which emphasizes their distinctness without providing any clue as to their relationships. According to the concepts developed in this work about the phyletic relationships and morphologic divergence within the Sapindales-Geraniales-Linales-Polygalales complex, the Limnanthaceae fit into the Geraniales rather than into any of the related orders. Among recent students, Hutchinson, Takhtajan, and Scholz (in the 12th edition of the Engler Syllabus) have agreed in placing the Limnanthaceae in the Geraniales. It may be significant that the Limnanthaceae and Tropaeolaceae are alike in having some unusual fatty acids (erucic, eicosenoic) in their seeds. No information is available about the seed fats of the other families of the order.

On a purely morphological basis, the Balsaminaceae might be accommodated in either the Geraniales or the Polygalales as here conceived. They would be wholly isolated in the Polygalales, however, whereas they do have certain similarities to the Tropaeolaceae in the Geraniales. The most obvious of these is the conspicuous retrorse spur on one of the sepals. The two families also have very similar pollen. On the other hand, they differ in so many other ways that one is tempted to consider the similarities as accidental. At our present level of knowledge it seems more appropriate to refer the Balsaminaceae to the Geraniales than to any other order.

There is nothing ecologically distinctive about the Geraniales. Nearly all of the species are herbs or soft shrubs, but this growth form is of course well known in diverse other orders. Many of them prefer moist or shady places, but *Erodium*, in the Geraniaceae, is a familiar vernal weed of dry regions. The curious, elastically dehiscent fruits of *Impatiens* (Balsaminaceae) are quite in contrast to the one-seeded mericarps of the Geraniaceae and Tropaeolaceae, but neither of these fruit types is notably efficient in dispersal nor obviously adapted to any particular habitat. The long-awned, spirally twisting mericarps of *Erodium* are evidently well adapted to animal transport and self-planting, but here we are dealing with a generic rather than a familial or ordinal feature.

SYNOPTICAL ARRANGEMENT OF THE FAMILIES OF GERANIALES

1. Flowers regular or slightly irregular, not spurred, or the spur inconspicuous and adnate to the pedicel; stamens 2 or 3 times as many as the sepals or petals, sometimes some of them staminodial; leaves mostly compound or deeply cleft

    2. Style not gynobasic; integuments 2; ovules, when few, pendulous; filaments basally connate; annual or more often perennial herbs, or seldom woody plants

        3. Fruit a beakless, loculicidal capsule or seldom a berry, with (1) 2–many seeds per carpel; styles free; endosperm well developed; pollen binucleate; leaves with or more often without stipules     **1. Oxalidaceae**

        3. Fruit a beaked, mostly septicidally dehiscent capsule or a schizocarp, the mericarps 1-seeded and mostly again dehiscent; styles usually united, at least below, into a prominent column; endosperm mostly scanty or none; pollen trinucleate; leaves with stipules     **2. Geraniaceae**

    2. Style gynobasic; integument 1; ovules 1 per carpel, erect; filaments free; annual herbs; seeds without endosperm
        **3. Limnanthaceae**

1. Flowers strongly irregular, one of the sepals with a conspicuous free spur; stamens 5 or 8, less than twice as many as the basic (unmodified) number of sepals or petals; leaves simple, not deeply lobed

    4. Leaves peltate, palmately nerved; stamens 8; carpels 3, each uniovulate, the fruit a schizocarp; plants with myrosin and without raphides     **4. Tropaeolaceae**

    4. Leaves not peltate, the venation pinnate; stamens 5; carpels 5, each with several or many ovules; fruit an elastically dehiscent capsule, or seldom a berry; plants with raphides and without myrosin     **5. Balsaminaceae**

SELECTED REFERENCES

Gibbs, R. D., A classical taxonomist's view of chemistry in taxonomy of higher plants. Lloydia 28: 279–299. 1965.

Maheshwari, P., and B. M. Johri, The morphology and embryology of *Floerkea proserpinacoides* Willd. with a discussion on the systematic position of the family Limnanthaceae. Bot. Mag. Tokyo 69: 410–423. 1956.

Mason, C. T., Jr., Development of the embryo-sac in the genus *Limnanthes*. Am. Jour. Bot. 38: 17–22. 1951.

Mathur, N., The embryology of *Limnanthes*. Phytomorph. 6: 41–51. 1956.

Narayana, L. L., Contributions to the embryology of Balsaminaceae. I. Jour. Indian Bot. Soc. 42: 102–109. 1963. II. Jour. Jap. Bot. 40: 104–116. 1965.

Warburg, E. F., Taxonomy and relationships in the Geraniales in the light of their cytology. New Phytol. 37: 130–159; 189–210. 1938.

ORDER 14. **Linales**

The order Linales as here defined consists of three families and about 700 species. More than half of the species belong to the Linaceae (450). The Erythroxylaceae are estimated at 200 species, and the Humiriaceae at about 50.

The Linales are here regarded as a simple-leaved offshoot of the Sapindales, parallel to the Geraniales and Polygalales. Further divergence from the Sapindales has involved reduction of cambial activity and loss of the nectary disk in most members of the order. The three families should be regarded as evolutionary siblings, no one of them ancestral to either of the others.

The Linaceae are a somewhat heterogeneous family from which several small segregates might with some reason be cut off. The core members of the family are alternate-leaved herbs or half-shubs without a floral disk and with a few-seeded, usually septicidal capsule. Three shrubby or arborescent segregates with indehiscent fruits are sometimes recognized. Of these, the Hugoniaceae (six genera, more than 50 species) lack a floral disk and have drupaceous fruits, the Ctenolophonaceae (one genus, four species) have opposite leaves, nutlike fruits, and an extrastaminal disk, and the Ixonanthaceae (two genera, 23 species) have an intrastaminal disk and drupaceous fruits. The latter two families or subfamilies, especially the Ixonanthaceae, are often considered to be transitional toward the Humiriaceae.

The Humiriaceae and Erythroxylaceae are relatively homogeneous and sharply limited. Both of them are generally regarded as closely allied to the Linaceae and have sometimes been included in that family. The Humiriaceae have also been compared to the Malpighiaceae, here regarded as the most primitive family of the Polygalales. Such a dual relationship of the Humiriaceae is entirely compatible with the concepts of phylogenetic relationships which have been set forth here. The wood of the Linales as a whole is also very much like that of the Polygalales.

The numerous (sometimes more than 100) stamens of *Vantanea*, of the Humiriaceae, are something of an embarrassment to the concept that the Linales are derived from the Sapindales, which rarely have more than two or three sets of stamens. However, *Peltostigma* in the Rutaceae, and some species of *Deinbollia*, in the Sapindaceae, have more than 20 stamens. All of these multistaminate genera also have a well developed nectary disk, which in this group of orders is usually interpreted as representing a reduced whorl of stamens.

The possibility must be considered that the number of stamens here reflects a secondary increase, rather than persistence of an original high number. We are so accustomed to thinking of evolutionary trends as unidirectional that the possibility of reversal is likely to be slow to come to mind and easy to put aside. Nevertheless, it is perfectly clear that many reversals in various evolutionary trends have occurred. Herbs with an active cambium have repeatedly reverted to the woody habit, and the multicarpellate species of *Nicotiana* (Solanaceae), for example, surely have a bicarpellate ancestry within the genus.

On the other hand, we must also consider the possibility that the low number of stamens in most species of this group of orders reflects many independent parallel reductions, and that the nearest common ancestor was multistaminate. In either case, it seems advisable to continue to associate the Sapindales, Linales, Geraniales, and Polygalales in the system, and to consider the Sapindales as the most primitive order in the group.

The Linales, even though they are one of the smaller and more narrowly defined orders, are ecologically diverse. They are trees, shrubs, or herbs, of tropical or temperate, moist or dry regions, with or without a nectary disk, and with capsular or drupaceous fruits. Indeed the single family Linaceae, if interpreted broadly, embraces all of these variations.

## SYNOPTICAL ARRANGEMENT OF THE FAMILIES OF LINALES

1. Plants woody, often arborescent; fruit drupaceous

    2. Petals exappendiculate; carpels 4–7, commonly 5; flowers with an intrastaminal disk; leaves without stipules
    **1. Humiriaceae**

    2. Petals internally appendiculate; carpels 3, 1 fertile; flowers without a disk; leaves with stipules     **2. Erythroxylaceae**

1. Plants herbaceous or occasionally shrubby, seldom arborescent; fruit mostly capsular, seldom drupaceous or nutlike; petals exappendiculate; leaves mostly stipulate; floral disk usually wanting     **3. Linaceae**

## SELECTED REFERENCES

Cuatrecasas, Jose, A taxonomic revision of the Humiriaceae. Contr. U. S. Nat. Herb. 35: 25–214. 1961.

Narayana, L. L., A contribution to the floral anatomy and embryology of Linaceae. Jour. Indian Bot. Soc. 43: 343–357. 1964.

Rao, D., Floral anatomy of Erythroxylaceae. Proc. Nat. Inst. Sci. India B35: 156–162. 1965.

Saad, S., Palynological studies in the Linaceae. Pollen et Spores 4: 65–82. 1962.

————, Pollen morphology of *Ctenolophon*. Bot. Notis. 115: 49–57. 1962.

ORDER 15. **Polygalales**

The order Polygalales as here defined consists of seven families and nearly 1900 species. The vast majority of the species belong to only three of the families, the Malpighiaceae (800), Polygalaceae (750), and Vochysiaceae (200). The remaining four families have only about 125 species amongst them.

The mutual affinity of five of these families has been widely accepted. These are the Polygalaceae, Xanthophyllaceae, Tremandraceae, Vochysiaceae, and Trigoniaceae. The Xanthophyllaceae, consisting of the single genus *Xanthophyllum* with about 40 species, have often been included in the Polygaceae as a separate tribe. The rank at which this distinctive small group should be received is a matter of opinion, scarcely subject to objective criteria. Its relationship to the Polygalaceae is generally admitted. The Tremandraceae (30 spp) are unusual in the order in having regular flowers. In other respects, however, including the wood anatomy and the poricidal stamens, they fit well into the Polygalales, and they do not appear to be allied to any other order.

The other two families, the Malpighiaceae and Krameriaceae, are more controversial. The Malpighiaceae have sometimes been associated with the families here grouped as the Linales. Anatomically the Linales and Polygalales as here defined are somewhat similar and share in common certain differences from the Sapindales, but each order also hangs together as a group, and the Malpighiaceae go with the Polygalales. The pollen morphology suggests some affinity of the Malpighiaceae to the Humiriaceae, Tremandraceae, Trigoniaceae, and Zygophyllaceae. These facts are all compatible with the concept of relationships here propounded, in which the Linales and Polygalales are regarded as parallel offshoots of the Sapindales. The Malpighiaceae are in some respects the most primitive family of the Polygalales. They cannot be considered as directly ancestral to the rest of the order, however, because their seeds lack endosperm, whereas the Polygalalaceae and some of the other families have endospermous seeds.

The Krameriaceae have by different authors been associated with the Polygalaceae or Leguminosae. The superficial appearance of the flower is that of the Leguminosae (Papilionoideae), but the actual structure of the perianth and androecium is more like that of the Polygalaceae. The wood anatomy is similar to that of the Polygalales, and differs from that of the Leguminosae. The seemingly unicarpellate pistil, with two pendulous apical ovules, is here regarded as pseudomonomerous; i.e., it is an ancestrally compound pistil in which all the carpels but one have been suppressed. It may be worth noting that *Xanthophyllum*, which is undoubtedly allied to the Polygalaceae, typically has a unilocular ovary and a subentire stigma.

The Polygalales are here regarded as a simple-leaved offshoot of the Sapindales, with mostly irregular flowers. Flowers of the more primitive families of Polygalales, such as the Malpighiaceae, are only slightly irregular and have the stamens dehiscent by longitudinal slits. It may also be significant that many of the Malpighiaceae have a jointed petiole, suggesting that the leaf blade is actually the terminal leaflet of a unifoliolate compound leaf. The more advanced families of the order, such as the Polygalaceae, have highly irregular flowers with the stamens opening by terminal pores. Primitively woody, the order has given rise to herbaceous forms in the Polygalaceae, Vochysiaceae, and Krameriaceae. These three families include both woody and herbaceous species, but only the Polygalaceae have a large proportion of herbs.

Ecologically the families of the Polygalales have little in common beyond the possible similarities in pollinating mechanisms as suggested by the structure of the flowers. The family Polygalaceae is indeed highly diversified ecologically, as it includes shrubs, ordinary herbs, twiners, and even a few nongreen root-parasites. The several families (and sometimes genera within the families) differ in number of carpels, number of ovules per locule, and presence or absence of stipules and endosperm, but these characters are difficult to assess in terms of ecological adaptation or survival value. The differences in habit and fruit type which mark some of the genera or families are apparently adaptive but are duplicated in many other groups of dicots.

## SYNOPTICAL ARRANGEMENT OF THE FAMILIES OF POLYGALALES

1. Carpels mostly 3, seldom 2–5; stamens mostly dehiscent by longitudinal slits, seldom by terminal pores; leaves usually stipulate

2. Fertile stamens several; filaments usually connate toward the base

   3. Petals 5, commonly fringed or toothed; fruit indehiscent; ovules 1 per carpel; stamens 10, some sterile
   **1. Malpighiaceae**

   3. Petals 3, entire; fruit capsular; ovules 1–many per carpel; stamens 3–12   **2. Trigoniaceae**

2. Fertile stamen 1, but staminodes also usually present; filaments free; fruit usually capsular; ovules 1–many per carpel
   **3. Vochysiaceae**

1. Carpels 2, or seemingly only 1; stamens dehiscent by terminal pores; leaves usually exstipulate

   4. Pistil more or less evidently bicarpellate, but not always bilocular; seeds with endosperm; stamens 8–10

      5. Stamens free

         6. Flowers regular; placentation axile in a bilocular ovary; heatherlike shrubs or subshrubs with small leaves
         **4. Tremandraceae**

         6. Flowers irregular; placentation apical or parietal in a unilocular ovary which may show traces of a partition; small trees with normal leaves   **5. Xanthophyllaceae**

      5. Stamens mostly monadelphous; flowers irregular; ovary evidently bilocular; herbs, vines, or shrubs   **6. Polygalaceae**

   4. Pistil pseudomonomerous and with a pair of apical ovules; seeds without endosperm; stamens 4 or seldom 3; low shrubs or seldom herbs   **7. Krameriaceae**

### SELECTED REFERENCES

Erdtman, G., The systematic position of the genus *Diclidanthera* Mart. Bot. Notis. 1944: 80–84.

Heimsch, C., Comparative anatomy of the secondary xylem in the "Gruinales" and "Terebinthales" of Wettstein with reference to taxonomic grouping. Lilloa 8: 83–198. 1942.

O'Donnell, C. A., Posicion sistematica de *"Diclidanthera"* Mart. Lilloa 6: 207–212. 1941.

Stafleu, F. A., A monograph of the Vochysiaceae. I-IV. Rec. Trav. Bot. Neerl. 41: 398–540. 1948. Acta Bot. Neerl. 1: 222–242. 1952. 2: 144–217. 1953. 3: 459–480. 1954.

Turner, B. L., Chromosome numbers in the genus *Krameria;* evidence of familial staus. Rhodora 60: 101–106. 1958.

ORDER 16. **Umbellales**

The order Umbellales as here defined consists of only two families, the Araliaceae, with about 700 species, and the Umbelliferae, with about 3000. The two families are generally admitted to be intimately related, with the Araliaceae being the more primitive. The Araliaceae are chiefly tropical, with a few temperate-zone species. The Umbelliferae, in contrast, occur chiefly in temperate regions.

The Araliaceae and Umbelliferae have traditionally been included in the same order with the Cornaceae and their immediate allies. However, the Cornales and Umbellales as here defined are coherent groups which differ from each other in so many respects that it seems unreasonable to keep them in the same order. More than 50 years ago Hoar pointed out a series of anatomical differences between the two groups and concluded that they belonged in different orders, but his work has been generally overlooked or ignored.

The most obvious morphological differences between the Umbellales and Cornales may be summarized as follows:

1. Leaves usually compound or conspicuously lobed or dissected, and either stipulate or with a broad, sheathing base; nodes multilacunar; pollen trinucleate; schizogenous secretory canals well developed; wood parenchyma vasicentric; vessels usually with a simple terminal pore; central axis of the gynoecium usually with vascular bundles **Umbellales**

1. Leaves always simple and entire or merely toothed, usually without stipules, not sheathing; nodes trilacunar; pollen binucleate; secretory canals usually wanting; wood parenchyma mostly diffuse or none; vessels mostly or all with scalariform perforations; central axis of the gynoecium without vascular tissue **Cornales**

The true relationships of the Umbellales appear to be with the Sapindales, in which compound leaves, stipules, multilacunar nodes, trinucleate pollen, schizogenous secretory canals, vasicentric parenchyma, vessels with a single terminal pore, and a centrally vasculated gynoecium are all well known. If the Araliaceae had the ovary superior instead of inferior they would be perfectly at home in the Sapindales. Anatomically they compare well with the Burseraceae except in having somewhat wider wood rays. The width of the rays probably reflects the tendency toward herbaceousness which begins in the Araliaceae and culminates in the Umbelliferae.

It may also be noted that on grounds of the chemical constituents Hegnauer (1964) favors separating the Umbellales from the Cornales. He compares the Umbellales to the Sapindales (sensu meo), and the Cornales to the Rosales (sensu meo).

A few of the Araliaceae have ten or more petals, stamens, and carpels in a regular, symmetrical arrangement. These polymerous types have sometimes been regarded as primitive within the family, but they more probably have undergone a secondary increase in the number of parts of each kind. Pentamerous flowers are here regarded as primitive in the family and order.

It may be noted for the record that, in contrast to the view here presented, Sophia Tamamschian, well known Russian student of the Umbelliferae, maintains in publication and in recent correspondence that the Umbellales are indeed closely related to the Cornales. She correctly points out that the anatomical characters differentiating the two orders as here conceived have been insufficiently studied, and that at least some of them are more subject to exception than might at first appear: for example, simply perforate vessels occur in *Alangium* (Cornales), and scalariform vessels in *Mydocarpus* (Araliaceae — see Rodriguez, p. 254); and the number of lacunae per node in the Umbelliferae tends to decrease upwards so that the upper nodes are often trilacunar. She further maintains that those Araliaceae with ten or more petals, stamens, and carpels are primitive in that respect, having inherited this feature from an eventual Rosalean or Myrtalean ancestry. She does admit, however, that the "Sapindales have some affinity to Umbellales."

There is nothing ecologically distinctive about the Umbellales or either of the included families. The adaptive significance of epigyny is dubious at best. The well developed secretory system may prove to be adaptive, but the fact remains to be demonstrated. It can scarcely be a consistently effective repellent to foraging animals, inasmuch as many species of Umbelliferae are well known to be palatable to livestock. The fruit of the Umbelliferae is morphologically unique but ecologically undistinguished. In a few genera it is beset with hooks or barbs which adapt it to distribution by animals; in other genera it is dispersed by wind, and in still others it has no very obvious means of dispersal.

SYNOPTICAL ARRANGEMENT OF THE FAMILIES OF UMBELLALES

1. Carpels 1–many, typically 5; fruit fleshy or dry, usually a berry, the carpels sometimes separating, but without a carpophore; trees and shrubs, seldom herbs or lianas **1. Araliaceae**

1. Carpels consistently 2; fruit a dry schizocarp, the mericarps separated by a usually persistent carpophore; herbs, rarely shrubs or trees **2. Umbelliferae**

## SELECTED REFERENCES

Eyde, R., The peculiar gynoecial vasculature of the Cornaceae and its systematic significance. Phytomorph. in press, 1967. (Manuscript available in advance of publication)

Hoar, C. L., A comparison of the stem anatomy of the cohort Umbelliflorae. Ann. Bot. Lond. 29: 55–63. 1915.

Mittall, S. P., Studies in the Umbellales. II. The vegetative anatomy. Jour. Indian Bot. Soc. 40: 424–443. 1961.

Rodriguez, R. L., Systematic anatomical studies on *Myrrhidendron* and other woody Umbellales. U. Calif. Pub. Bot. 29: 145–318. 1957.

Tamamschian, S., A tentative system of dicotyledons as exemplified by the order Umbellales. (In Russian) A paper delivered in 1964 at the Second Conference on Plant Phylogeny in Moscow, and now (1968) awaiting publication in the collective Proceedings of this and the Third (1967) Conference.

## Subclass VI. Asteridae

The subclass Asteridae as here defined consists of nine orders, 43 families, and about 56,000 species. About a third of all species of dicotyledons belong to this one subclass, which is second in size only to the Rosidae. A third of the species of the subclass belong to the order Asterales (19,000), with the single family Compositae (Asteraceae). The other orders, according to size, are the Scrophulariales (10,000), Lamiales (7800), Rubiales (6500), Gentianales (5600), Polemoniales (4600), Campanulales (2500), Dipsacales (1000), and Plantaginales (250).

The Asteridae are sympetalous (rarely polypetalous or apetalous) dicotyledons with unitegmic, tenuinucellate ovules, and with the stamens usually as many as or fewer than the corolla lobes and alternate with them. Most of them have two carpels, but a few have as many as five or even more carpels, and a few others are pseudomonomerous. The subclass is highly natural, and it is morphologically the best defined of the six subclasses of dicots. The step by step progression from the more primitive to the more advanced families of the subclass makes any grouping of families into orders difficult and seemingly precludes any reasonable possibility of deriving different parts of the group from diverse sources.

The ancestry of the Asteridae is not entirely certain. It is clear enough that they are derived eventually from the Magnoliidae, and equally clear that the connection to the Magnoliidae is through one of the other subclasses rather than direct. The Hamamelidae and Caryophyllidae are easily excluded as possible ancestors, but both the Rosidae and the Dilleniidae have the morphological characters from which those of the Asteridae could well have developed.

The morphological differences between the Asteridae and the sympetalous orders of the Dilleniidae are not very impressive. No one family is suggested as a possible direct ancestor, but the Ericales and Ebenales collectively have all that would be required for such an ancestor. Furthermore, the nectaries of those families of Asteridae with a superior ovary are very much like those of the Theales and the sympetalous orders of the Dilleniidae. Sympetaly itself is much more common in the Dilleniidae than in the Rosidae.

On the other hand, several different families or orders of the Rosidae have been compared to various members of the Asteridae in terms of a possible relationship. The status of the Pittosporaceae (Rosales) as possible collateral relatives of the Asteridae has already been noted. The order Celastrales is also a very possible starting point for the Asteridae, but no one family presents itself as a possible connecting link. Each of the other orders of the Rosidae is in its own way too specialized to be near-ancestral to the Asteridae. Neither the Cornales nor the Umbellales nor the Hydrangeaceae is a good starting point, for example, because they all have epigynous flowers, and the Cornales and Umbellales also have too few ovules per locule. Each of these groups has been compared to one or

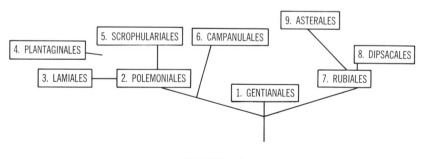

FIGURE 4.8

Probable relationships among the orders of Asteridae.

another member of the Asteridae, but any close similarity which may exist is probably secondary. The striking serological correspondence between *Cornus* and some of the Caprifoliaceae can scarcely outweigh the differences between the two groups, but it might lend some strength to the thought that the Asteridae are derived from the Rosidae.

The evidence from chemical characters is still scanty and inconclusive, but it is beginning to pile up on the side of a Rosidan rather than a Dilleniidan ancestry for the Asteridae. A series of simple chemical tests carried out by Gibbs suggests that most of the Rosidae and most of the Dilleniidae hang together as separate groups, and that the Asteridae probably go with the Rosidae. Not too much weight can yet be placed on this evidence, however, especially inasmuch as some individual families, such as the Rubiaceae, are heterogeneous in these same features. Recently Hegnauer (cited below, p. 544), using a different set of chemical data, has suggested, in effect, that the Asteridae are derived from the Rosidae. Here again some caution is called for, because within the Asteridae as here conceived he has two groups, separately derived from the Rosidae, which do not conform to the relationships within the Asteridae that are suggested by other information. He would, for example, put the Solanaceae in one of these groups, and the Scrophulariales in the other.

Even before the chemical information had been added to the scales, Takhtajan thought that the Asteridae were more probably related to the Rosidae than to the Dilleniidae, although I understand from personal conversation that he felt this to be one of the less certain parts of his system.

Contrary to my initial predilection, I am now prepared to accept the Rosidae as the more likely ancestors to the Asteridae. In a previous paper I deliberately left the present Asteridae as a group of uncertain ancestry, but in the linear sequence I put them after the present Dilleniidae. These subclasses were at that time still undefined and unnamed, but they can be recognized as groups in both Takhtajan's diagrams and my own. Similar thoughts had doubtless occurred to other botanists as well.

The sympetalous corolla and few stamens of the Asteridae presumably reflect specialization for advanced types of insect pollinators. The significance of the unitegmic, tenuinucellate condition of the ovules is more doubtful. The reduction of the nucellus may reflect merely an accumulation of loss mutations. I do not suggest that there is any survival value to the reduction of the nucellus per se, but rather that genes which have been incorporated into the system for other reasons may have carried nucellar reduction as an unimportant side-effect. The same sort of explanation may apply to the fusion of two integuments into one. It does not seem to be simple economy, for the fusion is not necessarily accompanied by any reduction in the total amount of tegumentary tissue in the ovule.

Indeed the massiveness of the single integument in many members of the subclass has often been noted. Furthermore, a change in the number of integumentary layers has no bearing on the amount of metabolic energy necessary to form the mature seed coat, which is derived from the integument(s). Simplification and reduction are common results of the operation of Finagle's law (anything that can go wrong, will go wrong) when it is not counterbalanced by selection.

SYNOPTICAL ARRANGEMENT OF THE ORDERS OF ASTERIDAE

1. Ovary chiefly superior

    2. Plants usually with well developed internal phloem; leaves opsite; endosperm nearly always nuclear; no integumentary tapetum; corolla regular, its lobes usually contorted in bud
    **1. Gentianales**

    2. Plants usually either without internal phloem, or with alternate leaves, or both; endosperm nuclear or more often cellular; integumentary tapetum often present; corolla regular or irregular, with various sorts of aestivation

        3. Fruit typically of 4 more or less distinct (or separating) nutlets, derived from a bicarpellate ovary that often has a gynobasic style    **3. Lamiales**

        3. Fruit otherwise; style rarely gynobasic

            4. Corolla scarious, persistent; flowers mostly anemophilous, tetramerous as to calyx, corolla, and androecium; leaves usually all basal    **4. Plantaginales**

            4. Corolla otherwise; flowers mostly entomophilous or ornithophilous, variously pentamerous or tetramerous, with isomerous or anisomerous stamens; leaves seldom all basal

                5. Flowers mostly regular and with as many functional stamens as corolla lobes, typically 5 of each
    **2. Polemoniales**

                5. Flowers mostly irregular and with fewer functional stamens than corolla lobes, or sometimes with regular, tetramerous corolla and 2 or 4 stamens, or the corolla rarely wanting    **5. Scrophulariales**

1. Ovary chiefly inferior

   6. Flowers borne in various sorts of inflorescences, but if borne in involucrate heads then the ovules either pendulous or more than one; endosperm present or absent

      7. Leaves nearly always alternate; stamens free from the corolla, or attached at the base of the tube; anthers in most families connivent or connate around the style, which pushes out the pollen; plants nearly always herbaceous
                            **6. Campanulales**

      7. Leaves usually opposite or whorled (notable exception: Calyceraceae); stamens attached to the corolla tube, generally well above the base; flowers mostly without a specialized pollen presentation mechanism (notable exceptions: Calyceraceae, some Rubiaceae); plants woody or herbaceous

         8. Stipules usually present and interpetiolar (sometimes reduced to a mere interpetiolar line, or enlarged into leaves); corolla typically regular and with isomerous stamens; endosperm nuclear; plants chiefly tropical and woody, but some members herbaceous, or of temperate regions, or both      **7. Rubiales**

         8. Stipules typically none, when present usually small and adnate to the petiole; corolla regular or often irregular, the stamens often fewer than the corolla lobes; endosperm cellular; plants woody or often herbaceous, of temperate or less often tropical regions   **8. Dipsacales**

   6. Flowers borne in involucrate, centripetally flowering heads; ovary 1-celled, with a solitary, erect ovule; anthers connate or connivent into a tube around the style, which pushes out the pollen; endosperm wanting from mature seeds
                             **9. Asterales**

## SELECTED REFERENCES

Cronquist, A., Outline of a new system of families and orders of dicotyledons. Bull. Jard. Bot. Brux. 27: 13–40. 1957.

Gibbs, R. D., Comparative chemistry of plants as applied to a problem of systematics: the Tubiflorae. Trans. Roy. Soc. Can., Ser. 3, Sect. 3. 46: 143–159. 1962.

————, A classical taxonomist's view of chemistry in taxonomy of higher plants. Lloydia 28: 279–299. 1965.

Hegnauer, R., Chemotaxonomic der Pflanzen. III. Dicotyledoneae: Acanthaceae–Cyrillaceae. Birkhäuser Verlag. Basel and Stuttgart. 1964.

Hillebrand, G. R., Phytoserological systematic investigation of the genus *Viburnum*. Ph.D. thesis, Rutgers University. 1966.

Leppik, E. E., Evolutionary relationship between entomophilous plants and anthophilous insects. Evolution 11: 466–481. 1957.

Takhtajan, A., Die Evolution der Angiospermen. Gustav Fischer Verlag. Jena. 1959.

ORDER 1. **Gentianales**

The order Gentianales as here defined consists of four families and about 5600 species. The Apocynaceae and Asclepiadaceae have about 2000 species each, the Gentianaceae about 1100, and the Loganiaceae about 500.

These several families are held together by a large number of characters to which there are but few exceptions. They have well developed internal phloem and opposite, simple, mostly entire leaves. The flowers are hypogynous, with a regular corolla, the lobes usually contorted in bud. There is a single cycle of stamens, alternate with the corolla lobes. The gynoecium is bicarpellate, with mostly numerous, anatropous ovules on axile to parietal placentae. The ovules are unitegmic and tenuinucellate, as in other members of the subclass, but they lack the integumentary tapetum found in most of the other orders. The endosperm is nearly always nuclear. Many of the genera contain alkaloids or glucosides.

The small genus *Desfontainea*, with about five species, is reported to lack internal phloem and has therefore sometimes been treated as a separate family. However, it is so much like the Loganiaceae in other respects that it is here included in that family.

The Buddlejaceae, Menyanthaceae, and Oleaceae, which have sometimes been included in the Gentianales, are here excluded. They lack internal phloem, and they have an integumentary tapetum and cellular endosperm. The Menyanthaceae further differ in having alternate leaves, and the Oleaceae usually have only two stamens. Tournay and Lawalrée have suggested that these three families be treated as forming a single order, but they differ so much inter se that the suggestion is not here adopted. The Menyanthaceae are here referred to the Polemoniales, and the Buddleiaceae and Oleaceae to the Scrophulariales.

The biological significance of the characters which distinguish the Gentianales as a group is doubtful. Diverse species of the order occur in a wide range of habitats which are also occupied by species of various other orders.

The principal characters marking the family Asclepiadaceae, on the other hand, are clearly related to pollination, ensuring that many pollen grains are transported in a group. The distinction between the Apocynaceae and Asclepiadaceae is much less sharp than the simple statement in the synoptical key would imply. Typical members of the two families are so different that it is easier to think of two families than of only one, but there is a step by step gradation in characters, so that the line between the two is only arbitrarily established. The significance of the progressive evolutionary separation of the carpels from the bottom upwards, beginning in the Apocynaceae and culminating in the Asclepiadaceae (in which the carpels are united only by the common stigma), is obscure.

### SYNOPTICAL ARRANGEMENT OF THE FAMILIES OF GENTIANALES

1.  Plants without a latex system; style not especially thickened and modified distally; carpels fully united, only the stigmas sometimes separate

    2.  Plants mostly woody and with more or less well developed stipules, often containing alkaloids, but lacking gentiopicrin; placentation axile                                    **1. Loganiaceae**

    2.  Plants mostly herbaceous, without stipules, containing the bitter glucoside gentiopicrin but lacking alkaloids; placentation mostly parietal                                      **2. Gentianaceae**

1.  Plants with a well developed latex system; style usually more or less thickened and modified at the tip; carpels typically united only by the style and/or stigma, but sometimes wholly united; stipules none; alkaloids often present, but gentiopicrin lacking

    3.  Androecium free from the stigma and without translators; pollen granular, not forming pollinia; carpels often united by part or all of the style as well as by the stigma, or even wholly united; corona mostly poorly developed or wanting; trees, shrubs, vines, or less often herbs
                                                                    **3. Apocynaceae**

    3.  Androecium concrescent to the stigma and provided with translators, the pollen grains borne in dense masses (pollinia); carpels united only by the stigma; corona more or less well developed; herbs, vines, or shrubs, rarely trees
                                                                    **4. Asclepiadaceae**

## SELECTED REFERENCES

Bissett, N. G., The occurrence of alkaloids in the Apocynaceae. Ann. Bogor. 3: 105–236. 1958. 4: 395–418. 1961.

Gopal Krishna, G., and V. Puri, Morphology of the flower of some Gentianaceae with special reference to placentation. Bot. Gaz. 124: 42–57. 1962.

Hegnauer, F., Chemotaxonomische Betrachtungen. V. Die systematische Bedeutung des Alkaloidmerkmales. Planta Medica 6: 1–34. 1958.

Korte, F., Die Beziehungen zwischen Inhaltstoffen und morphologischer Systematik in der Reihe Contortae unter besonderer Berücksichtigung der Bitterstoffe. Zeitschr. Naturforsch. 9b: 354–358. 1954.

Korte, F. and I., Über die Beziehungen zwischen morphologischer Systematik und chemischen Inhaltstoffen bei den Asclepiadaceen. Zeitschr. Naturforsch. 10b: 223–229. 1955.

Lindsey, A. A., Floral anatomy in the Gentianaceae. Am. Jour. Bot. 27: 640–652. 1940.

Moore, R. J., Cytotaxonomic studies in the Loganiaceae. I. Chromosome numbers and phylogeny in the Loganiaceae. Am. Jour. Bot. 34: 527–538. 1947.

Roy Tapadar, N. N., Cytotaxonomic studies in Apocynaceae and delineation of the different evolutionary tendencies operating within the family. Caryologia 17: 103–138. 1964.

Safwat, F. M., The floral morphology of Secamone and the evolution of the pollinating apparatus in the Asclepiadaceae. Ann. Mo. Bot. Gard. 49: 95–130. 1962.

Tournay, R., and A. Lawalrée, Une classification nouvelle des families appartenant aux ordres des Ligustrales et des Contortées. Bull. Soc. Bot. Fr. 99: 262–263. 1952.

Woodson, R. E., Jr., The North American Asclepiadaceae. I. Perspective of the genera. Ann. Mo. Bot. Gard. 28: 193–244. 1941.

Woodson, R. E., Jr., and J. A. Moore, The vascular anatomy and comparative morphology of apocynaceous flowers. Bull. Torrey Club 65: 135–166. 1938.

ORDER 2. **Polemoniales**

The order Polemoniales as here defined consists of eight families and about 4600 species. Two of the families, the Solanaceae (2300) and Convolvulaceae (1400), contain about four-fifths of the species. The Hydrophyllaceae and Polemoniaceae have about 300 species each, the Cuscutaceae about 150, the Nolanaceae nearly 100, the Menyanthaceae about 40, and the Lennoaceae only four.

Most of the 25 families here distributed among the orders Polemoniales,

Lamiales, and Scrophulariales are often treated as a single large order, usually under the irregular name Tubiflorae. The Tubiflorae as so conceived appear to be a largely natural group, but the group embraces so much diversity as to be mentally unwieldy. If so much is included in one group, there is little reason to maintain the family Plantaginaceae as an order, and even the status of the Gentianales might well be called into question. On the other hand, botanists continue to differ about how the Tubiflorae might best be divided into smaller groups. Even an agreed phylogeny might be presented in more than one taxonomic scheme. It is purely a matter of taste whether to maintain the larger and relatively amorphous concept of the group, or to recognize several smaller orders.

In general, the Polemoniales are sympetalous plants with regular or nearly regular, usually five-lobed corolla, and with isomerous stamens alternate with the corolla lobes. The pollen is binucleate except in the Cuscutaceae. The fruit is usually capsular or softly fleshy, only seldom drupaceous. Some members of the order show one or another expression of the tendency toward one-seeded, hemicarpellate nutlets that is more fully expressed in the derived order Lamiales. Most of the Polemoniales have internal phloem, as in the Gentianales, but five families, representing nearly a fifth of the species, lack it. Except for many of the Polemoniaceae, the leaves are nearly always alternate.

In addition to the alternate leaves, the order Polemoniales differs from the order Gentianales in its often cellular (rather than nuclear) endosperm and in often having an integumentary tapetum. The related order Lamiales differs from the Polemoniales in its characteristic gynoecium and fruit, and the order Scrophulariales differs in having a usually irregular corolla and usually fewer stamens than corolla lobes. These two orders represent the realization of tendencies which are evident but less fully developed in some of the Polemoniales.

It is widely agreed that the Nolanaceae, Solanaceae, and Convolvulaceae are closely related inter se, that the Polemoniaceae and Hydrophyllaceae form a pair, and that these two groups are related to each other. The relationship of the Cuscutaceae to the Convolvulaceae is also generally admitted; indeed the Cuscutaceae have often been submerged in the Convolvulaceae.

The Lennoaceae were at one time usually included in the Ericales, but more recent studies have shown that they lack the embryological peculiarities of the Ericales and are wholly out of place there. A relationship to the Hydrophyllaceae and Boraginaceae has been suggested instead. The large number (six to 14) of carpels in the Lennoaceae is usually regarded as reflecting a secondary increase rather than an original high number. A similar secondary increase in carpel number has evidently taken place in some of the half-wild species of *Nicotiana* (Solanaceae) in western North

America. The Lennoaceae resemble the Boraginaceae and other Lamiales in producing twice as many one-seeded nutlets as there are carpels, but they lack the gynobasic style. The similarity reflects parallelism from similar ancestors with similar potentialities, rather than direct inheritance from a common ancestor. The Lennoaceae are taxonomically somewhat isolated, but I believe they are best referred to the Polemoniales, somewhere near the point of origin of the Lamiales.

Perhaps the most startling addition to the Polemoniales, in the eyes of some botanists, will be the family Menyanthaceae. However, the Menyanthaceae are phenetically perfectly at home in the Polemoniales, whereas they differ from the Gentianales in having alternate leaves, cellular endosperm, and an integumentary tapetum, as well as in lacking internal phloem. They do have a bitter principle, meliatin, in common with certain Loganiaceae, but this can hardly outweigh the series of differences. No very close relationship of the Menyanthaceae to any other family of the Polemoniales is here suggested, but the ordinal assignment seems well justified nonetheless. Certainly the Menyanthaceae would be a highly discordant element in the Gentianales.

The Polemoniales are evidently related to the Gentianales. Aside from the general similarities in floral structure, the two orders are linked by the common possession of internal phloem in most species of Polemoniales and nearly all of the Gentianales. Except for these two orders and the taxonomically remote order Myrtales, only a few widely scattered groups of dicots have internal phloem. Embryologically the Gentianales are more primitive than the Polemoniales; so far as these characters are concerned, the one order might well be directly ancestral to the other. Although in the angiosperms as a whole alternate leaves are more primitive than opposite leaves, I will hazard the guess that in this pair of orders, as in the Asterales, opposite leaves are primitive and alternate leaves derived. However, any common ancestor of the Polemoniales and Gentianales would probably be more primitive than all of the Gentianales and the vast majority of the Polemoniales in having five carpels. The two orders should therefore be considered to have diverged from a common ancestor which combined the more primitive features of both.

The order Polemoniales as a whole is ecologically varied, and the adaptive significance of the characters which mark the group is at best debatable. Several of the individual families, on the other hand, clearly have ecological significance. The Cuscutaceae and Lennoaceae represent different, independently evolved adaptations to parasitism. The Menyanthaceae are adapted to aquatic life, but the formal characters which mark the family are not obviously related to the habitat. Both the Polemoniaceae and the Hydrophyllaceae have produced many desert and semidesert annuals, but they also occur in a wide range of other habitats, and the biological significance of the characters which distinguish the two families is still wholly unknown.

## SYNOPTICAL ARRANGEMENT OF THE FAMILIES OF POLEMONIALES

1. Plants with internal phloem, and always autotrophic

   2. Carpels 5; ovary with a terminal style and ripening into a schizo-carp, or with a gynobasic style and ripening into 5 or more 1- to several-seeded nutlets; endosperm cellular; more or less succulent herbs and shrubs, confined to the Pacific slope of S. Am. **1. Nolanaceae**

   2. Carpels 2 (–5); fruit not schizocarpic and style never gynobasic; plants widespread, seldom succulent

      3. Ovules and seeds mostly numerous, on axile placentae; cotyle-dons not plicate; style barely or scarcely lobed; endo-sperm mostly cellular **2. Solanaceae**

      3. Ovules 2 (rarely many) per carpel, basal, erect, the partition present or absent; cotyledons plicate; style entire or often cleft, or the styles sometimes wholly distinct; endosperm nuclear **3. Convolvulaceae**

1. Plants without internal phloem, and either autotrophic or parasitic

   4. Plants parasitic, without chlorophyll; embryo scarcely differen-tiated

      5. Carpels 2; fruit capsular; twining stem-parasites, not rooted in the ground at maturity; endosperm nuclear; pollen tri-nucleate **4. Cuscutaceae**

      5. Carpels 6–14; fruit of 1-seeded nutlets; erect root-parasites; endosperm probably cellular; pollen binucleate **8. Lennoaceae**

   4. Plants autotrophic; embryo normally developed

      6. Corolla lobes valvate or induplicate-valvate in bud; plants more or less aquatic; placentation parietal; endosperm cel-lular **5. Menyanthaceae**

6.  Corolla lobes convolute or imbricate in bud; plants terrestrial, though sometimes of wet places

7.  Carpels 3 (2–4); corolla lobes convolute in bud; placentation axile; endosperm nuclear          **6. Polemoniaceae**

7.  Carpels 2; corolla lobes imbricate or rarely convolute in bud; placentation parietal (often with intruded placentae) or occasionally axile; endosperm nuclear or cellular          **7. Hydrophyllaceae**

SELECTED REFERENCES

Copeland, H. F., The structure of the flower of *Pholisma arenariu*. Am. Jour. Bot. 22: 366–383. 1935.

Dawson, M. L., The floral morphology of the Polemoniaceae. Am. Jour. Bot. 23: 501–511. 1936.

Drugg, W. S., Pollen morphology of the Lennoaceae. Am. Jour. Bot. 49: 1027–1032. 1962.

Grant, V., Natural history of the Phlox family. Martinus Nijhoff. The Hague. 1959.

———, "The Diversity of Pollination Systems in the Phlox Family," in *Recent Advances in Botany* (Toronto: Univ. Toronto Press, 1961), pp. 55–60.

Johnston, I. M., A study of the Nolanaceae. Proc. Am. Acad. 71: 1–87. 1936.

Johri, B. M., and B. Tiagi, Floral morphology and seed formation in *Cuscuta reflexa* Roxb. Phytomorph. 2: 162–180. 1952.

Lindsey, A. A., Anatomical evidence for the Menyanthaceae. Am. Jour. Bot. 25: 480–485. 1938.

Tiagi, B., A contribution to the morphology and embryology of *Cuscuta hyalina* Roth and *C. planiflora* Tenore. Phytomorph. 1: 9–21. 1951.

Wilson, K., The genera of Hydrophyllaceae and Polemoniaceae in the southeastern United States. Jour. Arn. Arb. 41: 197–212. 1960.

———, The genera of Convolvulaceae in the southeastern United States. Jour. Arn. Arb. 41: 298–317. 1960.

ORDER 3. **Lamiales**

The order Lamiales as here defined consists of five families and more than 7800 species. The Labiatae (3200), Verbenaceae (2600), and Boraginaceae (2000) are all large families. The Callitrichaceae (25) and Phrymaceae (one) are much smaller. The Labiatae are also called the Lamiaceae.

The order is marked by its characteristic gynoecium, consisting of two (rarely four to five) biovulate carpels, each carpel divided between the ovules by a "false" partition, or the two halves of the carpel seemingly wholly separate. Except in *Phyrma* and some of the more primitive species of Boraginaceae and Verbenaceae, the fruit usually consists of four separate or separating nutlets. In the Labiatae and most of the Boraginaceae the four segments of the ovary are nearly or quite distinct from each other and are united chiefly by gynobasic style. The Lamiales have the typical unitegmic, tenuinucellate ovule of the subclass. The endosperm is chiefly cellular.

The relationship of the Verbenaceae to the Labiatae has been evident to most botanists for many years. Indeed it is difficult to know where to draw the line between them, and a case might be made for treating the present Verbenaceae and Labiatae as the more primitive and more advanced segments of a single family. The Labiatae relate to the Verbenaceae in the same way that the Asclepiadaceae relate to the Apocynaceae.

The Phrymaceae are clearly a reduced, pseudomonomerous type derived from the Verbenaceae. The single genus and species of Phrymaceae is habitally very much like *Verbena* and has often been included in the Verbenaceae. It is purely a matter of taste whether the evident relationship or the obviously different gynoecium should be emphasized in determining the status of *Phryma*.

The Boraginaceae stand somewhat apart from the three preceding families, but the primitive, tropical, woody, drupe-bearing species of the Boraginaceae and Verbenaceae are very much alike, differing mainly in the arrangement of the leaves. Botanists whose acquaintance with these families is restricted to the temperate-zone species might be surprised to discover that teakwood comes from a verbenaceous tree, and that *Cordia*, in the Boraginaceae, is also a tree.

In both the Boraginaceae and the Verbenaceae cum Labiatae the gynobasic style and characteristic four-nutlet fruit have evolved from the unlobed gynoecium that has a terminal style and ripens into a four-seeded drupe. Here we have another example of the fact that a character which helps to define a recognized group may actually have been developed independently by different members of the group from similar ancestors with similar potentialities. If the Verbenaceae and the more primitive genera of the Boraginaceae were to die out, the unique gynoecium of the Labiatae and the remaining Boraginaceae could easily be thought to have been inherited directly from a common ancestor.

The most aberrant family of the order is the Callitrichaceae. They have apetalous, unisexual flowers, and they have often been associated with the Haloragaceae or Euphorbiaceae. However, they have the characteristic unitegmic, tenuinucellate ovule of the Asteridae and higher Dilleniidae, and they have a fruit of four separating drupelets. These and other embryological features lead most current students of angiosperm phylogeny to associate the Callitrichaceae with the families here grouped in the Lamiales. It is perfectly obvious that the flowers of the Callitrichaceae are reduced, and the corolla that was lost could just as well have been sympetalous as polypetalous. Other secondarily apetalous types (e.g., *Besseya*, in the Scrophulariaceae) are well known in the subclass Asteridae. Even so, it should be noted that the Callitrichaceae differ from most of the Lamiales in having well developed endosperm and separate styles, both of which are primitive characters as compared to the characters of the other families of the order.

The Lamiales are evidently derived from the Polemoniales. If they were deprived of their special gynoecial character, the Boraginaceae and some of the Verbenaceae would fit comfortably into the Polemionales, and we have already noted that a tendency toward this type of gynoecium exists even within the Polemoniales. Many authors have noted a probable relationship between the Hydrophyllaceae (Polemoniales) and Boraginaceae.

The biological importance of the characters that mark the order Lamiales is doubtful. The structure of the mature gynoecium obviously influences the means of seed dispersal, but a comprehensive explanation of the competitive advantage to be achieved in the step by step progression from the one type of fruit to the other has not yet been presented. Within the order, only the aquatic family Callitrichaceae has obvious ecological significance. It is possible that the floral reduction in this family is attendant on the change in habitat. The role of the aromatic oils in the economy of the Labiatae is still obscure. In some cases they help to discourage grazing animals, but the generality of this function for the family as a whole is doubtful and in any case remains to be demonstrated.

### SYNOPTICAL ARRANGEMENT OF THE FAMILIES OF LAMIALES

1. Leaves mostly alternate, usually entire; flowers mostly regular and with 5 stamens; style typically but not always gynobasic, the fruit typically but not always of 4 distinct nutlets; stems not square **1. Boraginaceae**

1. Leaves mostly opposite (or whorled), and often toothed or cleft; flowers mostly more or less irregular (or apetalous) and with 1–4 stamens, the exceptions (regular corolla and/or 5 stamens) being found chiefly in the Verbenaceae; stems commonly square

2. Flowers apetalous, unisexual; styles 2, separate; stamen solitary; endosperm well developed; plants mostly aquatic, often free-floating **2. Callitrichaceae**

2. Flowers petaliferous, usually perfect; style usually solitary; stamens 2 or more; endosperm usually scanty or wanting; mostly terrestrial plants, never free-floating aquatics

   3. Style terminal, the ovary only slightly or not at all lobed; plants seldom aromatic

      4. Ovary with 2 (seldom more) usually biovulate carpels, with a false partition between the ovules of each carpel, ordinarily ripening into 4 separating nutlets or into a 4-stoned drupe **3. Verbenaceae**

      4. Ovary pseudomonomerous, with a single ovule and locule **4. Phrymaceae**

   3. Style mostly gynobasic, the ovary more or less deeply 4-cleft and ripening into 4 nutlets; plants mostly aromatic **5. Labiatae**

## SELECTED REFERENCES

Erdtman, G., Pollen morphology and plant taxonomy. IV. Labiatae, Verbenaceae and Avicenniaceae. Svensk. Bot. Tidsk. 39: 279–285. 1945.

Hillson, C. J., Comparative studies of floral morphology of the Labiatae. Am. Jour. Bot. 46: 451–459. 1959.

Johri, B. M., and I. K. Vasil, The embryology of *Ehretia laevis* Roxb. Phytomorph. 6: 134–143. 1956.

Junell, S., Zur Gynäceummmorphologie und Systematik der Verbenaceen und Labiaten. Symb. Bot. Upsal. 1 (4): 1–219. 1934.

————, Die Samenentwicklung bei einigen Labiaten. Svensk. Bot. Tidsk. 31: 67–110. 1937.

Lawrence, J. R., A correlation of the taxonomy and the floral anatomy of certain of the Boraginaceae. Am. Jour. Bot. 24: 433–444. 1937.

Maheshwari, P., *An Introduction to the Embryology of Angiosperms*. (New York: McGraw-Hill, 1950), pp. 366–368.

Rao, V. S., The floral anatomy of some Verbenaceae with special reference to the gynoecium. Jour. Indian Bot. Soc. 31: 297–315. 1952.

Risch, C., Die Pollenkörner der Labiaten. Willdenowia 1: 617–641. 1956.

ORDER 4. **Plantaginales**

The order Plantaginales consists of the single family Plantaginaceae, with about 250 species. The Plantaginales are Asteridae with small, chiefly wind-pollinated flowers that have a persistent, scarious, regular corolla. The flowers are hypogynous and typically have four calyx-lobes, four corolla-lobes, four (seldom fewer) stamens, and a bicarpellate (seldom pseudomonomerous) ovary. The plants are herbs or seldom half-shrubs with mostly basal, alternate leaves and without internal phloem. A few species have opposite, cauline leaves. Opinions differ as to whether opposite or alternate leaves are primitive within the family. The floral structure of some species suggests the possibility of a pentamerous ancestry. The endosperm is cellular and has both micropylar and chalazal haustoria.

It is clear enough that the Plantaginales belong to the Asteridae, but their precise relationships within the subclass are less certain. Deprived of their special features of anemophily and persitent, scarious corolla, the Plantaginaceae would fit reasonably well into the order Polemoniales, but no one family presents itself as a near ally. The presence of both micropylar and chalazal haustoria is more in accord with the Scrophulariales than the Polemoniales, but here again no one family is clearly ancestral. Derivation of the Plantaginales from the Scrophulariales would imply the same sort of reduction from an irregular, five-lobed corolla with four stamens to a regular, four-lobed corolla with four stamens that appears to have taken place in the Buddlejaceae. We have noted that the floral anatomy of some species lends some support to this hypothesis, but in the absence of known intermediate stages it is still largely speculative.

The anemophilous habit of the Plantaginales must be of considerable importance to the plants, but the reasons for the change are obscure. The Plantaginales grow in the same sort of places as ordinary entomophilous plants, and there is nothing in their structure, aside from the reduced corolla itself, that does not appear to be equally compatible with entomophily.

## SELECTED REFERENCES

Cooper, G. O., Development of the ovule and formation of the seed in *Plantago lanceolata*. Am. Jour. Bot. 29: 577–581. 1942.

Pilger, R., Plantaginaceae. Heft 102, Das Pflanzenreich. 1937.

ORDER 5. **Scrophulariales**

The order Scrophulariales, as here defined, consists of 12 families and nearly 10,000 species. Nearly three-fourths of the species belong to only three families, the Scrophulariaceae (2700), Acanthaceae (2600), and

Gesneriaceae (1800). The Bignoniaceae (800), Oleaceae (600), Globulariaceae (300), and Lentibulariaceae (300) are families of moderate size. The Myoporaceae (180), Buddlejaceae (160), and Orobanchaceae (150) are rather small families, and the Pedaliaceae (including Martyniaceae) and Hydrostachyaceae trail with only about 55 and 30 species, respectively.

The Scrophulariales are Asteridae with a usually superior ovary and generally either with an irregular corolla, or with fewer stamens than corolla lobes, or commonly both. They lack the specialized gynoecium of the Lamiales, and only a few of them have internal phloem. They uniformly lack stipules.

The Hydrostachyaceae, Buddlejaceae, and Oleaceae have not usually been referred to this alliance, but the other nine families are generally agreed to be related. The Hydrostachyaceae have usually been associated with the Podostemaceae, but recent careful studies especially of the embryological features have emphasized the differences between these two families and the similarity of the Hydrostachyaceae to the families here grouped in the Scrophulariales. The Hydrostachyaceae appear to be an aquatic, apetalous offshoot of the Scrophulariales, just as the Callitrichaceae are an aquatic, apetalous offshoot of the Lamiales.

The Buddlejaceae and Oleaceae have usually been referred to the Gentianales, where they are out of place in several respects as noted in the discussion of that order. The same features which make them anomalous in the Gentianales are highly compatible with the Scrophulariales, although they would not necessarily require placement in this order. The relationship of the Buddlejaceae to the Scrophulariales is further suggested by two genera (*Sanango*, *Peltanthera*) of the family, which have five corolla lobes but only four functional stamens. In *Sanango* the fifth stamen is represented by a staminode, as in many Scrophulariaceae. The tetramerous flowers of other members of the Buddlejaceae may have a pentamerous ancestry, just as does *Veronica* in the Scrophulariaceae. The Oleaceae are here carried along with the Buddlejaceae, with some reservation and for lack of a better alternative; the two families do have much in common.

The Columelliaceae, which have usually been associated with the families here grouped in the Scrophulariales, are here referred to the order Rosales. The wood of the Columelliaceae is more primitive than that of the Scrophulariales, and suggests that of the Rosales instead (unpublished studies of W. L. Stern). The sympetalous corolla and unitegmic, tenuinucellate ovules of the Columelliaceae are unusual in the Rosales, but these characters also occur in the Pittosporaceae, which are widely admitted to be of Rosalean affinity.

The Scrophulariales are related to and derived from the Polemoniales, which are typically more primitive in having a regular corolla with isomerous stamens. The relationship between the Solanaceae (Polemoniales) and Scrophulariaceae has long been noted.

Except for being herbaceous or merely frutescent, the Scrophulariaceae have the characters from which those of the other families of the order could have been derived. Several families, notably the Orobanchaceae, Gesneriaceae, Globulariaceae, Lentibulariaceae, and Acanthaceae, appear to have been derived directly from the Scrophulariaceae.

The relationship of the Orobanchaceae to the Scrophulariaceae is particularly close, and indeed there is no clear line between these two generally recognized families. Many of the Scrophulariaceae are partly parasitic, and a few of them (e.g., *Striga*) are wholly so. The Bignoniaceae are also closely related to the Scrophulariaceae, and the genus *Paulownia* has been shifted back and forth between the two. Its arborescent habit and winged seeds are anomalous in the Scrophulariaceae, but its copious endosperm is anomalous in the Bignoniaceae.

The Acanthaceae are likewise linked to the Scrophulariaceae by transitional forms. *Elytraria* and some related genera have been transferred back and forth between the two families by different authors. *Elytraria* differs from most of the Acanthaceae in having well developed endosperm, and its funiculus, though enlarged, is not developed into the typical jaculator which expels the seeds in most genera of the Acanthaceae.

The Globulariaceae are here defined to include the Selaginaceae. The two groups together form a single series marked by the progressive reduction of the gynoecium, the extreme form being pseudomonomerous and uniovulate. The Selaginaceae have often been included in the Scrophulariaceae, where they find their nearest relatives in the tribe Manuleae. The Manuleae, Selaginaceae, and Globulariaceae sens. strict. all have the anthers monothecal at maturity, in contrast to the persistently dithecal anthers of most Scrophulariaceae.

The characters which distinguish the Scrophulariales from other orders probably reflect adaptation to advanced kinds of insect pollinators which can distinguish particular floral patterns. Specialization for particular pollinators often marks individual genera and even species in the various families. Some of the individual families are ecologically specialized. Thus the Hydrostachyaceae are aquatics with reduced flowers, the Lentibulariaceae are insectivorous aquatics, and the Orobanchaceae are parasites. The Acanthaceae typically have explosively dehiscent fruits with the enlarged and thickened funiculus of each seed functioning as a jaculator. The value of the jaculator in its fully developed form is clear enough, but the selective significance of its evolutionary precursors remains to be clarified. *Elytraria*, mentioned above, is a case in point: Evolutionary changes in the funiculus were well under way before the propulsive function was acquired. Some other characters helping to mark various families of the order are even more difficult to interpret in terms of adaptation. The function of the unique mucilage hairs of the Pedaliaceae is still wholly unknown, as is the significance of the different types of placentation in different families.

SYNOPTICAL ARRANGEMENT OF THE FAMILIES OF SCROPHULARIALES

1. Flowers perfect or less often unisexual, almost always with a calyx and usually also with a corolla; stamens mostly 2–4 (5); habit and habitat various

   2. Placentation axile to parietal or apical; plants not insectivorous

      3. Corolla with mostly 4 (5) lobes and only slightly (or not at all) irregular, or sometimes wanting; woody plants (rarely herbs) with mostly opposite or whorled leaves

         4. Stamens 4, rarely 2 or 5; ovules mostly numerous in each locule       **1. Buddlejaceae**

         4. Stamens 2, rarely 4; ovules mostly 2 (seldom 4–many) in each locule       **2. Oleaceae**

      3. Corolla with mostly 5 (seldom 4) lobes, or 2-lipped with scarcely lobed lips; habit and leaves various

         5. Mature seeds usually with a well developed endosperm

            6. Placentation mostly axile or apical

               7. Ovules 2 or more in each locule; anthers (1) 2–locular

                  8. Fruit dry and mostly dehiscent, rarely fleshy; herbs, rarely shrubs; ovules mostly more or less numerous in each locule, seldom only 2; leaves opposite or alternate       **3. Scrophulariaceae**

                  8. Fruit drupaceous, indehiscent; woody plants, sometimes arborescent; ovules 2 (4–8) in each locule; leaves mostly alternate       **4. Myoporaceae**

               7. Ovules mostly solitary in each locule, sometimes only one locule developed; anthers unilocular at least

at maturity; herbs or shrubs, mostly with alternate leaves; fruit small, indehiscent, the carpels sometimes separating     **5. Globulariaceae**

6. Placentation mostly parietal

9. Plants chlorophyllous, not parasitic, with well developed, mostly opposite or whorled leaves; ovary often more or less inferior; embryo well developed     **6. Gesneriaceae**

9. Plants parasitic, without chlorophyll, the leaves reduced and alternate; ovary superior; embryo minute, scarcely differentiated     **7. Orobanchaceae**

5. Mature seeds mostly with little or no endosperm; leaves chiefly opposite or whorled

10. Trees or large shrubs, seldom herbs; leaves mostly compound, seeds usually winged     **8. Bignoniaceae**

10. Herbs, rarely shrubs or trees; leaves mostly simple; seeds not winged

11. Seeds with an enlarged and specialized funiculus; plants without specialized mucilage hairs; fruits fleshy to more often capsular, with axile placentation     **9. Acanthaceae**

11. Seeds with normal, unspecialized funiculus; plants with a unique type of mucilage hair; fruits capsular or drupaceous to more often nutlike, often with conspicuous appendages     **10. Pedaliaceae**

2. Placentation free-central; herbs, aquatic or of wet places, mostly insectivorous     **11. Lentibulariaceae**

1. Flowers unisexual, without perianth; stamen solitary; aquatic herbs     **12. Hydrostachyaceae**

SELECTED  REFERENCES

Boeshore, I., Morphological continuity of Scrophulariaceae and Orobanchaceae. Contr. Bot. Lab. Univ. Penn. 5: 139–177. 1920.

Bremekamp, C. E. B., Delimitation and subdivision of the Acanthaceae. Bull. Bot. Surv. India 7: 21–30. 1965.

Bunting, G., and J. A. Duke, *Sanango*: New Amazonian genus of Loganiaceae. Ann. Mo. Bot. Gard. 48: 269–274. 1961.

Campbell, D. H., The relationships of *Paulownia*. Bull. Torrey Club 57: 47–50. 1930.

Casper, S. J., "Systematisch massgebende" Merkemale für die Einordnung der Lentibulariaceen in das system. Österr. Bot. Zeits. 110: 108–131. 1963.

Crété, P., L'application de certaines données embryologiques à la systématique des Orobanchacées et de quelques familles voisins. Phytomorph. 5: 422–435. 1955.

Hartl, D., Die Bezeihungen zwischen den Plazenten der Lentibulariaceen und Scrophulariaceen nebst einem Exkurz über die Spezialisationsrichtungen der Plazentation. Beitr. Biol. Pfl. 32: 471–490. 1956.

Jäger-Zürn, I., Zur Frage der systematischen Stellung der Hydrostachyaceae auf Grund ihrer Embryologie, Blüten- und infloreszenzmorphologie. Österr. Bot. Zeits. 112: 621–639. 1965.

Johri, B. M., and H. Singh, The morphology, embryology and systematic position of *Elytraria acaulis* (Linn. F.) Lindau. Bot. Notis. 112: 227–251. 1959.

Junell, S., Ovarian morphology and taxonomical position of Selagineae. Svensk. Bot. Tidsk. 55: 168–192. 1961.

Khan, R., A contribution to the embryology of *Utricularia flexuosa* Vahl. Phytomorph. 4: 80–117. 1954.

Li, H. L., Trapellaceae, a familial segregate from the Asiatic flora. Jour. Wash. Acad. Sci. 44: 11–13. 1954.

Mohan Ram, H. Y., and M. Wahdi, Embryology and the delimitation of the Acanthaceae. Phytomorph. 15: 201–205. 1965.

Neubauer, H. F., Die Entwicklungsgeschichte des Bignoniazeenblattes und die Entwicklungslinien innerhalb dieser Familie. Ber. Deuts. Bot. Ges. 72. 299–307. 1959.

Raj, B., Pollen morphological studies in the Acanthaceae. Grana Palynologica 3: 3–108. 1961.

Straka, H., and H. D. Ihlenfeldt, Pollenmorphologie und Systematik der Pedaliaceae R. Br. Beitr. Biol. Pfl. 41: 175–207. 1965.

Tiagi, B., A contribution to the embryology of *Striga orobanchoides* Benth. and *Striga euphrasioides* Benth. Bull. Torrey Club 83: 154–170. 1956.

———, Studies in the family Orobanchaceae. IV. Embryology of *Boschniaka himalaica* Hook. and *B. tuberosa* (Hook.) Jepson, with remarks on the evolution of the family. Bot. Notis. 116: 81–93. 1963.

Westfall, J. J., Cytological and embryological evidence for the reclassification of *Paulownia*. Am. Jour. Bot. 36: 805. 1949.

Wilson, K., and C. E. Wood, Jr., The genera of Oleaceae in the southeastern United States. Jour. Arn. Arb. 40: 369–384. 1959.

## ORDER 6. Campanulales

The order Campanulales as here defined consists of six families and about 2500 species. About four-fifths of the species belong to the single family Campanulaceae (2000), which is here broadly interpreted to include the Lobeliaceae, and most of the remainder belong to the Goodeniaceae (320). The Stylidiaceae have about 140 species, the Pentaphragmataceae about 25, the Sphenocleaceae only one or two, and the Brunoniaceae only one. These last three families have only a single genus each.

The Campanulales are Asteridae with a typically inferior ovary, mostly alternate leaves, and with the filaments free from the corolla or attached at the base of the corolla tube. Nearly all of the species are herbaceous. Three of the families and more than nine-tenths of the species have more or less connivent or connate anthers which dehisce around the young style. The style, in turn, commonly has collecting hairs or some other kind of specialized appendage which catches the pollen and carries it upward (by growth) to a position where it is exposed to visiting insects. The stigmas in most species are concealed during the growth of the style, not becoming exposed and receptive to pollen until later. One small family (Brunoniaceae) of the Campanulales has uniovulate flowers in involucrate heads as in the Compositae, but it also has a superior ovary and is in other respects an unlikely choice for an ancestor to the Compositae. We shall see that in spite of the similarity in pollen presentation mechanism between the Campanulales and the Asterales, the latter order is probably more closely related to the Dipsacales and Rubiales than to the Campanulales.

There is some question whether all of the families here assigned to the Campanulales belong together, and every family appears to be sharply set off from all the others. All hands agree that the Brunoniaceae are closely allied to the Goodeniaceae, and indeed *Brunonia* has often been included in the Goodeniaceae, but even so there are some notable differences as indicated in the key. The stylar indusium of the Goodeniaceae and Brunoniaceae is apparently not homologous with the collecting hairs of the Campanulaceae, and these two groups possibly do not have a common ancestor short of the Polemoniales. It may be significant that some few members of the Campanulaceae, as well as the genus *Brunonia*, have the ovary essentially superior.

The Sphenocleaceae and Pentaphragmataceae have usually been included in the Campanulaceae, where they are aberrant in their undivided, hairless styles and in some other features. *Sphenoclea* is habitally reminiscent of *Phytolacca*, in the Caryophyllidae, but the embryological and other differences between these two genera are so overwhelming that the habital similarity must be purely concidental. *Sphenoclea* is embryologically and palynologically much like the Campanulaceae, and its relationship here seems reasonably certain. *Pentaphragma* is a little more doubtful. It has a unique type of pollen and relatively primitive wood. Embryologically it is much like the Campanulaceae, yet with some features of its own. For the present, at least, the genus is best held as a separate family within the order Campanulales.

The Stylidiaceae lack the specialized pollen-presentation mechanism of the other larger families of the order, and their relationship has sometimes been questioned on that account. In other respects they fit well enough into the Campanulales, however, and there is no other obvious place to put them. It may be significant that the pollen of most of the Stylidiaceae is very like that of some of the Campanulaceae.

The Calyceraceae, which are often referred to the Campanulales, are here referred with some hesitation to the Dipsacales.

The ancestry of the Campanulales is to be sought in or near the Polemoniales. If all the more primitive characters shown in various members of the Campanulales (regular corolla, isomerous stamens, superior, five-carpellate ovary, and no specialized mechanism of pollen presentation) were combined into one plant, only the number of carpels would be unusual in the Polemoniales. The present members of the Campanulales appear to have undergone more or less parallel changes from similar ancestors of Polemonialean affinity.

The specialized pollen presentation mechanism of the Campanulales would seem a priori to be of great importance to the plant, and differences in the nature of this mechanism furnish important taxonomic characters in the delimitation of families within the order. On the other hand, it is not easy to see how this complex mechanism is a real improvement over more ordinary arrangements. A study of the kinds of pollinators in the order as a whole might be instructive. The herbaceous habit of the order is of course significant ecologically, but it is also far from unique. The other characters which distinguish the Campanulales from the other orders of the Asteridae (alternate leaves, inferior ovary, stamens free from the corolla or attached at the base of the tube) are even more difficult to interpret in terms of survival values.

## SYNOPTICAL ARRANGEMENT OF THE FAMILIES OF CAMPANULALES

1. Style without an indusium, but often with collecting hairs

    2. Stamens as many as the corolla lobes, typically 5, free or connate only by their anthers; anthers introrse

3. Style glabrous, with a solitary stigma; plants apparently without a latex system

    4. Petals valvate; fruit a berry; leaves asymmetric; pollen binucleate         **1. Pentaphragmataceae**

    4. Petals imbricate; fruit a circumscissle capsule; leaves not notably asymmetric; pollen trinucleate
        **2. Sphenocleaceae**

3. Style with well developed collecting hairs below the usually separate stigmas; plants with a well developed latex system         **3. Campanulaceae**

2. Stamens 2 or 3, the filaments adnate to the style or connate into a tube around it; anthers extrorse; no latex system
        **4. Stylidiaceae**

1. Style with a more or less cupulate indusium just beneath the stigma(s), but without collecting hairs; no latex system

5. Flowers regular or nearly so, borne in involucrate heads; endosperm none; ovary superior, unilocular, with a single basal ovule         **5. Brunoniaceae**

5. Flowers mostly irregular, not borne in involucrate heads; endosperm well developed; ovary generally inferior, 1–2-locular, with 2 or more ovules         **6. Goodeniaceae**

## SELECTED REFERENCES

Airy-Shaw, H. K., Sphenocleaceae and Pentaphragmataceae, in Flora Malesiana 4: 27–20; 517–528. 1948; 1954.

Carolin, R. C., Floral structure and anatomy in the family Goodeniaceae Dumort. Proc. Linn. Soc. N. S. Wales 84: 242–255. 1959.

————, Floral structure and anatomy in the family Stylidiaceae. Ibid. 95: 189–196. 1960.

————, The structures involved in the presentation of pollen to visiting insects in the order Campanales. Ibid. 85: 197–207. 1960.

————, The concept of the inflorescence in the order Campanulales. Proc. Linn. Soc. N.S. Wales 92: 7–26. 1967.

Crété, R., Contribution a l'étude de l'albumen et de l'embryon chez les Campanulacées et les Lobeliacées. Bull. Soc. Bot. Fr. 103: 446–454. 1956.

Gupta, D. P., Vascular anatomy of the flower of *Sphenoclea zeylanica* Gaertn. and some other related species. Proc. Nat. Inst. Sci. India B25: 55–64. 1959.

Kapil, R. N., and M. R. Vijayaraghavan, Embryology of *Pentaphragma horsfieldii* (Miq.) Airy-Shaw with a discussion on the systematic position of the genus. Phytomorph. 15: 93–102. 1965.

Maheshwari, P., The embryology of angiosperms, a retrospect and prospect. Curr. Sci. (India) 25: 106–110. 1956. (For position of *Sphenoclea*)

Rosén, W., Further notes on the morphology of the Goodeniaceae. Acta. Hort. Gotob. 16: 235–249. 1946.

————, Endosperm development in Campanulaceae and closely related families. Bot. Notis. 1949: 137–147.

Subramanyan, K., A contribution to our knowledge of the systematic position of Sphenocleaceae. Proc. Indian Acad. Sci. B31: 60–65. 1950.

————, A morphological study of *Stylidium graminifolium*. Lloydia 14: 65–81. 1951.

ORDER 7. **Rubiales**

The order Rubiales as here defined consists of the single large family Rubiaceae, with about 6500 species. The Rubiales are Asteridae with inferior ovary, regular or nearly regular corolla, stamens as many as and alternate with the corolla lobes, and opposite leaves with interpetiolar stipules or whorled leaves without stipules. The endosperm is, so far as known, nuclear. The species that are familiar in north temperate regions are mostly herbs, but the vast majority of the species of the family are tropical, woody plants, many of them good-sized trees. The origin of whorled leaves in *Galium* by enlargement of interpetiolar stipules has been discussed in an earlier chapter.

The small genera *Henriquezia* and *Platycarpum*, which have recently been segregated as a family Henriqueziaceae, are here tentatively retained in the Rubiaceae. At this stage of knowledge I do not believe that the admitted peculiarities (no endosperm, no colleters, filaments bent at the base) of these genera require familial segregation, and I am not convinced by comparisons with families of the "Tubiflorae" as possible allies. It may be noted that *Gleasonia*, which Bremekamp keeps in the Rubiaceae, also lacks endosperm, and that there are certain other resemblances between *Gleasonia*, *Henriquezia*, and *Platycarpum* which have caused these genera to be associated in the past.

The Rubiaceae form a connecting link between the Gentianales and the Dipsacales, and would be an aberrant element in either order. Each of these orders is relatively homogeneous and well defined without the Rubiaceae. The characters of the groups have been discussed at some length by Wagenitz, who concludes that the Rubiaceae should be included in the Gentianales and that the resemblance of the Rubiaceae to the Caprifoliaceae (Dipsacales) is due to convergence rather than close relationship. I see no need to deny the one relationship in affirming the other. In my opinion the Loganiaceae (Gentianales) are near-ancestral to the Rubiaceae, which in turn are near-ancestral to the Caprifoliaceae. The other families of the Dipsacales appear to be derived from the Caprifoliaceae.

Characters which link the Rubiaceae to the Gentianales as a whole or specifically to the Loganiaceae are: (1) nuclear endosperm; (2) well developed stipules; (3) specialized glandular trichomes, called colleters, on the inner surface of the stipules; (4) the wall structure of the pollen; and (5) the frequent presence of complex alkaloids of the tryptophane family, which are known only in the Loganiaceae, Apocynaceae, and Rubiaceae. There are also two genera, *Gaertnera* and *Pagamea*, which are generally referred to the Rubiaceae but which have the ovary superior and have on that account sometimes been referred to the Loganiaceae instead. The pollen of *Gaertnera* and *Pagamea* is typically rubiaceous, and differs from that of the Loganiaceae (Walter Lewis, personal communication).

The Rubiaceae differ sharply from the Gentianales in having an inferior ovary (with the exceptions noted above) and in lacking internal phloem. These characters tend to ally the Rubiaceae with the Dipsacales. Within the Dipsacales, the Rubiaceae are most nearly like the Caprifoliaceae, which as we have noted are the most primitive family of the order. Botanists have repeatedly expressed doubt that the Caprifoliaceae should be maintained as a family separate from the Rubiaceae.

The Caprifoliaceae and other Dipsacales differ from the Rubiales, however, in having a cellular endosperm, in being almost entirely without alkaloids, and in lacking colleters. The Dipsacales are sometimes said to lack stipules, but this is an overstatement. The Caprifoliaceae often have small stipules, and occasionally the stipules are relatively well developed and conspicuous, as in *Viburnum ellipticum*. Such stipules as there are in the Caprifoliaceae are usually adnate to the petiole, however, instead of being interpetiolar as in the Rubiaceae. I am not convinced by efforts to interpret all stipules in the Caprifoliaceae as mere pseudostipules, of different nature than true stipules.

It should be noted that although the classical taxonomic characters have been fairly well observed, the more recondite characters such as the structure of the pollen, the arrangement of vascular tissue, the development of the endosperm, and the distribution of various chemical substances are still very inadequately known in the Rubiaceae and most other families. Examination of more species is likely to bring surprises. If the position of the ovary required specialized equipment and hours or days of time to determine, we might well be still unaware that the ovary in *Gaertnera* and

*Pagamea* is superior, instead of inferior as in other members of the family. According to how one weighs the evidence, the Rubiaceae could be referred to the Gentianales, or to the Dipsacales, or held as a separate, unifamilial order. If the family consisted of only a handful of species, it could easily be included in either order as a peripheral group, transitional to the other order. In actual fact, however, there are about as many species of Rubiaceae as of Gentianales and Dipsacales combined, and the Rubiaceae would dominate either order if they were included therein. It therefore seems most useful to maintain the Rubiaceae in an order by themselves.

## SELECTED REFERENCES

Bremekamp, C. E. B., On the position of *Platycarpum* Humb. et Bonpl., *Henriquezia* Spruce ex Bth. and *Gleasonia* Standl. Meddel. Bot. Mus. Herb. Rijksuniv. Utrecht 141: 361–377. 1957.

————, Remarks on the position, the delimitation and the subdivision of the Rubiaceae. Acta. Bot. Neerl. 15: 1–33. 1966.

Hegnauer, R., Chemotaxonomische Betrachtungen. V. Die systematische Bedeutung des Alkaloidmerkmales. Planta Medica 6: 1–34. 1958.

Verdcourt, B., Remarks on the classification of the Rubiaceae. Bull. Jard. Bot. Brux. 28: 209–290. 1958.

Wagenitz, G., Die systematische Stellung der Rubiaceae. Bot. Jahrb. 79: 17–35. 1959.

ORDER 8. **Dipsacales**

The order Dipsacales as here defined consists of five families and about 1100 species. The Caprifoliaceae (400), Valerianaceae (360), and Dipsacaceae (270) are all of comparable size, but the Calyceraceae (60) are a much smaller group, and the Adoxaceae consist of only a single species.

The Dipsacales are Asteridae with an inferior ovary, usually opposite leaves, and (except in the Calyceraceae) without the specialized pollen presentation mechanism found in the Campanulales and Asterales. They are mostly herbs or shrubs, seldom small trees. Except for some members of the Caprifoliaceae, they lack stipules. Many of them have an irregular corolla, or fewer stamens than corolla lobes, or both. The filaments are attached to the corolla tube, commonly well above the base. The pollen is reported to be trinucleate in all families except the Calyceraceae, which are binucleate. All of the families except the Calyceraceae are best developed in temperate or boreal regions, in contrast to the chiefly tropical Rubiales.

The Caprifoliaceae are clearly the most primitive family of the order, from which all the others (except perhaps the Calyceraceae) have been derived. Several genera of the Caprifoliaceae, belonging to the tribe Linaeae, are transitional toward the Valerianaceae, and some authors have submerged the Valerianaceae in the Caprifoliaceae. The Dipsacaceae have a highly specialized inflorescence, but their close relationship to the Valerianaceae has been obvious to all. The Adoxaceae are a specialized, herbaceous offshoot of the Caprifoliaceae marked by the lengthwise division of each stamen into two. Earlier suggestions that the Adoxaceae might be related to the Saxifragaceae instead of to the Caprifoliaceae do not withstand reexamination. It may be noteworthy that the pollen of *Adoxa* is much like that of *Sambucus*.

*Sambucus* and *Viburnum*, in the Caprifoliaceae, stand somewhat apart from the rest of the family and from each other, and have been the subject of much comment and phylogenetic speculation. *Sambucus* is the more distinctive of the two, but no other family presents itself as a likely close relative. Both the Hydrangeaceae and the Cornaceae have been suggested as possible near-ancestors of *Viburnum*, and the similarity of some species of *Hydrangea* to species of *Viburnum* in both vegetative and floral aspect is striking. More detailed comparisons reveal so many differences that any close relationship seems most unlikely. *Cornus* differs from *Viburnum* in its tetramerous flowers, separate petals, binucleate pollen, bilocular ovary with an ovule in each locule, two-celled endocarp, and other features. *Hydrangea* differs from *Viburnum* in having raphide sacs and in its separate petals, diplostemonous flowers, binucleate pollen, bilocular ovary, separate styles, capsular fruits with numerous seeds, and other features. There is no more reason to remove *Viburnum* from the Caprifoliaceae and associate it with *Hydrangea* or *Cornus* than to associate a part of *Euphorbia* with *Cereus* (Cactaceae).

Recent serological studies by Hillebrand would seem to indicate a close relationship of the Caprifoliaceae as a whole to the Cornaceae. *Kolkwitzia* (Caprifoliaceae) reacts strongly with some species of *Cornus, Lonicera,* and *Diervilla,* and somewhat less strongly with *Nyssa* and *Sambucus;* and *Sambucus* reacts well with some species of *Viburnum.*

Admitting as always the possibility of error, I am reluctant to place decisive weight on the serological evidence, especially inasmuch as we are still wholly unaware of what other seeming affinities might be uncovered by more extensive application of the same methods. How would the Caprifoliaceae react with the Compositae, or the Rubiaceae, or the Hydrangeaeceae, or the Grossulariaceae, or the Magnoliaceae? We do not yet know. Strong serological reactions between otherwise very different groups are frequent enough so that Moritz and his associates have sought to devise methods to deal with the problems created by these "antisystematic reactions." The wide differences in reactions by different species of the same genus also indicate a need for caution. Thus, on a scale of 100 for reaction with itself, *Cornus florida* gives an 84 percent reaction with *C. kousa*, 58 percent with *C. stolonifera*, and only 20 percent with *C. racemosa*.[9] I understand from Marion Johnson that a rabbit can be so strongly sensitized that its blood serum will react to almost anything. At the higher levels of sensitivity, which are required in order to show more or less distant relationships, it would seem that the possibility of an "antisystematic" reaction might be substantially increased. Serological information, like other taxonomic data, needs to be evaluated in the context of the rest of the information about the group.

If Hillebrand's results are to be taken at face value, the whole family Caprifoliaceae (not just *Viburnum*) must be associated with the Cornales and specifically with *Cornus*. Here we bump into the problem that the Asteridae appear to be a highly natural group, in which the Caprifoliaceae are not primitive. Even if we attempt to extricate the Caprifoliaceae or the Dipsacales as a whole from the rest of the Asteridae, we must still seek an ancestor with capsular fruits and numerous seeds. Such an ancestor would be more primitive than anything in the Cornales, and the two groups could not have a common ancestor short of the Rosales. Inasmuch as the Asteridae as a whole are here considered to be of Rosalean origin, such an interpretation of the significance of Hillebrand's studies would seem to be compatible with the phyletic concepts which have been established on other grounds. We would even without the serological evidence conclude that the nearest common ancestry of the Caprifoliaceae and Cornaceae would be in the Rosales. It may thus be possible to reconcile Hillebrand's results with phyletic concepts which have been established on other grounds.

Two small genera (*Carlemannia, Silvanthus*) which have sometimes been included in the Rubiaceae are here referred with some hesitation to the Caprifoliaceae. Their exstipulate, toothed leaves and anisomerous androecium would make them anomalous in the Rubiaceae but are more compatible with their inclusion in the Caprifoliaceae. I do not believe that the presently available evidence requires the separation of these two genera as a family Carlemanniaceae, as suggested by Airy-Shaw. The fact that their pollen is rather different from that of other Caprifoliaceae indicates the need for some caution, however. I am prepared to change my mind if future studies call for it.

The Calyceraceae have variously been referred to the Campanulales, or to the Dipsacales, or more recently (by Takhtajan, 1966) to a separate order of their own. Each of these possibilities has something to recom-

---

[9] D. E. Fairbrothers and M. A. Johnson, "Comparative Serological Studies Within the Families Cornaeceae (Dogwood) and Nyssaceae (Sour Gum)," in Ch. Leone, ed., *Taxonomic Biochemistry and Serology* (New York: The Ronald Press, 1964), pp. 305–318.

mend it, as well as certain difficulties. The pollen of the Calyceraceae is much like that of the Brunoniaceae and Goodeniaceae, and the anthers are connivent around the style and introrsely dehiscent as in typical members of the Campanulales. The mechanism of pollen presentation may not be so important, taxonomically, as it once seemed, however, especially inasmuch as Bremekamp has recently pointed out that the subfamily Ixoroideae in the Rubiaceae has a similar arrangement. The filaments of the Calyceraceae are attached well up in the corolla tube, as in the Rubiales and Dipsacales, instead of being free or attached at the base of the corolla tube as in the Campanulales. Furthermore, the Calyceraceae resemble the Rubiales and Dipsacales, and differ from the Campanulales, in certain embryological features. Poddubnaja-Arnoldi[10] comments that inclusion of the Dipsacaceae and Calyceraceae in the Rubiales would be in accord with the embryological data, and further that from an embryological point of view the families Calyceraceae, Dipsacaceae, and Compositae are closely related. The resemblance of the Calyceraceae to the Compositae in general floral morphology and inflorescence has often been remarked, but the relationship can be no more than collateral at best, inasmuch as the ovule is apical in the Calyceraceae and basal in the Compositae. The binucleate pollen, alternate leaves, and centripetal (racemose) inflorescence of the Calyceraceae are all out of harmony with the rest of the Dipsacales. One might sharpen the characters and definition of both the Campanulales and the Dipsacales by holding the Calyceraceae as a distinct order, but I am reluctant to accord ordinal status to such a small and undistinguished group. Believing that the relationship of the Calyceraceae is with the Dipsacales and eventually the Rubiales, rather than with the Campanulales, I prefer to include this family in the Dipsacales as an admittedly somewhat aberrant group.

The adaptive significance of most of the characters that distinguish the Dipsacales from the other orders of Asteridae is doubtful. Within the order, there is a progression from woody to herbaceous habit, from regular to irregular corolla, from isomerous to anisomerous stamens, from several fertile locules to only one, from numerous seeds to only one, and from endospermous to nonendospermous seeds. Each of these changes has also taken place in other orders of angiosperms, and there is nothing ecologically distinctive about the order or any of its families (save for the specialized pollen presentation mechanism of the Calyceraceae). The presence of an involucel surrounding the individual flowers of the Dipsacaceae suggests that the heads in this family may actually be second-order heads composed

---

[10] P. 376 in *The General Embryology of Angiosperms* (Moscow: 1964). (In Russian)

of primary heads that have been reduced to a single flower each, but an interpretation in terms of survival value still eludes us.

## SYNOPTICAL ARRANGEMENT OF THE FAMILIES OF DIPSACALES

1. Filaments and anthers distinct; leaves nearly always opposite; pollen trinucleate; inflorescence various

   2. Plants mostly woody, seldom herbaceous; stamens mostly as many as the corolla lobes, typically 5, seldom only 2; ovules often more than 1 per locule; endosperm well developed; fruits diverse, often with several or many ovules per locule
   **1. Caprifoliaceae**

   2. Plants herbaceous or nearly so; stamens seldom (except some Dipsacaceae) of the same number as the corolla lobes; ovules pendulous, not more than 1 per locule

      3. Stamens 8–10, twice as many as the corolla lobes, paired in the sinuses of the corolla, each with only a single pollen sac; fruit a dry drupe with several stones; endosperm well developed; flowers in compact heads
      **2. Adoxaceae**

      3. Stamens 1–4 (rarely 5), as many as or usually fewer than the corolla lobes, each with 2 pollen sacs; fruit dry, 1-seeded; endosperm scanty or none

         4. Ovaries not individually enclosed or subtended by a gamophyllous involucel; flowers not in compact, involucrate heads; ovary mostly trilocular, with 1 fertile and 2 sterile locules; corolla mostly 5-lobed
         **3. Valerianaceae**

         4. Ovaries individually enclosed or subtended by a gamophyllous involucel; flowers mostly borne in compact, involucrate heads; ovary strictly unilocular; corolla mostly 4–5-lobed
         **4. Dipsacaceae**

1. Filaments and anthers connate into a tube around the style; leaves alternate; pollen binucleate; flowers in involucrate, centripetally flowering heads, with regular corolla **5. Calyceraceae**

## SELECTED REFERENCES

Airy-Shaw, H. K., On a new species of the genus *Silvianthus* Hook. f., and on the family Carlemanniaceae. Kew Bull. 19: 507–512. 1965.

Bremekamp, C. E. B., On the position of the genera *Carlemannia* Benth. and *Sylviantus* Hook. f. Rec. Trav. Bot. Neerl. 36: 372. 1939.

————, Remarks on the position, the delimitation and the subdivision of the Rubiaceae. Acta. Bot. Neerl. 15: 1–33. 1966.

Ferguson, I. K., The genera of Valerianaceae and Dipsaceae in the southeastern United States. Jour. Arn. Arb. 46: 218–231. 1965.

Hillebrand, G. R., Phytoserological systematic investigation of the genus *Viburnum*. Ph.D. thesis. Rutgers University. 1966.

Moissl, E., Vergleichende embryologische Studien über die Familie der Caprifoliaceae. Österr. Bot. Zeits. 90: 153–212. 1941.

Moritz, O., "Some Special Features of Serobotanical Work," in Ch. Leone, ed., *Taxonomic Biochemistry and Serology* (New York: The Ronald Press, 1964), pp. 275–290.

Sprague, T. A., The morphology and taxonomic position of the Adoxaceae. Jour. Linn. Soc. 47: 471–487. 1927.

Troll, W., and F. Weberling, Die Infloreszenzen der Caprifoliaceae und ihre systematische Bedeutung. Abh. Akad. Wiss. Lit. Mainz, math.-nat. Kl. 1966; 459–605.

Weberling, F., Morphologische Untersuchungen zur Systematik der Caprifolaceen. Abh. Akad. Wiss. Lit. Mainz, math.-nat. Kl. 1957: No. 1.

————, Die Infloreszenzen der Valerianaceen und ihre systematische Bedeutung. Ibid. 1961: 151–281.

Wilkinson, A. M., Floral anatomy and morphology of *Triosteum* and of the Caprifoliaceae in general. Am. Jour. Bot. 36: 481–489. 1949.

### ORDER 9. **Asterales**

The order Asterales consists of the single family Compositae (Asteraceae). One of the largest families — perhaps the largest family — of flowering plants, the Compositae have been estimated to contain about 19,000 species.

The Asterales are Asteridae with an inferior ovary, a single basal ovule, and introrsely dehiscent anthers that are more or less connate (or at least connivent) around the style, which grows up through the anther-tube and pushes the pollen out into the open. The flowers are borne in involucrate, centripetally flowering heads, but the sequence of flowering among the heads is usually cymose. The pollen is trinucleate, and the seeds lack endosperm. The calyx, when present at all, is modified into a set of usually scale-like or bristle-like or hair-like structures which are collectively called the pappus.

The pollen presentation mechanism of the Compositae is similar to and seemingly homologous with that of the Campanulales. For this and other reasons the Compositae have often been included in that order. Before the evolutionary trends in the family were understood, and before Breme-kamp pointed out the existence of a similar pollen presentation mechanism in some of the Rubiaceae, that seemed to be a reasonable disposition of the group. More recent information, however, makes it seem highly improbable.

Comparative morphological studies of genera and species within the Compositae indicate clearly that the ancestral prototype must have been a woody-plant — a large shrub or perhaps a small tree — with opposite leaves. At one time I thought that the family was primitively herbaceous and that the woody types were derived from herbaceous ancestors within the family. Further study convinces me that although some of the woody Compositae are drived from herbs, others — notably some of the woody Heliantheae — are primitively woody. On the basis of extensive anatomical studies, Carlquist has also concluded that the family is primitively woody. Inasmuch as the Campanulales are herbaceous or nearly so and have chiefly alternate leaves, the likelihood that they are ancestral to the Com-positae is minimal. Only the Rubiales-Dipsacales complex has the charac-ters necessary for a near ancestor of the Compositae.

Evidence from some other sources also points to the Rubiales-Dipsacales complex as more likely ancestors than the Campanulales. Phenolic com-pounds are widespread in the Compositae and are also present in many Rubiales and Dipsacales, whereas they are unknown in the Campanulales. We have noted on page 308 the embryological similarity of the Calycera-ceae, Dipsacaceae, and Compositae. For the other side of the coin, Crété has noted that "Les Campanales d'une part, les Asterales, d'autre part, représentent plutôt deux séries évolutives, parfaitement individualisées, dont les ancêtres communs nous demeurent encore inconnus."

If my interpretation of the evolutionary history of the inflorescence in the group is correct, the ancestors of the Compositae had a cymose in-florescence, like that of the Rubiales and most of the Dipsacales. The primitive head in the ancestors of the modern Compositae was probably dichasially cymose, in keeping with the opposite arrangement of the leaves. Once the inflorescence had been condensed into a bracteate, cymose head, the head was converted into a racemose head with spirally arranged bracts and flowers. The change in arrangement of the leaves (bracts), once estab-lished in the head, progressed downward, phyletically, on the stem, so that more and more — eventually all — of the leaves were alternate instead of opposite. Many members of the primitive tribe Heliantheae, including *Helianthus* itself, now have the lower leaves opposite and the upper ones alternate. Even though the sequence of flowering within the heads changed from racemose to cymose, the cymose sequence *among* the heads was retained in most species; i.e., the terminal head blooms first, followed by the terminal heads of the major branches, and the sequence progresses downwards on each branch. In some members of the family the sequence of flowering among the heads has become mixed instead of strictly determinate, and in a very few — e.g., *Eupatorium capillifolium* — the arrangement of heads has become strictly racemose, without taint of determinate sequence. The evolutionary force which could drive such a change in inflorescence and leaf arrangement is wholly unknown, but no more so than the forces governing recognized changes in inflorescence and leaf arrangement in other groups of angiosperms.

*Cephalaria*, in the Dipsacaceae, seems to present an early stage in a similar change from a cymose to a racemose head. In *Cephalaria* the terminal (central) flower of the head blooms first, followed by a ring of the outermost flowers. Thereafter the sequence appears on external inspection to be mixed and confused. If the terminal flower were to abort, so that the outermost flowers were the first to bloom, *Cephalaria* would be well on the way toward a racemose sequence of flowering within the head. The Dipsacaceae are too advanced in their own way to be ancestral to the Compositae, but they seem to have undergone some of the same changes which led to the origin of the Compositae from common ancestors in the Rubiales-Dipsacales complex.

The heads of the Compositae are pseudanthia which often closely simulate a large individual flower. The nature of the evolutionary advantage of such a structure over a large individual flower with separate pistils, as in *Rosa* or *Ranunculus*, is uncertain, but the actual success of the group is unmistakable. Comparing more nearly related groups, one can see that the Compositae head is usually more conspicuous and probably less dependent on particular kinds of pollinators than the individual flowers of members of the Rubiales and Dipsacales. The pollen is exposed to all comers, and no unusual equipment is needed to get at the nectar. The heads of any one species are likely to be visited by a wide range of pollinators, and I well remember the value of *Chrysothamnus* as a source of insects for my collection in high school biology. It is not hard to believe that the condensation of a more open inflorescence into a large and showy pseudanthium had real survival value, in coordination with a change in pollinators.

The relationship between this presumably adaptive change and the generally recognized tendency *within* the family toward aggregation and reduction in the inflorescence is more speculative, and the adaptive significance of composite composites, such as *Echinops* and *Lagascea*, still defies analysis. These genera have one-flowered, separately involucrate heads aggregated into a secondary head with a common involucre. In contemplating plants such as these it is hard to avoid the thought that the nature of the supply of mutations may have more evolutionary importance than is generally realized.

Stebbins has rightly pointed out that the number of flowers in a Compositae head may increase (phyletically) as well as decrease. Increases usually occur in phylads with enough flowers in each head so that actual numbers are rarely counted and even more rarely given in formal descriptions. As we have pointed out in an earlier chapter, the same thing is true of the number of stamens or other parts of a flower: When the number is indefinite and fairly large, it may go up or down. Once the number has been stabilized and the parts (or whole flowers) set in a definite pattern, however, a further evolutionary change leads much more often to decrease than to increase. To borrow an analogy which Stebbins has used in another context, evolutionary increase in parts of relatively small and definite number follows a goatpath, whereas reduction follows a paved highway.

## SELECTED REFERENCES

Bate-Smith, E. C., The phenolic constituents of plants and their taxonomic significance. Jour. Linn. Soc. 58: 95–173. 1962.

Bremekamp, C. E. B., Remarks on the position, the delimitation and the subdivision of the Rubiaceae. Acta Bot. Neerl. 15: 1–33. 1966.

Carlquist, S., Wood anatomy of Compositae: A summary, with comments on factors controlling wood evolution. Aliso 6 (2): 25–44. 1966.

Cronquist, A., Phylogeny and taxonomy of the Compositae. Am. Midl. Nat. 53: 478–511. 1955.

Crété, R., Contribution a l'étude de l'albumen et de l'embryon chez les Campanulacées et les Lobéliacees. Bull. Soc. Bot. Fr. 103: 446–454. 1956.

Philipson, W. R., The relationships of the Compositae particularly as illustrated by the morphology of the inflorescence in the Rubiales and Campanulatae. Phytomorph. 3: 391–404. 1953.

Stebbins, G. L., "Adaptive Radiation and Trends of Evolution in Higher Plants," in Th. Dobzhansky, M. K. Hecht, and W. C. Steere, eds., Evolutionary Biology, Vol. I (New York: Appleton-Century-Crofts, 1967), pp. 101–142.

Zohary, M., Evolutionary trends in the fruiting head of Compositae. Evolution 4: 103–109. 1950.

## Chapter 5

Moccasin flower, Cypripedium acaule, a member of the orchid family. The Orchidaceae are one of the largest families of flowering plants, notable for diverse and often bizarre adaptations to specific pollinators, but they are not among the most successful families in terms of biomass or number of individuals.

# The Subclasses, Orders, and Families of Monocotyledons

## Origin and Characteristics of the Monocots

The class Liliatae (monocots) consists of four subclasses, 18 orders, 61 families, and roughly 55,000 species. As we have noted on p. 128, it is now generally agreed that the Liliatae as a whole are derived from primitive Magnoliatae (dicots). The differences between the two classes are summarized on p. 128.

The dicots which gave rise to the monocots must have had apocarpous flowers with a fairly ordinary (not highly specialized) perianth, and with uniaperturate pollen. They must also have been herbs without a very active cambium, and they presumably had laminar placentation. The only order of dicots which fits these specifications is the Nymphaeales. It is not here suggested that the Nymphaeales are directly ancestral to the monocots as a whole, but rather than the premonocotyledonous dicots were probably something like the modern Nymphaeales. It is noteworthy that the Nymphaeales are aquatic, that they lack vessels, and that they show tendencies toward the fusion of two cotyledons into one. The vesselless condition of the Nymphaeales is here regarded as probably secondary and consequent on the aquatic habitat, as explained on p. 149.

The monocots are here considered to be of aquatic ancestry, and the typical parallel-veined leaf is considered to be a modified, bladeless petiole. This morphological interpretation of the monocot leaf was proposed more than a century ago by deCandolle[1] and was further elaborated in evolutionary terms by Arber[2] in 1925. It is the only hypothesis known to me

[1] A. P. deCandolle, Organographie Vegetale, 2 vols. Paris, 1827.
[2] Agnes Arber, Monocotyledons. Cambridge: Cambridge University Press, 1925.

which permits all the information about monocots to fall into place and make sense.

Several species of *Sagittaria* show all transitions, within a single local population, between a normal leaf with blade and petiole, and a typical monocotyledonous leaf with parallel veins. The transitional series, which was reported by deCandolle and which I have observed in another species in Minnesota, is conspicuous and impressive. Depending mainly on the depth of the water, the blade is well developed or progressively reduced and lost, with a concomitant change in the structure of the petiole. Here the parallel-veined, submerged leaf is clearly nothing but a modified, blade-less petiole. The blade of *Sagittaria* is itself probably only a secondarily expanded petiole-tip, not strictly homologous with the blade of dicots, but the present structure is nonetheless that of a petiolate leaf with a well defined, palmately veined blade. It is perfectly clear that in *Sagittaria* plants with the genetic potentiality to produce normal leaves with blade and petiole can be induced to produce instead typical monocotyledonous leaves which are really flattened petioles. Genetic (and eventual evolutionary) fixation of a character which first appeared as a direct response to the environment is amply provided for in modern evolutionary theory.

Whether the interpretation that fits *Sagittaria* can be extended to the whole class Liliatae is of course another question. I believe that it can and should be so extended. Under this concept terrestrial monocots with a well defined, net-veined blade are considered to be derived from ancestors with narrow, parallel-veined leaves without a well defined blade, and indeed all transitional stages can be seen in several families. An attempt to read the system the other way means that we must start with broad, more or less net-veined leaves in diverse groups of monocots having nothing to do with each other, and have these all converge in both floral and vegetative characters into a hopelessly polyphyletic core of typical monocots. The resulting system, if it could be called that, would be shot full of internal contradictions.

There are two principal ways by which the typical monocot leaf can become broad and more or less net-veined. One way is to spread the main veins farther apart near the middle of the blade and amplify the cross-connections among them. Subsequently the main veins can fade out before reaching the leaf tip, so that a more or less palmate venation is established. *Alisma, Sagittaria, Dioscorea, Smilax,* and *Trillium* exemplify this type of change. The other way is for each of many closely set parallel veins to diverge in turn toward the margin, the outermost ones first, those nearest the midrib last. The result is a pinnately veined leaf with numerous, closely parallel primary lateral veins. The Zingiberales reflect this sort of change. The palms have their own special variation of this second type, involving the intercalation of new tissue between the veins during the early growth of the leaf.

An interesting difference between monocots and dicots is that whereas in dicots the vessels appear first (phyletically) in the secondary wood of the stem and spread thence to other tissues and organs, in monocots they appear first in the roots. This fact has led Cheadle[3] and others to the conclusion that vessels originated independently in the two classes. They therefore consider that the evolutionary divergence of the two classes preceded the origin of vessels.

On the same facts, plus some others, I come to a different conclusion. I suggest that in the ancestral premonocots, as in their probable relatives the Nymphaeales, vessels were phyletically lost in association with the aquatic habitat. Loss of the cambium eliminated at one stroke all vessels which had not worked their way (phyletically) into the primary tissues. The same factors which operated to produce a primitive, vesselless xylem structure in the Nymphaeales (q.v.) operated in the monocots to leave vestigial, seemingly primitive vessels in the roots. It was only after returning to a land habitat that some of the monocots developed an effective vessel system in the shoot. In this as in other features, the aquatic ancestry is the key to understanding the monocotyledons.

If the interpretation here presented is correct, the aquatic ancestry of the Liliatae has had a profound influence on the subsequent evolutionary history of the group. Deprived of cambium, deprived of a normal leaf blade, and virtually deprived of vessels (except for some vestiges in the roots), the early monocots might seem to have faced a bleak evolutionary future. Nevertheless, they managed to return to terrestrial habitats, and some of them evolved one or more of these features: woody habit, a vessel system, a normal leaf blade, and a sort of secondary thickening. These changes, however, were not easy; it is always easier to lose something than to regain it. Evolution of woody monocots from their herbaceous ancestors might better be compared with the rare reversion of dicotyledonous herbs to woody plants than to the common and easy origin of herbs from woody ancestors. Even among those monocots which have evolved a broad, more or less net-veined leaf blade, traces of the ancestral parallel-veined pattern usually persist. It should not be surprising that habital differences often turn out to be taxonomically more important among monocots than among dicots.

Although the monocots are here considered to have an aquatic ancestry, the situation is not simple. It appears that terrestrial monocots, derived

[3] V. I. Cheadle, Independent origin of vessels in the monocotyledons and dicotyledons. Phytomorph. 5: 399–411. 1953.

from aquatic premonocots, have themselves repeatedly given rise to groups which have returned to the water. Among the modern Alismatidae there appears to be a progressive adaptation to an aquatic and eventually even marine habitat. In the subclass Arecidae, the terrestrial family Araceae has some secondarily aquatic forms which in turn gave rise to the thalloid, aquatic family Lemnaceae. In a third subclass, the Commelinidae, such aquatic families as the Mayacaceae, Sparganiaceae, and Typhaceae appear to be derived from terrestrial ancestors within the group. The aquatic habitat of the Pontederiaceae, in the subclass Liliidae, may likewise prove to be secondary.

The adventitious, fibrous root system of monocots is a consequence of the absence of cambium. Having no adequate means of secondary thickening, the individual roots cannot persist, enlarge, and ramify. The roots of some monocots do manage to penetrate deeply into the soil, but the largest single family, the Orchidaceae, is shallow-rooted and mycorhizal, and the next largest family, the Gramineae, tends to exploit mainly the upper part of the soil, often forming a dense turf. Another large family, the Liliaceae, often has contractile roots which pull the bulb progressively deeper into the soil as years go by. Creeping rhizomes, which may penetrate to any depth, serve as rootstocks for many of the Gramineae, Cyperaceae, Liliaceae, and other monocots.

The nature of the single cotyledon in the monocots has occasioned much study and controversy, and is not yet fully resolved. Like the foliage leaves, the cotyledon typically has a basal sheath surmounted on one side by a limb which may or may not be divided into blade and petiole. Typically the sheath is closed and tubular, at least near the base. The vascular supply typically consists of two near-median bundles, as in the individual cotyledons of dicots. This implies that the monocot cotyledon is equivalent to a single leaf, and is not a double structure as has sometimes been supposed. The sheathing base of the cotyledon is thus left unexplained, except in that it is comparable to the sheathing base of a foliage leaf.

Drawing on the evidence from the living members of the Nymphaeales (dicots) I suggest another alternative: Two ancestral cotyledons have become connate by their margins toward the base, forming a bilobed, basally tubular, compound cotyledon. One of the lobes has subsequently been reduced and lost, and its vascular supply suppressed, so that the embryo has in effect a single cotyledon with a sheathing, tubular base. I further suggest that this modified cotyledonary structure has so firmly impressed itself on the growth pattern of the embryo that subsequent leaves are also built on the same plan. The sheathing base of monocot leaves is therefore a reflection of cotyledonary structure, rather than the reverse. I have no idea of the causes or survival value of these evolutionary changes.

Regardless of the morphological nature of the single cotyledon, it appears that throughout the Liliatae it is basically the same organ. Highly modified though the cotyledon may be in such plants as grasses, there is no reason or need to assume that the monocotyledonous condition of the Liliatae is of more than one origin. A more ample discussion of the nature and possible evolutionary history of the single cotyledon of the Liliatae can be found in the books by Arber and by Eames that have previously been cited.

The unique septal nectary of many Liliatae is an interesting and neglected feature which helps not only to unify the class but also to strengthen the concept that the Alismatidae are near-basal. This structure is apparently unknown in the Magnoliatae.

Septal nectaries

> occur in the septa between two carpels and represent places where the adjacent walls of the two carpels have not fused. They discharge nectar to the outside by means of small openings. They are such complicated structures that they would seem to indicate a relationship between all plants having them. . . . The openings of septal glands may be near the base of the ovary, but often they are at the top of the ovary or even in the style. Where the glands are at the top of the ovary, it would seem that they would not interfere with the fusion of the ovary and torus. This may be connected with the fact that the higher monocotyledons frequently have inferior ovaries.[4]

Septal nectaries are characteristic of those Palmae which are nectariferous, and of the Liliales, Bromeliales, and Zingiberales. Not every genus in every family of these orders has septal nectaries, but they are common enough so that their absence is exceptional rather than typical. (The Smilacaceae and the tribe Tulipeae of the Liliaceae may be mentioned among the more notable exceptions.) The complex, external nectaries of some of the Zingiberales are evidently derived from septal nectaries. It will be noted that septal nectaries occur in all three of the subclasses of Liliatae which are typically syncarpous.

Septal nectaries probably take their origin in the mostly apocarpous subclass Alismatidae. As Brown has pointed out, *Sagittaria* and other *Alismataceae* have nectaries between the petals and staminodes and between and around the staminodes and lower carpels. *Alisma* itself, with a single whorl of separate carpels, has a nectary at the base of the slit between any two adjacent carpels. The palms, which range from apocarpous to syncarpous, have correspondingly alismatoid to septal nectaries. Presumably a

[4] W. H. Brown, Proc. Am. Phil. Soc. 79: 551–553. 1938.

similar change has occurred in the line(s) leading to the Liliidae and Commelinidae.

## The Subclasses

Takhtajan's four subclasses of Liliatae are here adopted, but with a major change in the contents of the Liliidae and Commelinidae. The Bromeliales, Zingiberales, Juncales, and Cyperaceae, which refers to the Liliidae, are here referred to the Commelinidae instead. The Typhaceae and Sparganiaceae, which he includes in the Arecidae, are here referred to the Commelinidae as an order Typhales, and the Alismatidae are modified only by the addition of the small family Petrosaviaceae to the order Triuridales.

Each of the four subclasses of Liliatae tends to exploit a different ecological niche or set of niches, although with much overlapping. (This is in contrast to the Magnoliatae, in which only one of the subclasses has some obvious ecological coherence.) The Alismatidae are chiefly aquatic, whereas the other subclasses are chiefly terrestrial. The Arecidae typically have large, often petiolate leaves and usually are either arborescent or have the flowers crowded into a spadix. The Commelinidae have intensively

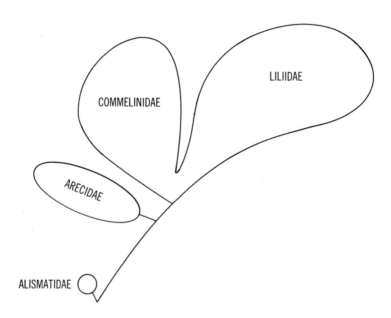

FIGURE 5.1

Probable relationships among the subclasses of Liliatae. The size of the balloon for each subclass is proportional to the number of species.

exploited the avenue of floral reduction and wind pollination. The Liliidae have intensively exploited insect pollination, and they also have bulbs, tubers, or corms much more commonly than do the other subclasses.

It may be of some interest to follow the partly divergent and partly parallel series of advances in the Commelinidae and Liliidae. The hypothetical common ancestor of these two groups may be assumed to have been a narrow-leaved herb with the vessels confined to the roots, with several subsidiary cells associated with each stomate, with the sepals and petals well differentiated from each other, with a superior ovary, septal nectaries, and a starchy endosperm. Such a plant would probably be referred to the Commelinidae if we had it, but no such plant exists today.

Following the Commelinid line, we find first an early divergence of the Zingiberales, leading to broad, veiny leaves and epigynous flowers, although there is no obvious reason why these two features should be linked. No other group appears to be derived from the Zingiberales, which are so distinctive that they might with some reason be treated as a separate subclass. Returning to the main Commelinid line, we next see the development of a vessel system in all vegetative organs. When this development was nearing completion the Bromeliales diverged into their specialized ecologic niche as xerophytes and then epiphytes, with an inferior ovary. The Bromeliales are a terminal group, which has given rise to nothing else. The main Commelinid line then loses its nectaries, and we reach the order Commelinales. Several lines of floral reduction then lead from the Commelinales to the terminal groups Eriocaulales, Restionales, Juncales, Cyperales, and Typhales. We do not know why the Commelinales lost the capacity to produce nectar, but the loss evidently set the stage for floral reduction associated with wind pollination. In at least some of the Commelinales (e.g., Xyris) insect visitors to the flowers collect pollen instead of nectar.[5] It is easy to suppose that the need to produce more pollen to offset the losses to marauding insects might also prepare the way for anemophily.

Picking up the Liliid line at its pre-Commelinid base, we find an immediate critical change in the perianth, so that both the sepals and the petals are petaloid. Concomitantly with this change, so far as present information shows, the number of subsidiary cells associated with each stomate was reduced to two. At this stage the Pontederiaceae diverged. It is still an open question whether the aquatic habitat of the Pontederiaceae is phyletically continuous with that of the primitive monocots, or whether this family has returned to the water after a terrestrial interlude. In any case the remainder of the Liliidae are basically land plants (or

---

[5] Robert Kral, Xyris (Xyridaceae) of the continental United States and Canada. Sida 2: 177–260. 1966. Also personal communication.

epiphytes). The next step in the main Liliid line is the substitution of other food reserves (especially cellulose or oils) for starch in the endosperm, followed or accompanied by the loss of subsidiary cells from about the stomates. We are now in the midst of the order Liliales. Three of the characters which have been attained by the Liliid line at this stage are nearly or quite without parallel in the Commelinidae: (1) petaloid sepals; (2) nonstarchy endosperm; (3) stomates without subsidiary cells. Some of the Liliaceae now develop vessels in the stem and leaves, usually in association with an expansion of the leaf surface, and some of these vesseliferous lines develop an inferior ovary. The larger group of typical Liliidae, however, moves separately into epigyny and thence into mycotrophy and the development of highly complex pollinating mechanisms, as in the Orchidaceae.

### SYNOPTICAL ARRANGEMENT OF THE SUBCLASSES OF LILIATAE

1.  Flowers mostly apocarpous, the syncarpous forms with a modified type of laminar placentation; plants mostly aquatic or subaquatic, always herbaceous; pollen consistently trinucleate; endosperm mostly wanting, not starchy when present; subsidiary cells mostly 2      **I. Alismatidae**

1.  Flowers mostly syncarpous, the apocarpous forms borne on terrestrial, woody plants; plants terrestrial or less often aquatic; pollen binucleate or trinucleate; endosperm present or absent, often starchy; subsidiary cells none to several

    2.  Seeds mostly with a starchy endosperm; flowers either with well differentiated sepals and petals (the sepals mostly not petaloid), or more or less reduced and with chaffy or bristly (or no) perianth, not aggregated into a spadix
         **II. Commelinidae**

    2.  Seeds mostly with a nonstarchy endosperm (notable exceptions: Araceae, Pontederiaceae, Philydraceae) or without endosperm; flowers usually either with the petals and sepals much alike (commonly both petaloid), or reduced, collectively subtended by a spathe, and aggregated into a spadix

        3.  Flowers usually numerous, small, subtended by a prominent spathe, and often aggregated into a spadix; plants often arborescent; leaves in most species broad, petiolate, and not with typical parallel venation; supporting cells typi-

cally 4, less often 3 or 2; ovary superior (sometimes sunken in the axis of the spadix)  **III. Arecidae**

3. Flowers few to numerous and small to large, but generally not subtended by a spathe and never aggregated into a spadix; plants seldom arborescent; leaves narrow and parallel-veined, or less often broader and net-veined; supporting cells typically none, less often 2, rarely more; ovary superior or more often inferior  **IV. Liliidae**

## SELECTED REFERENCES

Arber, Agnes, *Monocotyledons: A morphological study.* Cambridge: Cambridge University Press, 1925.

Brown, W. H., The bearing of nectaries on the phylogeny of flowering plants. Proc. Am. Phil. Soc. 79: 549–595. 1938.

Cheadle, V. I., The occurrence and types of vessels in the various organs of the plant in the monocotyledons. Am. Jour. Bot. 29: 441–450. 1942.

————, The origin and certain trends of specialization of the vessel in the Monocotyledoneae. Am. Jour. Bot. 30: 484–490. 1943.

————, Independent origin of vessels in the monocotyledons and dicotyledons. Phytomorph. 3: 23–44. 1953.

Holttum, R. E., Growth-habits of monocotyledons — variations on a theme. Phytomorph. 5: 399–411. 1955.

Lowe, J., The phylogeny of monocotyledons. New Phytol. 60: 355–387. 1961.

Stebbins, G. L., and G. S. Khush, Variation in the organization of the stomatal complex in the leaf epidermis of monocotyledons and its bearing on their phylogeny. Am. Jour. Bot. 48: 51–59. 1961.

## Subclass I. Alismatidae

The subclass Alismatidae as here defined consists of four orders, 14 families, and scarcely 500 species. The Najadales, with a little more than 200 species, are the largest order of the group. The Hydrocharitales have about a hundred species, and the Alismatales and Triuridales have about 80 species each. The Alismatidae have often been treated as a single order under the name Helobiae. The rank at which the group is received is perhaps less important than the fact that the group itself is widely recognized.

The Alismatidae are mostly aquatic or semiaquatic, except for the aberrant order Triuridales, and mostly apocarpous except for the Hydrocharitales. They all or nearly all have trinucleate pollen, and except for the Triuridales they all lack endosperm in the mature seed. Nearly all of them have stomates with two subsidiary cells, so far as present investigations show, but *Scheuchzeria*, with four subsidiary cells, is a notable exception. The development of the young endosperm usually follows the helobial pattern, which indeed takes its name from this group.

The Alismatidae are often considered to be the most primitive group of Liliatae. They can scarcely be on the main line of evolution of the class, however, because a primitive monocot should have binucleate pollen and endospermous seeds. They are here considered to be a near-basal side-branch, a relictual group which has retained a number of primitive characters. The apocarpous gynoecium of most members of the subclass, and the numerous stamens of some members, combined with the chiefly uni-aperturate pollen of the Liliatae as a whole, indicate that any connection of the Liliatae to the Magnoliatae must be to the primitive subclass Magnoliidae. Within the Magnoliidae the aquatic order Nymphaeales presents the closest approach to the Alismatidae. The concept that the Liliatae as a whole are of aquatic origin has already been discussed.

The trinucleate pollen of the Alismatidae may well relate to their aquatic habitat. As we have noted in Chapter 3, trinucleate pollen typically re-quires specialized conditions for germination, and a disproportionately high number of aquatic angiosperms have trinucleate pollen. However, judging by the presently extant members of the subclass, it appears that trinucleate pollen in the group may have antedated subsurface pollination. The trinucleate condition should probably therefore be regarded as pre-adapted to subsurface pollination, rather than as evolved in response to it.

The biological significance of the helobial pattern of endosperm de-velopment is wholly obscure, as is also the usual disappearance of the endosperm before maturity of the seed.


SYNOPTICAL ARRANGEMENT OF THE ORDERS OF ALISMATIDAE


1. Seeds without endosperm at maturity; plants mostly aquatic or semiaquatic, autotrophic

    2. Perianth differentiated into evident sepals and petals; flowers often bracteate

        3. Flowers hypogynous; carpels free or nearly so; embryo sac, except in Butomaceae, bisporic        **1. Alismatales**

        3. Flowers epigynous, with compound ovary and modified lami-nar placentation; embryo sac monosporic
        **2. Hydrocharitales**

2. Perianth, when present, not differentiated into sepals and petals; bracts usually wanting, except in Scheuchzeriaceae, but the perianth sometimes consisting of only a single, bractlike tepal; embryo sac monosporic      **3. Najadales**

1. Seeds with well developed endosperm; plants terrestrial, mycotrophic, without chlorophyll      **4. Triuridales**

ORDER 1. **Alismatales**

The order Alismatales as here defined consists of three families and less than a hundred species. Most of the species belong to the single family Alismataceae (70), and most of the remainder to the Limnocharitaceae (12). The family Butomaceae consists of a single species.

The Alismatales are Alismatidae with a well developed, biseriate perianth that is usually differentiated into three sepals and three petals, and with a gynoecium of several or many, more or less separate carpels. Each flower is usually subtended by a bract. The three families are amply distinct, although the Limnocharitaceae have often been submerged in the Butomaceae. The Butomaceae and Limnocharitaceae are interesting especially because of their laminar placentation, a presumably primitive type otherwise known (aside from the modified form in the Hydrocharitaceae) only in a few families of the subclass Magnoliidae, including those of the order Nymphaeales.

The adaptive significance of most of the characters which mark the families of the Alismatales is obscure. Placentation, the number of pores in a pollen grain, the number of pollen grains in a group, the shape of the embryo, the presence or absence of secretory canals, and the presence or absence of a ligule at the base of the leaf blade are all difficult to interpret in terms of survival value.

SYNOPTICAL ARRANGEMENT OF THE FAMILIES OF ALISMATALES

1. Plants without secretory canals; pollen uniaperturate; embryo sac monosporic; embryo straight; placentation laminar; leaves linear, not differentiated into blade and petiole

     **1. Butomaceae**

1. Plants with schizogenous secretory canals; pollen multiaperturate; embryo sac bisporic; embryo horseshoe-shaped; placentation and leaves various

   2. Ovules several or many, scattered over the inner surface of the carpel (i.e., the placentation laminar)    **2. Limnocharitaceae**

2. Ovules 1 or 2, seldom more, basal or on a marginal placenta
**3. Alismataceae**

## SELECTED REFERENCES

Dauman, E., Zur Morphologie der Blüte von *Alisma plantago-aquatica* L. Preslia 36: 226–239. 1964.

Kaul, R. B., Development and vasculature of the androecium in the Butomaceae. Am. Jour. Bot. 52: 624. 1965. (Abstract)

Markgraf, F., Blütenbau und Verwandtschaft bei den einfachsten Helobiae. Ber. Deuts. Bot. Ges. 54: 191–229. 1936.

Rao, Y. S., Karyosystematic studies in Helobiales. I. Butomaceae. Proc. Nat. Inst. Sci. India 19: 563–581. 1953.

Roper, P. B., The embryo sac of *Butomus umbellatus* L. Phytomorph. 2: 61–74. 1952.

Singh, V., Morphological and anatomical studies in the Helobiae. VIII. Vascular anatomy of the flower of *Butomus umbellatus* L. Proc. Indian Acad. Sci. B63: 313–320. 1966.

Stant, M. Y., Anatomy of the Alismataceae. Jour. Linn. Soc. 59: 1–42. 1964.

————, Anatomy of Butomaceae. Jour. Linn. Soc. 60: 31–60. 1967.

Troll, W., Beiträge zur Morphologie des Gynaeceums. II. Über das Gynaeceum von *Limnocharis* Humb. & Bonpl. Planta 17: 453–460. 1932.

ORDER 2. **Hydrocharitales**

The order Hydrocharitales consists of the single family Hydrocharitaceae, with about 100 species. The Hydrocharitaceae differ from all other Alismatidae in their inferior, compound ovary. The placentation is usually said to be parietal, with the placentae often more or less deeply intruded. It is in fact, however, basically laminar, with the ovules scattered over the walls of the individual carpels, which are weakly connate to form partial partitions (intruded placentae) in the ovary. Laminar placentation is otherwise known only in the Butomaceae, Limnocharitaceae, and a few families of the Magnoliidae. The Hydrocharitales are all aquatic, variously submerged or emergent. Pollination is by insects, wind, or water, according to the genus. The detached, free-floating male flowers of *Vallisneria* are famous as part of an ingeniously complex pollinating mechanism in this genus. The adaptive significance of epigyny in the Hydrocharitaceae is obscure.

The Hydrocharitales evidently constitute a special side-branch from the Alismatales. It is not difficult to imagine that they may have been derived from an ancestor similar to the Butomaceae.

## SELECTED REFERENCES

Govindappa, D. A., and T. R. B. Naidu, The embryo sac and endosperm of *Blyxa oryzetorum* Hook. f. Phytomorph. 35: 417–422. 1956.

Islam, A. S., A contribution to the life history of *Ottelia alismoides* Pers. Jour. Indian Bot. Soc. 29: 79–91. 1950.

Lakshmanan, K. K., Embryological studies in the Hydrocharitaceae. III. *Nechamandra alternifolia.* Phyton 20: 49–58. 1963.

Rangasamy, K., A morphological study of the flower of *Blyxa echinosperma* Hook. f. Jour. Indian Bot. Soc. 20: 123–133. 1941.

Troll, W., Beiträge zur Morphologie des Gynaeceums. I. Über das Gynaeceum der Hydrocharitaceen. Planta 14: 1–18. 1931.

ORDER 3. **Najadales**

The order Najadales (Potamogetonales) as here defined consists of eight families and a little over 200 species. The Potamogetonaceae, consisting of the single genus *Potamogeton*, are the largest family of the order, with about 90 species. The Aponogetonaceae have about 40 species, all in the genus *Aponogeton*; the Najadaceae have about 35 species, all in the genus *Najas*; the Zosteraceae have about 30 species in several genera, some of which have sometimes been parceled out into the separate families Cymodoceaceae and Posidoniaceae; the Juncaginaceae have about 18 species in four genera (including *Lilaea*, sometimes treated as a separate family); the Zannichelliaceae have about seven species in two genera, the Ruppiaceae have two species in the single genus *Ruppia*, and the Scheuchzeriaceae have only a single species, *Scheuchzeria palustris.*

The Najadales are Alismatidae in which the perianth, when present, is not differentiated into evident sepals and petals. Except in *Scheuchzeria* the flowers are not individually subtended by bracts, but the perianth sometimes consists of a single bractlike tepal. They all have ordinary, eight-nucleate, monosporic embryo sacs, so far as known.

The tepals of the Juncaginaceae and Potamogetonaceae have sometimes been interpreted as enlarged appendages of the connective of the anther, but I see no need for such an interpretation. The Juncaginaceae appear to be merely the next step in reduction beyond *Scheuchzeria*, which has extrorse anthers like those of the Juncaginaceae and Potamogetonaceae but is considered to have a normal perianth of six tepals. Making the anthers sessile and adnate to the tepals does not turn the tepals into appendages of the connective. The interpretation of the flowers of *Potamogeton* as pseudanthia has even less to recommend it.

*Scheuchzeria* forms an interesting connecting link between the Alismatales and Najadales. It resembles the Alismatales in having each flower subtended by a bract, but otherwise it appears to be a relatively primitive member of the Najadales. The Aponogetonaceae stand somewhat apart from the rest of the order and may perhaps represent a separate reduction from the Alismatales. The Najadaceae, although also isolated, are pretty clearly allied to the core families of the order. The Potamogetonaceae, Ruppiaceae, Zannichelliaceae, and Zosteraceae appear to be related inter se, and they have often been treated as a single family.

The evolutionary history of the Najadales is in large part a story of floral reduction associated with progressive adaptation to aquatic and eventually marine habitats. The Scheuchzeriaceae and Juncaginaceae are typically emergent plants of marshes. The Aponogetonaceae and Potamogetonaceae are fresh-water aquatics with submerged or floating leaves, but often with emergent inflorescences. The Najadaceae, Ruppiaceae, and Zannichelliaceae are submerged aquatics of fresh or brackish water, and the Zosteraceae are submerged marine plants.

The reduction of the perianth in the Najadales doubtless reflects abandonment of insect pollination, but the concomitant reduction in number of stamens, carpels, and ovules is more difficult to interpret in ecological terms. The threadlike pollen grains of the Zosteraceae may well be related to the submersed, marine habitat of the group. It is not hard to believe that in moving water a long thread has more chance of brushing against a stigma than does a little ball. It is interesting to note that two submersed marine genera of the Hydrocharitaceae have the pollen grains united into threadlike chains — apparently a different means of achieving a similar result.

## SYNOPTICAL ARRANGEMENT OF THE FAMILIES OF NAJADALES

1. Ovules 2 or more, basal, anatropous; stamens 6 or more, free, often more numerous than the tepals; fruits dehiscent, follicular; endosperm helobial

2. Tepals 0–3, sometimes petaloid; inflorescence a simple or basally forking spike, bractless, although the perianth sometimes consists of a single bractlike tepal; pollen uniaperturate, the grains borne singly; aquatic plants with submerged or floating leaves                 **1. Aponogetonaceae**

2. Tepals 6, never petaloid; inflorescence a raceme, each pedicel subtended by a bract; pollen nonaperturate, in dyads; emergent marsh-plants                 **2. Scheuchzeriaceae**

1. Ovule solitary; stamens either fewer than 6, or as many as and adnate to the tepals, or both; fruits indehiscent or seldom schizocarpic or irregularly dehiscent; tepals 0–6, never petaloid; pollen mostly nonaperturate

   3. Ovule basal, erect, anatropous; endosperm nuclear

      4. More or less emergent marsh-plants, scapose or subscapose, with chiefly or wholly basal leaves and a terminal raceme or spike; carpels and stamens usually more than 1
               **3. Juncaginaceae**

      4. Submerged aquatics with branching stems and opposite or whorled leaves, the flowers solitary in the axils; carpel 1; stamen 1
               **4. Najadaceae**

   3. Ovule apical or lateral, pendulous, more or less orthotropous

      5. Carpels usually 2 or more, separate; pollen not threadlike; endosperm chiefly helobial; plants not truly marine

         6. Flowers in spikes or racemes, perfect or less often unisexual; habit various

            7. Tepals 4; stamens 4; spikes axillary; fruiting carpels sessile; pollen ellipsoid or spheroid; plants of fresh water
               **5. Potamogetonaceae**

            7. Tepals 0; stamens 2; spikes terminal; fruiting carpels stipitate; pollen of a unique, bilateral type; plants of brackish water
               **6. Ruppiaceae**

         6. Flowers in axillary cymes, or solitary in the axils, unisexual; submerged plants of fresh or brackish water
               **7. Zannichelliaceae**

      5. Carpel 1; pollen threadlike; endosperm often or usually nuclear; submersed marine plants
               **8. Zosteraceae**

## SELECTED REFERENCES

Agrawal, J. S., The embryology of *Lilaea subulata* H.B.K. with a discussion on its systematic position. Phytomorph. 2: 15–29. 1952.

Daumann, E., Zur Frage nach dem Ursprung der Hydrogamie. Zugleich ein Beitrag zur Blütenökologie von *Potamogeton*. Preslia 35: 23–30. 1963.

Markgraf, F., Blütenbau und Verwandtschaft bei den einfachsten Helobiae. Ber. Deuts. Bot. Ges. 54: 191–229. 1936.

Reinecke, P., A contribution to the morphology of *Zannichellia aschersoniana* Graebn. Jour. So. Afr. Bot. 30: 93–101. 1964.

Sane, Y. K., A contribution to the embryology of the Aponogetonaceae. Jour. Indian Bot. Soc. 18: 79–91. 1939.

Sattler, R., Perianth development of *Potamogeton richardsonii*. Am. Jour. Bot. 52: 35–41. 1965.

Singh, V., Morphological and anatomical studies in Helobiae. II. Vascular anatomy of the flower of Potamogetonaceae. Bot. Gaz. 126: 137–144. 1965. III. Vascular anatomy of the node and flower of the Najadaceae. Proc. Indian Acad. Sci. B61: 98–108. 1965. IV. Vegetative and floral anatomy of Aponogetonaceae. Ibid., pp. 147–159. V. Vascular anatomy of the flower of *Lilaea scilloides* (Poir.) Hamm. Ibid., pp. 316–325.

Stenar, H., Embryologische Beobachtungen über *Scheuchzeria palustris* L. Bot. Notis. 1935: 78–86.

Swamy, B. G. L., and K. K. Lakshmanan, Contributions to the embryology of the Najadaceae. Jour. Indian Bot. Soc. 41: 247–267. 1962.

## ORDER 4. **Triuridales**

The order Triuridales consists of two families, the Triuridaceae, with about 70 species, and the Petrosaviaceae, with only about four species. The Petrosaviaceae have usually been submerged in the Liliaceae, where they are anomalous in their free or nearly free carpels and mycotrophic, nongreen habit. In both of these respects they resemble the Triuridaceae. It remains to be seen whether or not they have trinucleate pollen like the Triuridaceae and the rest of the Alismatidae.

Because of their apocarpous flowers and trinucleate pollen, the Triuridaceae are commonly associated with the Alismatidae. They differ from the other orders of Alismatides, however, in having well developed endosperm in the mature seeds, and in being terrestrial, mycotrophic plants without chlorophyll. The undifferentiated condition of the embryo is doubtless merely a reflection of mycotrophy. These two conditions are associated too often for mere happenstance. Any phyletic connection between the Triuridales and the other orders of the Alismatidae must presumably antedate the loss of endosperm by the other orders.

SYNOPTICAL ARRANGEMENT OF THE FAMILIES OF TRIURIDALES

1. Flowers perfect; carpels 3; seeds numerous      **1. Petrosaviaceae**

1. Flowers unisexual; carpels several; seeds 1 per carpel

**2. Triuridaceae**

SELECTED REFERENCE

Green, P. S., and O. Solbrig, *Sciaphila dolichostyla* (Triuridaceae). Jour. Arn. Arb. 47: 266–269. 1966.

## Subclass II. Commelinidae

The subclass Commelinidae as here defined consists of eight orders, 25 families, and nearly 19,000 species. The Cyperales, with about 12,000 species, are by far the largest order of the group. The Zingiberales, with about 2100 species, are the next largest order, followed by the Bromeliales (1700), Eriocaulales (1200), Commelinales (1000), Restionales (500), and Juncales (300). The Typhales, with only about 35 species, are the smallest order of the subclass.

The Commelinidae are syncarpous (or pseudomonomerous) monocots with a usually starchy endosperm and with the perianth either well differentiated into sepals and petals or more or less reduced and not petaloid. The stomates have two or more subsidiary cells, the pollen is either binucleate, or more often trinucleate, and the endosperm is nuclear or less often helobial. Except for the Zingiberales and some of the Bromeliales, they usually have vessels in all vegetative organs. The vast majority of the species are terrestrial herbs, although often of moist places. A smaller number are more or less aquatic, and only a few are arborescent. Although the more primitive (as well as some of the more advanced) members of the subclass have well developed petals and are pollinated by insects or birds, several lines within the group have taken the path of floral reduction

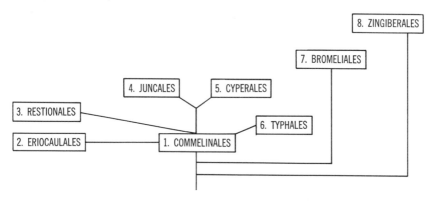

FIGURE 5.2

Probable relationships among the orders of Commelinidae.

and wind pollination. Floral reduction in the Commelinidae, unlike that in the Arecidae, has not led to the formation of a spadix.

Although the Commelinidae pretty clearly hang together as a natural taxonomic group, they are ecologically diverse. They have more fully exploited the opportunity for wind pollination associated with floral reduction than any of the other subclasses of Liliatae, but they also include highly specialized entomophilous and ornithophilous types. The arborescent habit and the bulbous habit, which have been exploited by the Arecidae and the Liliidae, respectively, are not much developed in the Commelinidae, but otherwise they run the gamut from ordinary mesophytes to hydrophytes, xerophytes, and epiphytes. Although the aquatic habitat is probably primitive for the Liliatae as a whole, it appears to be strictly secondary in the Commelinidae, which are basically terrestrial.

## SYNOPTICAL ARRANGEMENT OF THE ORDERS OF COMMELINIDAE

1.  Flowers without nectaries or nectar; ovary superior; vessels well developed in all vegetative organs

    2.  Flowers with more or less showy petals that are well differentiated from the sepals, nearly always perfect; pollination mostly by insects **1. Commelinales**

    2.  Flowers perfect or often unisexual, without showy petals, the perianth sometimes in 2 series, but the 2 series much alike and not petaloid

        3.  Ovules various, but never at once solitary, pendulous from the apex of the ovary, and anatropous; pollen mostly trinucleate; embryo embedded or more often peripheral to the endosperm; plants terrestrial or less often aquatic; flowers perfect or unisexual, borne in various sorts of inflorescences

            4.  Ovary with 1–3 fertile locules and as many stigmas, each fertile locule with a single, apical, pendulous, orthotropous ovule; embryo peripheral to the endosperm; flowers in most species and families unisexual

                5.  Flowers aggregated into dense, pseudanthial, involucrate heads, often pollinated by insects; ovules tenuinucellate; anthers usually bilocular; leaves all basal **2. Eriocaulales**

                5.  Flowers seldom aggregated into dense heads, normally

wind-pollinated or self-pollinated; ovules crassinucellate; anthers often unilocular; leaves cauline or
(in some small families) all basal    **3. Restionales**

4. Ovary otherwise, either with more stigmas than locules,
or with more than one ovule per locule, the ovules of
various form and position; embryo peripheral (Gramineae) or embedded (Juncales, Cyperaceae) in the endosperm; flowers perfect or unisexual

6. Ovary with 1–3 locules and 3–many ovules; fruit capsular;
pollen in tetrads; flowers in open or compact inflorescences unlike those of the Cyperales
**4. Juncales**

6. Ovary with a single locule and ovule; fruit indehiscent;
pollen in monads; flowers in characteristic spikes
or spikelets    **5. Cyperales**

3. Ovules solitary (or solitary in each locule), pendulous from
the apex of the ovary, anatropous; pollen binucleate;
embryo embedded in the endosperm; flowers unisexual,
borne in dense spikes or heads; plants aquatic
**6. Typhales**

1. Flowers with septal or septal-derived nectaries; ovary inferior
except in some Bromeliaceae; vessels chiefly confined to the
roots, except mainly in some Bromeliaceae

7. Stamens 6; flowers regular or sometimes somewhat irregular;
xerophytes and epiphytes with firm, narrow, often spiny-
margined leaves    **7. Bromeliales**

7. Stamens fewer than 6 (6 only in *Ravenala,* of the Strelitziaceae);
flowers distinctly irregular; mesophytes with unarmed, pinnately veined, often broad leaves    **8. Zingiberales**

ORDER 1. **Commelinales**

The order Commelinales as here defined consists of four families and
nearly 1000 species. The Commelinaceae (600) and Xyridaceae (270) together make up the bulk of the order. The Rapateaceae have about 80
species and the Mayacaceae only four.

The   Commelinales are Commelinidae with flowers that ordinarily
have a biseriate perianth well differentiated into sepals and petals, but do

not have nectaries or nectar. The ovary is consistently superior and the fruit capsular. Nearly all of the species have either three or six stamens (one number or the other, not both).

Within the Commelinales, the close relationship of the Xyridaceae to the Rapateaceae has been evident to all. Most authors have considered these two families to be related to the Commelinaceae, a view with which I agree. The Mayacaceae are also generally admitted to be allied to the Commelinaceae. The possibility that certain genera of the Xyridaceae should be segregated as a distinct family Abolbodaceae merits further study.

Ecologically, the Commelinales are characterized by having relatively simple adaptations favoring pollination by pollen-gathering insects. The other orders of the subclass either have nectariferous (and often much more complex) flowers, or have reduced flowers that are in most families wind-pollinated or self-pollinated. Aside from pollination, the Commelinales are ecologically rather diverse. They are all essentially herbaceous, but they occur in a wide range of habitats. The Mayacaceae are strictly aquatic. The Xyridaceae mostly occur in wet or marshy places but are rarely truly aquatic. The Rapateaceae and Commelinaceae range from ordinary mesophytes to marsh plants or occasional xerophytes.

SYNOPTICAL ARRANGEMENT OF THE FAMILIES OF COMMELINALES

1.  Leaf sheath open, often not well differentiated from the blade

2.  Plants mostly terrestrial, often of wet or marshy places, but seldom truly aquatic; leaves generally all clustered at the base, the flowers in a compact inflorescence terminating a long scape or peduncle

3.  Stamens 6, opening by terminal pores; inflorescence a head of spikelets, each spikelet with several bracts subtending its single flower                                          **1. Rapateaceae**

3.  Stamens 3, often accompanied by 3 staminodes; anthers opening by longitudinal slits; inflorescence a simple racemose head                                                         **2. Xyridaceae**

2.  Plants aquatic; leaves numerous, linear or threadlike, distributed along the stem; stamens 3, the anthers opening by terminal pores or short slits                                   **3. Mayacaceae**

1.  Leaves differentiated into a closed sheath and a well defined, commonly somewhat succulent blade        **4. Commelinaceae**

## SELECTED REFERENCES

Arber, Agnes, Leaves of the Farinosae. Bot. Gaz. 74: 80–94. 1922.

Carlquist, S., Anatomy of Guyana Xyridaceae: *Abolboda, Orectanthe* and *Achlyphila*. Mem. N. Y. Bot. Gard. 10 (2): 65–117. 1960.

————, Pollen morphology of Rapateaceae. Aliso 5: 39–66. 1961.

Hamann, U., Merkmalsbestand und Verwandtschaftsbeziehungen der Farinosae. Willdenowia 2: 643–768. 1961.

————, Weiteres über Merkmalsbestand und Verwandtschaftsbeziehungen der "Farinosae." Willdenowia 3: 169–207. 1962.

Maguire, B., J. Wurdack, et al., pp. 11–15 in *Botany of the Guyana Highland* — IV. Mem. N. Y. Bot. Gard. 10 (2). 1960.

Maheshwari, S. C., and B. Balder, A contribution to the morphology and embryology of *Commelina forskalaei* Vahl. Phytomorph. 8: 277–298. 1958.

## ORDER 2. **Eriocaulales**

The order Eriocaulales as here defined consists of the single family Eriocaulaceae, with about 1200 species. The Eriocaulales are Commelinidae with reduced (or no) perianth and with the unisexual flowers aggregated into a dense, involucrate head which typically functions as a pseudanthium, after the manner of the Compositae. Although the individual flowers are small and inconspicuous, the heads are more or less showy, and pollination is usually by insects. The perianth typically consists of two series of similar, white-hyaline tepals, three members in each series. Less often there are only two tepals in a series, or the outer series is missing. Like the Restionales, a parallel group with reduced flowers, the Erioacaulales have a single, pendulous, orthotropous ovule in each locule of the ovary. Unlike the Restionales, however, the Eriocaulales have tenuinucellate ovules.

Habitally the Eriocaulaceae are much like the Xyridaceae, with clustered, basal leaves and a terminal, racemose head on a long peduncle or scape. There is no very obvious reason why the Eriocaulaceae might not have been derived directly from the Xyridaceae or from similar common ancestors. It may be noted that the ovules of *Xyris* are reported to be "almost tenuinucellate."

The relationships of the Eriocaulaceae to the families here grouped in the Commelinales and Restionales might be expressed in several different systematic arrangements, each with its own advantages and disadvantages. One might include them all in a single order, but the order would then be so heterogeneous that it would be hard to exclude the Juncales. One might include the Eriocaulaceae in the Restionales, but this would probably overemphasize the likenesses. The several differences between the Eriocaulaceae and the Restionales as here defined suggest that these two groups probably take their origin in different parts of the Commelinales. Numerically the Eriocaulaceae would dominate the Restionales, becoming the tail that wags the dog. One might group the Eriocaulaceae with the Xyridaceae and Rapateaceae, on the basis of the habit and inflorescence. This would raise questions about some small families here included in the Restionales, which resemble the Eriocaulaceae in habit but are otherwise apparently at home in the Restionales. Finally, one might recognize the Eriocaulaceae as an order distinct from the Commelinales and Restionales. This last alternative is selected as most in accord with the combined phenetic and phyletic criteria.

## SELECTED REFERENCES

Hamann, U., Merkmalsbestand und Verwandtschaftsbeziehungen der Farinosae. Willdenowia 2: 639–768. 1961.

————, Weiteres über Merkmalsbestand und Verwantschaftsbeziehungen der "Farinosae." Willdenowia 3: 169–207. 1962.

Hare, C. L., The structure and development of *Eriocaulon septangulare* With. Jour. Linn. Soc. 53: 422–448. 1950.

### ORDER 3. **Restionales**

The order Restionales as here defined consists of five families and fewer than 500 species. The Restionaceae, with 400 or more species, make up the bulk of the order. The Centrolepidaceae have about 35 species, the Flagellariaceae six or seven, the Anarthriaceae five, and the Ecdeioleaceae only one.

The Restionales are wind-pollinated or self-pollinated Commelinidae with reduced flowers and a single, pendulous, orthotropous ovule in each locule of the ovary. The ovules are always, so far as known, bitegmic and crassinucellate. Except as noted below, the pollen in all five families is much alike and also much like that of the Gramineae. As in most of the families of the Commelinidae, the embryo lies alongside the endosperm, rather than being surrounded by it.

The Flagellariaceae, with perfect flowers and closed sheaths, are evidently allied to ancestral or near-ancestral Commelinaceae as well as to the Restionaceae. The Centrolepidaceae appear on classical morphological grounds to be an advanced but not highly successful offshoot of the Restionaceae. On grounds of pollen morphology, however, the Centrolepidaceae are the more primitive of the two families. Chanda sees a palynological series, indicating possible relationship, from the Centrolepidaceae to the Restionaceae to the Flagellariaceae to the Gramineae, with a possible connection of the Cyperaceae (via *Mapania* and its allies) to the Flagellariaceae. The Anarthriaceae and Ecdeiocoleaceae have often been included in the Restionaceae, but they are anatomically so distinctive that it seems better to exclude them.

Two small genera (*Trithuria* and *Hydatella*) of submerged aquatics from Australia and New Zealand are not provided for in the following key. In most respects they resemble the Centrolepidaceae, and they have usually been referred to that family. However, they have bilocular instead of unilocular anthers, and a unique type of pollen, said by Erdtman (Pollenmorphologische Notizen über einige Blütenpflanzen incertae sedis. Bot. Notis. 119: 160–168. 1966) to be difficult to describe. It is possible that these two genera should be treated as a distinct family, but no such family is here proposed.

## SYNOPTICAL ARRANGEMENT OF THE FAMILIES OF RESTIONALES

1.  Flowers perfect; stamens 6; leaves chiefly or wholly cauline, with closed sheath                                             **1. Flagellariaceae**

1.  Flowers unisexual; stamens 1–3; leaves cauline or basal, the sheath mostly open or poorly developed

    2.  Stamens (1) 2–3; tepals usually present; mostly coarse perennials

        3.  Leaves chiefly or wholly basal; anthers bilocular, free

            4.  Inflorescence open, cymose; flowers bracteate; ovary trilocular; plants mostly dioecious     **2. Anarthriaceae**

            4.  Inflorescence a dense spike suggesting that of *Xyris;* flowers ebracteate; ovary bilocular; plants monoecious
                                                            **3. Ecdeiocoleaceae**

        3.  Leaves wholly cauline; anthers unilocular, or rarely bilocular but then laterally connate; inflorescence more or less open                                             **4. Restionaceae**

    2.  Stamen 1; tepals none; diminutive, annual herbs with chiefly basal leaves; anthers unilocular      **5. Centrolepidaceae**

## SELECTED REFERENCES

Chanda, S., On the pollen morphology of the Centrolepidaceae, Restionaceae and Flagellariaceae, with special reference to taxonomy. Grana Palyn. 6: 355–415. 1966.

Cutler, D. F., Anatomy and taxonomy of the Restionaceae. Jodrell Lab. Notes 4: 1–25. 1966.

Cutler, D. F., ad H. K. Airy-Shaw, Anarthriaceae and Ecdeiocoleacae: two new monocotyledonous families, separated from the Restionaceae. Kew Bull. 19: 489–499. 1961.

Hamann, U., Merkmalsbestand und Verwandtschaftsbeziehungen der Farinosae. Willdenowia 2: 639–768. 1961.

————, Weiteres über Merkmalsbestand und Verwandtschaftsbeziehungen der "Farinosae." Willdenowia 3: 169–207. 1962.

Krupko, S., Embryological and cytological investigations in *Hypodiscus aristatus* Nees (Restionaceae). Jour. So. Afr. Bot. 28: 21–44. 1962.

## ORDER 4. **Juncales**

The order Juncales as here defined consists of two families, the Juncaceae, with about 300 species, and the Thurniaceae, with only three. The probable error of estimation of the number of species of Juncaceae is greater than the number of species in the Thurniaceae, and the order Juncales may be said to have about 300 (rather than 303) species.

The Juncales are Commelinidae with reduced, mostly wind-pollinated flowers and capsular fruits with one to many anatropous ovules per carpel. They have six tepals arranged in two more or less similar whorls, both sets chaffy and usually brown or green. The ovary is tricarpellate, with axile or parietal placentation. When there is only one ovule per carpel it is basal and ascending, rather than apical and pendulous as in the Restionales. The pollen grains are borne in tetrads, and the embryo is surrounded by endosperm.

The small family Thurniaceae has often been included in the Juncaceae. Indeed there is little in the classical morphological characters, or in the pollen, to suggest that they should be separated. In their habit and especially in their anatomy, however, they are so distinctive as to be a highly discordant element in the family. It may be significant that the characteristic silica bodies in the leaf epidermis of the Thurniaceae resemble the silica bodies found in the Rapateaceae, Restionaceae, and some of the Cyperaceae. I follow recent opinion in maintaining the Thurniaceae as distinct.

It has become customary in recent decades to consider the Juncaceae as florally reduced derivatives of the Liliales, often specifically from the Liliaceae. This simple view may have been tenable thirty years ago, but it is much more difficult when recent information about the stomatal apparatus and the distribution of vessels is considered. The Juncaceae characteristically have vessels throughout the plant, and they have stomates with two subsidiary cells. In both of these respects they resemble the Restionaceae, Cyperaceae, and Gramineae and differ sharply from the

Liliaceae, which have no subsidiary cells and in which the vessels are confined chiefly to the roots. The stomatal apparatus is particularly significant, because as Stebbins and Khush[6] have pointed out, the liliaceous type with no subsidiary cells is clearly advanced and is unlikely to be ancestral to the type with two subsidiary cells. Some few members of the Liliales, notably the Haemodoraceae, have two subsidiary cells and have vessels in all parts of the plant, but these plants are not otherwise much like the Juncaceae. It may be further noted that the Juncaceae have starchy endosperm, which is standard in the Commelinidae but rare in the Liliales. Any connection of the Juncaceae to the Liliaceae seems likely to be remote and tenuous, rather than close and direct. A search for other relatives and a different ancestry is therefore appropriate and a different scheme emerges.

There is no good reason, on the basis of present evidence, why the Juncaceae should not be associated with the Cyperaceae, Gramineae, and Restionales, which they resemble in a number of respects. The critical association here is with the Restionales, which as we have noted appear to be florally reduced derivatives of the Commelinales. The Cyperaceae and Gramineae are more advanced, and figure only indirectly in speculation on the ancestry of the Juncaceae.

If the Juncaceae are derived from the Commelinales, parallel to the Restionaceae, then there is nothing incompatible in the interlocking set of resemblances among the Juncaceae, Cyperaceae, Gramineae, and Restionales. If, on the other hand, the Juncaceae are derived from the Liliales, then it is difficult to make a coherent scheme. The Cyperaceae resemble both the Juncaceae and the Gramineae, and the Gramineae also appear to be allied to the Restionales, which in turn are allied to the Commelinales. A more direct connection between the Juncaceae, Cyperaceae, and Gramineae, on the one hand, and the Restionales on the other seems unlikely, especially inasmuch as the orthotropus ovules of the Restionales could not well be ancestral to the anatropous ovules found in the Juncaceae, Cyperaceae, and some of the Gramineae. Neither can the Juncales, with tetardinous pollen, be directly ancestral to the Gramineae and Restionales.

If there is anything ecologically distinctive about the Juncales, as compared to the other orders of Commelinidae with reduced, wind-pollinated flowers, the fact has escaped my attention. The adaptive significance of the known differences between the Juncaceae and Cyperaceae, for example, is obscure. The reasons why the Juncaceae are less successful than the more abundant and varied Cyperaceae remain to be discovered.

[6] G. L. Stebbins and G. S. Khush, Variation in the organization of the stomatal complex in the leaf epidermis of monocotyledons and its bearing on their phylogeny. Am. Jour. Bot. 48: 51–59. 1961.

## SYNOPTICAL ARRANGEMENT OF THE FAMILIES OF JUNCALES

1. Inflorescence of diverse sorts, but not as in the Thurniaceae; vascular bundles of ordinary nature, with abaxial phloem, not as in the Thurniaceae; cells without silica bodies

**1. Juncaceae**

1. Inflorescence of one or more dense heads subtended by spreading, leafy bracts; vascular bundles of the leaf in vertical pairs, the lower (and smaller) bundle of a pair with the phloem on top (adaxial), facing the phloem of the upper bundle; some cells of the leaf epidermis containing silica bodies

**2. Thurniaceae**

### SELECTED REFERENCES

Cutler, D. F., Vegetative anatomy of Thurniaceae. Kew Bull. 19: 431–441. 1965.

Hamann, U., Merkmalsbestand und Verwandtschaftsbeziehungen der Farinosae. Willdenowia 2: 639–768. 1961.

Wulff, H. D., Die Pollenentwicklung der Juncaceen. . . . Jahrb. wiss. Bot. 97: 533–556. 1939.

Order 5. **Cyperales**

The order Cyperales as here defined consists of two families, the Gramineae (Poaceae), with about 8000 species, and the Cyperaceae, with nearly 4000. The order has also been called Glumiflorae, Graminales, and Poales. The name Cyperales (1901) is older than Poales (1930).

The Cyperales are Commelinidae with reduced, mostly wind-pollinated or self-pollinated flowers that have a unilocular, two-to-three-carpellate ovary bearing a single ovule. The flowers are arranged in characteristic spikes or spikelets which represent reduced inflorescences, the condensation and reduction having taken somewhat different courses in the two families. The perianth is represented only by a set of bristles or tiny scales, or is completely wanting. The leaves are ordinarily differentiated into a well defined sheath and narrow blade, often with a small adaxial appendage, the ligule, at the junction of the two. The two families of the order are alike, and differ from most other monocots (as well as dicots) in having the stomates in straight files or rows of one or two and all oriented in the same direction. Like the Juncales and Restionales, the Cyperales have stomates with two supporting cells and have vessels in all vegetative organs. The pollen in the Cyperales and Juncales is uniformly trinucleate, as is that of the single species of Restionaceae that has been studied.

Most genera in both families have perfect flowers. The perfect flowers of the Cyperaceae have sometimes been interpreted as pseudanthia, but I do not believe the evidence calls for such an interpretation. I see no need or reasonable possibility to treat the perfect flower of Scirpus, for example, as anything but a flower. If the pseudanthial interpretation were accepted, the Cyperaceae would hang on an evolutionary skyhook, without evident relatives — unless one were to accept also the even more difficult interpretation (which has not been seriously proposed) that the individual grass flower is a pseudanthium. Occam's razor has its uses.

Data on the chemical constituents of the Cyperales and their relatives are only now becoming sufficiently abundant to warrant their serious consideration in broad-scale taxonomic studies. Hegnauer (1963, pp. 224–5, 267) finds that the admittedly still scanty chemical data are fully compatible with a relationship among the Juncaceae, Cyperaceae, Gramineae, and Restionaceae. Furthermore, although the Cyperaceae differ from the Gramineae in several respects, the chemical similarities between them, when considered in the context of the more classical characters, are in my opinion more suggestive of phyletic unity than convergence.

In recent years many authors have taken the Gramineae and Cyperaceae each to represent a separate, unifamilial order. This view has been conditioned partly by the pseudanthial interpretation of the cyperaceous flower, and partly also by the seemingly incompatible relationships of the two families — the Cyperaceae to the Juncales, and the Gramineae to the Restionales. As we have noted in the discussion of the Juncales, however, these orders are all closely related, and the Juncales probably have no close connection with the Liliales.

Most authors have considered the Cyperaceae to be related to the Juncaceae, a view with which I concur. However, for reasons developed in the following paragraphs, I believe the relationship is probably more nearly collateral than filial.

The seemingly single pollen grains of the Cyperaceae are generally considered to be cryptotetrads, in which three pollen grains are obsolete. This condition has often been compared to the truly tetradinous condition of the pollen in the Juncaceae, and has been adduced as evidence that the Cyperaceae are derived from the Juncaceae. This obvious interpretation may prove to be correct, but at present it does not seem to provide for all the facts. The Gramineae are generally considered to have truly monadinous pollen, and the tetradinous pollen of the Juncaceae is an unlikely antecedent for it. Mapania, one of the more primitive genera of the Cyperaceae, has pollen grains much like those of the Gramineae and Restionaceae, differing (aside from being cryptotetradinous) mainly in the subreticulate rather than papillate ornamentation of the exine (teste Koyama). If the Cyperaceae and Gramineae are evolutionary siblings, as here suggested, then it seems unlikely that the Cyperaceae inherited tetradinous pollen from some prior group.

One of the interesting links between the Cyperaceae and the Juncaceae is the occurrence of diffuse centromeres in *Luzula, Juncus,* and certain genera (e.g., *Carex*) of the Cyperaceae. Although the Cyperaceae are generally conceded to be the more advanced of the two families, only some of the cyperaceous genera have diffuse centromeres, whereas others have the more standard point centromeres.[7] It seems very probable, therefore, that the occurrence of the diffuse centromeres in these two families reflects parallelism rather than inheritance from a common ancestor. We have previously noted that many of the similarities between related taxa of plants often turn out on close inspection to be due to genetically favored parallelism rather than direct inheritance.

Most recent authors have considered the Gramineae to be related to the Restionales or to a broadly defined order Commelinales in which the Restionales are submerged. Takhtajan has emphasized the significance of *Streptochaeta* as a primitive grass which suggests a close relationship to the Flagellariaceae, and this view has also been adopted by Butzin. I concur that the families are related, but I suspect that the relationship is collateral, and that both have a common ancestry in the Commelinales. Some of the grasses, with open leaf sheaths and more or less anatropous ovules, are more primitive in these respects than anything now known in the Flagellariaceae.

Two features which militate against grouping the Gramineae and Cyperaceae in a single order need to be considered. The Cyperaceae have chiefly anatropous ovules, as in the Juncaceae, whereas the Gramineae have chiefly orthotropous ovules, as in the Restionales. Furthermore, the Cyperaceae have the embryo embedded in the endosperm, whereas the Gramineae have the embryo peripheral, as in the Restionales and most families of the Commelinales.

Although the grass ovule is typically apical, pendulous, and orthotropous, as in the Restionales (including Flagellariaceae), the attachment ranges from apical through lateral to basal, and the structure from orthotropous to hemianatropous (or anatropous, according to interpretation by different students). Butzin suggests that the hemianatropous type is derived within the family from the orthotropous type, but this is contrary to the usual sequence of ovular evolution. The Commelinaceae, which are closely allied to the Flagellariaceae, have the ovules variously anatropous to hemianatropous to orthotropus. I suggest that in the grasses, as well as in

---

[7] Very recently (August, 1967) L. J. Harms has maintained in conversation that diffuse centromeres occur throughout the Cyperaceae, even in *Fimbristylis,* which Sharma and Bal (cited below) had reported to have typical chromosomes with point centromeres. Further investigation is obviously in order.

the Commelinales and Restionales, the orthotropous ovule is the most modified type. Thus it is perfectly possible to consider that the Gramineae, Cyperaceae, Juncaceae, and the families of the Restionales are derived from a common ancestry in the Commelinales near the Commelinaceae, and that whereas the ovules have remained anatropous in the Juncaceae and Cyperaceae, they have become orthotropous in the Restionales and (separately) most of the Gramineae.

Both the Restionales and the Commelinales generally have the embryo peripheral to endosperm, as in the grasses. However, some few of the Bromeliaceae, in a related order, have the embryo embedded in the endosperm, in contrast to the peripheral embryo of most members of the family. This variation has caused no great concern to bromeliologists. As we have previously noted, characters are as important as they prove to be in each individual instance, and I see no reason why this one should be assigned such great a priori importance as to cause the dissolution of an order.

In the anemophilous families of the Commelinidae, as in some other groups, a consideration of the overlapping sets of similarities among related taxa leads us into a thicket with no established exit. If the similarity between A and B in character x is considered to be inherited from a common ancestry, then the similarity between B and C in character y must be due to close parallelism, and vice versa. This sort of parallelism, however, reflects a genetic similarity of the ancestors which conditioned them to undergo similar evolutionary changes. At this point we must back out of the thicket, or cut our way through it, and put together those things which are now most alike. For strictly phylogenetic purposes it is important to distinguish similarities due to close parallelism from similarities inherited from a common ancestry, but for taxonomic purposes this distinction is less significant. We have noted in an earlier chapter that taxonomy can present no more than a muddy reflection of phylogeny, and that many similarities which on first inspection seem to be inherited from a common ancestry turn out on more careful consideration to reflect close parallelism instead. (I here distinguish between close parallelism, in which similar changes occur in groups that are initially rather similar, and convergence, in which certain similarities are imposed on basically different groups. There is of course every gradation from the one to the other, but the extremes are different enough.)

On the basis of all the available information, I believe it is most useful to group the Gramineae and Cyperaceae in a single order Cyperales. The two families have too much in common to be divorced from each other. They are lines diverging at an angle from a common source.

The Cyperales are to be distinguished from the Juncales on the one hand and the Restionales on the other. All three of these orders take their origin in the Commelinales. The putative phylogenetic relationships shown in Fig. 5.2 represent only one of the several possibilities. All that is really required is that the Commelinales be ancestral to the other three orders.

The Cyperaceae and Gramineae are both large and successful families with reduced flowers adapted to wind pollination. The Gramineae are notably the larger and more successful of the two, in terms of both the number of species and the number of individuals. If the formal morphological characters usually cited as distinguishing the two families have any great adaptive significance, the fact remains to be established. One may reasonably speculate instead that it is the intercalary meristem of the grass leaf which plays the major role in the greater success of the grasses. The leaves of monocots in general usually mature from the tip downwards; in the grasses the base of the leaf blade remains more or less permanently immature and meristematic. Some of the Cyperaceae, notably species of *Carex*, share this same feature (teste Koyama), but in general it is much less developed in the Cyperaceae than in the Gramineae. Anyone who has pushed a lawnmower should recognize the significance of the intercalary meristem in permitting a plant to withstand grazing.

### SYNOPTICAL ARRANGEMENT OF THE FAMILIES OF CYPERALES

1. Flowers spirally or less often distichously arranged on the axis of the spike or spikelet, usually each flower seemingly or actually subtended by only a single bract, without an evident bract between the flower and the axis; seed coat generally free from the pericarp; leaf sheath usually closed, stem usually solid, often triangular; flowers often with a perianth of evident bristles; carpels 3 or less often 2; embryo embedded in the endosperm                                    **1. Cyperaceae**

1. Flowers distichously arranged on the axis of the spikelet (or only one per spikelet), each flower ordinarily subtended by a pair of bracts (lemma and palea), the palea inserted between the flower and the axis; seed coat generally adnate to the pericarp; leaf sheath usually open; stem usually hollow, never triangular; flowers without a perianth, unless the lodicules are so interpreted; carpels 2, rarely 3; embryo peripheral to the endosperm                                    **2. Gramineae**

### SELECTED REFERENCES

Artschwager, E., E. W. Brandes, and R. C. Starret, Development of flower and seed of some varieties of Sugar Cane. Jour. Agric. Res. 39: 1–30. 1929.

Butzin, F., Neue Untersuchungen über die Blüte der Gramineae. Inaug. Diss. Fr. Univ. Berlin 1–183. 1965.

Chandra, N., Morphological studies in the Gramineae. I. Vascular anatomy of the spikelet in the Pooidae. Proc. Nat. Inst. Sci. India B28: 545–562. 1962. II. Vascular anatomy of the spikelets in the Paniceae. Proc. Indian Acad. Sci. B56: 217–231. 1962. V. Vascular anatomy of the spike and spikelets in the Andropogoneae. (With N. P. Saxena) Proc. Indian Acad. Sci. B59: 1–23. 1964.

Cheadle, V. I., The taxonomic use of specialization of vessels in the metaxylem of Gramineae, Cyperaceae, Juncaceae and Restionaceae. Jour. Arn. Arb. 36: 141–157. 1955.

Clifford, H. T., Floral evolution in the family Gramineae. Evolution 15: 455–460. 1961.

Dunn, D. B., G. P. Sharma, and C. C. Campbell, Stomatal patterns of dicotyledons and monocotyledons. Am. Midl. Nat. 74: 185–195. 1965.

Koyama, T., Classification of the family Cyperaceae (1). Jour. Fac. Sci. Univ. Tokyo, Sect. III. 8: 37–148. 1961.

Metcalfe, C. R., Anatomy of the Monocotyledons, I: Gramineae. Oxford: The Clarendon Press, 1960.

Schultze-Motel, W., Entwicklungsgeschichtlichte und vergleichend-morphologische Untersuchungen im Blütenbereich der Cyperaceae. Bot. Jahrb. 78: 129–170. 1959.

Sharma, A. K., and A. K. Bal, A cytological investigation of some members of the Cyperaceae. Φyton 6: 7–22. 1956.

## ORDER 6. Typhales

The order Typhales as here defined consists of two closely related small families, the Typhaceae and Sparganiaceae. The Typhaceae, with only the genus *Typha*, have about 15 species, and the Sparganiaceae, with only the genus *Sparganium*, have about 20 species. Authors are agreed that the two genera are related. Were it not for the long historical tradition, they might well be accommodated in a single family.

The Typhales are Commelinidae with reduced, unisexual, wind-pollinated flowers and a single ovule in a pseudomonomerous ovary, or seldom with one ovule in each of the two or more locules of a compound ovary. In either case the ovule is apical, pendulous, and anatropous. The stomates have two subsidiary cells. They are all marsh or aquatic plants with emergent or floating stems and leaves. Like other Commelinidae, they usually have vessels in all vegetative organs. The Sparganiaceae have starchy endosperm like that of most other Commelinidae, but the Typhaceae contain more oil and protein (as well as some starch). The pollen of the two families is binucleate and much alike, and differs to some extent from that of other orders. It is interesting that both tetrads and monads occur in Typha.

Although the two genera have nearly the same number of species, *Typha* is much the more abundant. The difference may reflect the emergent and more vigorously rhizomatous habit of *Typha*, which tends to occur in dense stands from which other vegetation is largely excluded. The very numerous, windborne fruits of *Typha* may also be more effectively distributed than the fewer and larger fruits of *Sparganium*.

The Typhales have often been associated with the Pandanaceae in an expanded order Pandanales, but Eckardt indicates some uneasiness about this grouping in his treatment for the current Engler Syllabus. Some other authors, notably Hutchinson, have considered the Typhales to be related to the Liliaceae. The assignment of the Typhales to the Commelinidae in the present treatment follows the same arguments of stomatal organization and vessel distribution which have been partly responsible for my transfer of the Juncales and Cyperaceae to the Commelinidae. Floral reduction associated with wind-pollination has occurred repeatedly in the Commelinidae, but only minimally in the Liliidae. The Pandanaceae are in their own way too specialized to be likely ancestors to the Typhales, and I see no reason to assume that the Typhales have a woody ancestry within the monocotyledons.

The characters of the Typhales do not suggest that they are directly related to any of the other members of the Commelinidae that have reduced, wind-pollinated flowers. I take them to be a separate line from a generalized commelinalean ancestry, parallel in some respects to the other groups with reduced flowers.

### SYNOPTICAL ARRANGEMENT OF THE FAMILIES OF TYPHALES

1. Inflorescence of globose heads; vestigial perianth usually present; achenes sessile or nearly so, not wind-distributed
**1. Sparganiaceae**

1. Inflorescence of dense, cylindrical spikes; perianth apparently none; achenes slenderly long-stipitate, with long hairs on the stipe, wind-distributed **2. Typhaceae**

### SELECTED REFERENCE

Graef, P., Ovule and embryo sac development in *Typha latifolia* and *Typha angustifolia*. Am. Jour. Bot. 42: 806–809. 1955.

ORDER 7. **Bromeliales**

The order Bromeliales consists of the single family Bromeliaceae, with more than 1700 species. They are firm-leaved, terrestrial xerophytes or more often epiphytes with regular or somewhat irregular flowers that usually have septal nectaries and an inferior ovary. Although the flowers are mostly pollinated by insects (or birds), some few genera, such as Navia, are wind-pollinated.

The subfamily Pitcairnoideae, with about a third of the species, consists mainly of terrestrial xerophytes with the ovary superior. This is clearly the most primitive subfamily in the group. It appears that in the Bromeliaceae xeromorphism has opened the way to epiphytism.

Some authors have associated the Bromeliaceae with the families here grouped in the subclass Liliidae, instead of with the Commelinales. It is true that some of the Bromeliaceae have somewhat petaloid sepals suggesting those of the Liliidae, but most of them have the sepals and petals well differentiated as in the Commelinales. Five species in as many genera of the Bromeliaceae examined by Stebbins and Khush (Am. Jour. Bot. 48: 51–59. 1961) were alike in having several subsidiary cells associated with each stomate, a condition which is common in the Commelinidae and unknown in the Liliidae. Furthermore, the Bromeliaceae have a starchy endosperm as in the Commelinidae, whereas in the Liliidae the endosperm (when present) is mostly cellulosic or fatty. I believe that the Bromeliaceae are properly referred to the Commelinidae rather than to the Liliidae.

Although the Bromeliales are a rather isolated group, they are probably best associated with another distinctive order, the Zingiberales. The two orders are alike, and differ from the rest of the subclass, in having retained the floral nectaries and in having achieved epigyny. It is clear, however, that the epigynous condition has been independently achieved in each group. The hypogynous members of the Bromeliaceae are already too xeromorphic to be likely ancestors of the mesophytic Zingiberales.

ORDER 8. **Zingiberales**

The order Zingiberales consists of eight families and about 2100 species. Nearly two-thirds of the species belong to the Zingiberaceae (1300), and most of the rest belong to the three families Marantaceae (350), Costaceae (200), and Heliconiaceae (150). The Musaceae (70), Cannaceae (30), Strelitziaceae (eight), and Lowiaceae (five) have only a little more than a hundred species amongst them. The ordinal name Scitamineae has often been used instead of Zingiberales.

The Zingiberales are Commelinidae with irregular flowers, an inferior ovary, and pinnately veined leaves. Nearly all of them have either one or five functional stamens. Only the monotypic genus *Ravenala* (Strelitziaceae) has six stamens. Except for a few of the arborescent species, vessels are confined chiefly to the roots.

Authors are agreed that the Zingiberales constitute a well defined, taxonomically isolated order. Indeed they are so distinct from all other groups that their relationships are in dispute. For example, Takhtajan refers them to Liliidae, whereas Hutchinson associates them with the Bromeliales and Commelinales.

The Zingiberales are here referred to the Commelinidae because (1) they have sepals well differentiated from the petals; (2) they have a starchy endosperm; and (3) they have two or more (usually more) subsidiary cells associated with each stomate. All of these characters would be anomalous (although not necessarily unknown) in the Liliidae. The only respect in which the Zingiberales are more like the Liliidae than like the Commelinidae is that the vessels are confined to the roots instead of occurring in all vegetative organs. This implies that the Zingiberales must have diverged from the main line of the Commelinidae before that line had achieved a general distribution of vessels throughout the plant body.

The floral structure of the Zingiberales as a whole, and some of the differences between individual families, appear to be intimately related to the mechanism of pollination. A detailed correlation of pollinators with the families and genera of the order remains to be undertaken, however. The adaptive significance of many of the characters which distinguish the individual families remains to be established. Among such characters of presently unknown or doubtful significance are the arrangement of leaves, and the presence or absence of raphide sacs, laticifers, mucilage ducts, and oil cells.

### SYNOPTICAL ARRANGEMENT OF THE FAMILIES OF ZINGIBERALES

1 Functional stamens 5 or rarely 6, each with 2 pollen sacs; plants with raphide sacs; guard cells symmetrical except in Lowiaceae

2. Flowers perfect; leaves and bracts distichously arranged; plants without laticifers; fruit capsular or schizocarpic

3. Ovules numerous in each locule; fruit capsular; seeds arillate; stigma trifid

4. Flowers without an evident hypanthium; leaves obviously penniveined; guard cells symmetrical; most species caulescent or even arborescent     **1. Strelitziaceae**

4. Flowers with a long hypanthium much surpassing the ovary; leaves seemingly parallel-veined (actually penniveined with narrowly ascending veins); guard cells asymmetrical; acaulescent herbs    **2. Lowiaceae**

   3. Ovules solitary in each locule; fruit schizocarpic; seeds not arillate; stigma capitate    **3. Heliconiaceae**

  2. Flowers unisexual; leaves and bracts spirally arranged; plants with laticifers; fruit fleshy, indehiscent, the seeds not arillate    **4. Musaceae**

1. Functional stamen 1; plants without raphide sacs; guard cells asymmetrical except in Cannaceae

   5. Stamen with 2 pollen sacs, not petaloid; endosperm helobial

     6. Leaves and bracts distichously arranged; sheaths open; plants aromatic, with abundant oil cells    **5. Zingiberaceae**

     6. Leaves and bracts spirally arranged; sheaths closed; plants without oil cells    **6. Costaceae**

   5. Stamen with a single functional pollen sac, the other sac modified and petaloid; endosperm nuclear

     7. Ovules numerous in each of the 1–3 locules; stem with mucilage canals; embryo straight; seeds not arillate    **7. Cannaceae**

     7. Ovules solitary in each of the (1–2–) 3 locules; stem without mucilage canals; embryo folded; seeds mostly arillate    **8. Marantaceae**

## SELECTED REFERENCES

Lane, I. E., Genera and generic relationships in the Musaceae. Mitt. Staatssamml. München 2: 114–141. 1955.

Nakai, T., Notulae ad plantas Asiae orientalis (XVI). Jour. Jap. Bot. 17: 189–210. 1941. (For the first organization of the order into the families here recognized)

Tilak, V. D., and R. M. Pai, Studies in the floral morphology of the Marantaceae. I. Vascular anatomy of the flower of Schumannianthus virgatus Rolfe, with special reference to the labellum. Can. Jour. Bot. 44: 1365–1370. 1966.

Tomlinson, B., Phylogeny of the Scitamineae — morphological and anatomical considerations. Evolution 16: 192–213. 1962.

## Subclass III. Arecidae

The subclass Arecidae as here defined consists of four orders, five families, and about 6400 species. The Arecales (palms), with about 3500 species, make up more than half of the group. The Arales (aroids and duckweeds) have about 1800 species, the Pandanales (screw-pines) about 900, and the Cyclanthales less than 200.

The Arecidae are monocots with an inflorescence of usually numerous, small flowers generally subtended by a prominent spathe and often aggregated into a spadix. The endosperm, when present, variously contains fats, oils, proteins, and/or hemicelluloses, but only rarely (except in the Araceae) starch. Except for the Lemnaceae (which have reduced stomates without supporting cells), the stomates of the Arecidae usually have two or more (typically four) supporting cells. None of these characters is by itself definitive, but in combination they set the Arecidae apart from the rest of the Liliatae.

Ecologically, the Arecidae are characterized by several prominent tendencies which, however, are not shared by all members. They tend to be arborescent (more than half of the species), they tend to have broad, petiolate leaves (more than ⅘ of the species) that do not have the typical parallel venation commonly associated with monocots, and they tend toward floral reduction and eventually (about ⅖ of the species) the formation of a spadix. Except for the Arales, they usually have vessels in all vegetative organs. The arborescent habit, broad leaves, and well distributed vessels would seem a priori to form an integrated, mutually adaptive character-complex. It should not be surprising that the palms, the only group to combine all of these features, are the largest family and largest order of the subclass.

### SYNOPTICAL ARRANGEMENT OF THE ORDERS OF ARECIDAE

1. Flowers well developed (though small), with an evident, biseriate perianth of 6 members, not crowded into a spadix; carpels free or united; plants very often arborescent and with a terminal crown of large leaves, the leaves in any case with sheath, petiole, and expanded, plicate blade; endosperm nuclear
**1. Arecales**

1. Flowers more or less reduced and very often crowded into a spadix; carpels united to form a compound ovary; plants

of various habit, but never at once arborescent and with a terminal crown of broad leaves; leaves not plicate except in some Cyclanthaceae

2. Stamens mostly 10–numerous; vessels generally present in leaves and stems as well as in roots; endosperm nuclear or helobial; flowers unisexual; leaves usually either with bifid blade or not divided into blade and petiole

    3. Leaves with blade, petiole, and basal sheath, the blade usually bifid; plants monoecious; endosperm helobial; neotropical herbs, seldom woody     **2. Cyclanthales**

    3. Leaves elongate and narrow, somewhat sheathing at the base, but not divided into blade and petiole, not bifid; plants dioecious; endosperm nuclear; paleotropical woody plants     **3. Pandanales**

2. Stamens mostly 1–8 (flowers rarely produced in Lemnaceae); vessels confined to the roots, or wholly wanting; endosperm cellular; flowers perfect or unisexual; leaves (when present) often divided into blade and petiole; herbs, or seldom woody climbers     **4. Arales**

ORDER 1. **Arecales**

The order Arecales consists of the single family Palmae (Arecaceae), with about 3500 species. The order has also been called Principes and Palmales.

> It is generally recognized that the palms are morphologically a unique and isolated group of monocotyledons. Palms also possess distinctive anatomical features by which they can be recognized. The Cyclanthaceae is the one monocotyledonous family that appears to be most closely related to the Palmae; in particular, some members of this family share the peculiar method of development of a compound leaf that has been evolved by palms.[8]

Indeed the seedling leaves of some palms, such as *Hypospathe*, are remarkably like the leaves of mature plants of the Cyclanthaceae.

Palms are mostly trees with an unbranched trunk and a terminal crown of large leaves. Less common types are scandent, with scattered leaves, or

[8] Tomlinson, cited below.

more or less acaulescent, with the leaves arising from the ground. The leaf has a blade, petiole, and sheath; the sheath is open or closed, but in any case it fully encircles the stem at the base. The blade as a whole is usually pinnately compound, less often palmately compound or merely lobed. It is likewise pinnately or less often palmately veined, but the individual pinnae or segments have more or less parallel veins and are plicate between the veins. The plicate structure of all palm leaves relates to a complex and unique ontogeny shared only by some of the Cyclanthaceae. As to venation alone, the hypothetical archtype of the palm leaf is much like the characteristic leaf of the Zingiberales, with a strong midvein which gives rise to numerous closely parallel primary branches running toward the margin and connected by secondary cross-veins. I do not suggest that the two groups are very closely related.

As with other monocotyledonous trees, the woodiness of the stem of palms derives not from the presence of great amounts of xylem, but from lignification of other tissues. In the context of the monocotyledons as a whole and the angiosperms in general, woodiness in palms is clearly secondary, but it does not necessarily follow that the acaulescent types are primitive within the family. These may well be reduced from arborescent ancestors.

The palms have the most primitive type of flower and inflorescence in the subclass Arecidae, but they are more advanced in these regards than most of the Alismatidae. Like most other members of the subclass, palms have numerous, rather small flowers collectively subtended by a large spathe, but the inflorescence is more or less branched instead of contracted into a spadix. The flowers have three sepals and three petals that are not strongly differentiated from each other, usually six stamens, and typically three separate or more or less united carpels, one or two of the carpels sometimes being reduced and infertile.

### SELECTED  REFERENCES

Arber, A., On the development and morphology of the leaves of palms. Proc. Roy. Soc. Lond. B93: 249–261. 1922.

Eames, A., Neglected morphology of the palm leaf. Phytomorph. 3: 172–189. 1953.

Tomlinson, P. B., Seedling leaves in palms and their morphological significance. Jour. Arn. Arb. 41: 414–428. 1960.

————, Anatomy of the Monocotyledons, II: Palmae. Oxford: The Clarendon Press, 1961.

### Order 2.  Cyclanthales

The order Cyclanthales consists of the single tropical American family

Cyclanthaceae, with about 180 species. The Cyclanthaceae are herbs or seldom woody plants, with characteristic leaves and numerous, small, unisexual flowers that are crowded into a spadix. The whole inflorescence acts as a sort of pseudanthium, and pollination is commonly by insects. The leaves have a basal sheath, a petiole, and an expanded, usually bifid blade, which in at least some species is plicate and follows the otherwise unique ontogeny of palm leaves.

Most authors have agreed that the Cyclanthaceae are related to the palms, aroids, and Pandanaceae. In spite of their probable relationship, these families are so different inter se that it does not seem useful to include any two of them in the same order. It seems likely that the similarity of the Cyclanthales to the palms in leaf ontogeny and structure reflects an independent realization of similar potentialities, rather than direct inheritance from a common ancestor. Neither group could have given rise to the other.

The Cyclanthales are typically herbs of the forest floor, less often lianas or plants of open places. The significance of their morphology in fitting them to this habitat is still largely speculative. One may legitimately speculate that the shady habitat placed a premium on the expansion of the photosynthetic surface, and that the available supply of mutations determined the particular pathway by which this expansion was achieved. The type of flower and inflorescence does not appear to be closely governed by the habitat. The Zingiberales, occupying similar habitats, found a different method of leaf expansion and a wholly different sort of adaptation to insect pollination.

SELECTED REFERENCE

Harling, G., Monograph of the Cyclanthaceae. Acta Hort. Berg. 18: 1–428. 1948.

ORDER 3. **Pandanales**

The order Pandanales as here defined consists of the single paleotropical family Pandanaceae, with nearly 900 species. The Typhaceae and Sparganiaceae, which have often been included in the Pandanales, are here treated as an order Typhales in the subclass Commelinidae.

The Pandanales are woody Arecidae with numerous, firm, narrow, parallel-veined leaves that are usually spiny-margined. The flowers are small, very numerous, unisexual, and usually crowded into a dense spadix. The large genus *Pandanus*, with more than 600 species, is a characteristic and familiar component of the vegetation in the Old-World tropics, especially on the Pacific Islands and along continental coasts.

Most authors have related the Pandanales to the Cyclanthales, Arecales, and Arales. I concur. The relationship, however, is still rather distant, and it is necessary to hold the several orders as distinct. It is noteworthy that *Sararanga* (two spp) has pedicellate flowers in a branching inflorescence, indicating that the characteristic spadix of the other genera has evolved within the family rather than being phyletically linked with that of the Cyclanthaceae and Araceae. Here again the phenotypic similarity reflects independent realization of similar potentialities rather than direct inheritance from a common ancestor.

The firm, xeromorphic leaves of the Pandanales may well be related to the sunny, often maritime sites in which the plants grow. Otherwise the adaptive significance of the characters which mark the group is obscure. It may be noted that in spite of the relatively stereotyped inflorescence, pollination is effected in different species by such diverse agents as wind, insects, birds, and bats.

## SELECTED REFERENCES

Fagerlind, F., Stempelbau und Embryosackentwicklung bei einigen Pandanazeen. Ann. Jard. Bot. Buitenzorg 49: 55–78. 1940.

Strömberg, B., Embryo-sac development of the genus *Freycinetia*. Svensk. Bot. Tidskr. 50: 129–134. 1956.

van der Pijl, L., Remarks on pollination by bats in *Freycinetia*, *Duabanda* and *Haplophragma*, and on chiroptery in general. Acta Bot. Neerl. 5: 135–144. 1956.

ORDER 4. **Arales**

The order Arales (Spathiflorae) as here defined consists of two families and about 1800 species. The Lemnaceae, with only about 30 species, have fewer species than the margin for error in estimating the number of species of Araceae (1800). A mere addition of the estimates for the two families would give a misleading figure for the order as a whole, because it would imply greater precision than actually exists.

The Arales are herbaceous or seldom somewhat woody Arecidae with few (one to eight) stamens and without the special features which separately characterize the leaves of the other orders of the subclass. Very often the leaves have an expanded, pinnately net-veined blade of dicotyledonous appearance. The endosperm is mostly cellular.

The Araceae are, like the Cyclanthaceae and the Zingiberales, mostly tropical and subtropical herbs of the forest floor. Some few are woody climbers, and some others extend well into temperate or even subarctic regions. Like the Palmae, Cyclanthaceae, and Pandanaceae, they present their own combination of primitive and advanced characters. One respect in which they are more primitive than these other families is that their vessels are chiefly or wholly confined to the roots, instead of being well distributed throughout the plant. There is no obvious reason why the Araceae have been more successful than the Cyclanthaceae, which occur in the same sort of habitat, but the fact of their greater success is evident both in terms of number of species and number of individuals, as well as in geographic limits.

The Lemnaceae are free-floating, thalloid aquatics which hardly extend above the surface of the water. They occupy a distinctive ecological niche which has scarcely been entered by other angiosperms. Their whole structure reflects their habitat, and they are so highly reduced and specialized, that they could easily be treated as a distinct order Lemnales. On the other hand, it is now widely agreed that the Lemnaceae are related to and derived from the Araceae. *Pistia*, a free-floating aquatic aroid with a relatively small and few-flowered spadix, is seen as pointing the way toward *Spirodela*, the least reduced genus of Lemnaceae. In the absence of precedent, I would be inclined to treat the Lemnaceae as a separate order, and indeed at one time I did so. However, as a contribution toward consensus in a matter not subject to objective determination, I here follow Hutchinson, Melchior (in the 12th Engler Syllabus), and Takhtajan in associating the Lemnaceae and Araceae in a single order.

### SYNOPTICAL ARRANGEMENT OF THE FAMILIES OF ARALES

1. Plants with roots, stems, and leaves, terrestrial or sometimes more or less aquatic, but only rarely free-floating; inflorescence a spadix; plants usually with vessels in the roots and tracheids in the roots, stems, and leaves  **1. Araceae**

1. Plants thalloid, free-floating, with or without 1–several short, slender roots; flowers (rarely produced) not forming a definite spadix; plants lacking both vessels and tracheids  **2. Lemnaceae**

### SELECTED REFERENCES

Daubs, E., A monograph of the Lemnaceae. Ill. Biol. Monog. 34: 1–118. 1965.

Maheshwari, S., The endosperm and embryo of *Lemna* and systematic position of the Lemnaceae. Phytomorph. 6: 51–55. 1956.

————, *Spirodela polyrhiza*: the link between the aroids and the duckweeds. Nature 181: 1745–1746. 1958.

Maheshwari, S., and R. Kapil, Morphological and embryological studies on the Lemnaceae. I. The floral structure and gametophytes of *Lemna paucicostata*. Am. Jour. Bot. 50: 677–686. 1963.

Maheshwari, S., and P. P. Khanna, The embryology of Arisaema wallichianum Hook. f. and the systematic position of the Araceae. Phytomorph. 6: 379–388. 1956.

Wilson, K., The genera of the Arales in the southeastern United States. Jour. Arn. Arb. 41: 47–72. 1960.

## Subclass IV. Liliidae

The subclass Liliidae as here defined consists of two orders, 17 families, and nearly 28,000 species. The single family Orchidaceae, constituting the bulk of the order Orchidales, includes about 20,000 species.

The Liliidae are syncarpous monocots with both the sepals and the petals usually petaloid. The seeds are either nonendospermous or have an endosperm with various sorts of reserve foods such as cellulose, fats, or protein, only seldom starch. The stomates are without subsidiary cells, or (in a few small families) have two subsidiary cells. The pollen is mostly binucleate, and the young endosperm is variously helobial, nuclear, or cellular. Probably more than $\frac{9}{10}$ of the species have the vessels chiefly or wholly confined to the roots. The flowers generally have more or less well developed nectaries of one or another sort, and pollination is usually by insects or other animals.

### SYNOPTICAL ARRANGEMENT OF THE ORDERS OF LILIIDAE

1. Plants not obviously mycotrophic; seeds of ordinary number and structure, usually with a well differentiated embryo and well developed endosperm; most families and genera with septal nectaries, sometimes with other kinds of nectaries in addition or instead; ovary superior or inferior          **1. Liliales**

1. Plants strongly mycotrophic, sometimes without chorophyll; seeds very numerous and tiny, with undifferentiated embryo and very little or no endosperm; nectaries diverse, but not (or at least not usually) septal; ovary inferior          **2. Orchidales**

### ORDER 1. Liliales

The order Liliales as here defined consists of 13 families and nearly 7700 species. Well over half of the species belong to the single family Liliaceae (4200), and most of the rest belong to only four more: the Iridaceae (1500), Dioscoreaceae (650), Agavaceae (550), and Smilacaceae (300). The remaining eight families have fewer than 500 species amongst them: Velloziaceae (200), Haemodoraceae (120), Xanthorrhoeaceae (50),

Pontederiaceae (30), Stemonaceae (30), Taccaceae (30), Philydraceae (five), and Cyanastraceae (five). It may be noted that the Amaryllidaceae are here included in the Liliaceae.

The Pontederiaceae and Philydraceae stand off sharply from the rest of the Liliales and from each other. Both families appear to represent early and relatively unsuccessful offshoots from the primitive Liliid stock, when its differentiation from the Commelinid stock was not yet complete. *Philydrum* is indeed much like the Commelinaceae in general appearance, and like the Commelinaceae it has a starchy endosperm and apparently lacks nectaries. On the other hand, the wholly corolloid perianth and the absence of vessels from the shoot of the Philydraceae are more compatible with the Liliales, to which they are referred by Takhtajan and by Hamann in the current Engler Syllabus. At least until studies now underway in Berlin are completed, the Philydraceae may remain as a somewhat anomalous member of the Liliales.[9]

The arrangement of the genera of Liliales into families has been the subject of much controversy. All of the other families except the Philydraceae and Pontederiaceae appear to be derived from the highly diverse and loosely knit family Liliaceae. There is room for considerable difference of opinion both as to how many terminal groups should be chopped off from the Liliaceae and as to where to draw the line between the ancestral and the derived group. I make no claim to expertise in this alliance. The treatment here presented is necessarily tentative — the more so because I have emphasized some characters (distribution of vessels and number of subsidiary cells associated with each stomate) which appear to be taxonomically important but which have been investigated in only a scattering of species.

For many years it was customary to distinguish the Amaryllidaceae from the Liliaceae by the single character of position of the ovary — inferior in Amaryllidaceae, superior in Liliaceae. This distinction broke down more than three decades ago, when it was discovered that *Yucca* (with superior ovary), *Agave* (with inferior ovary), and certain other genera form a coherent natural group marked by cytological as well as habital features. With these several genera removed to a separate family Agavaceae, the traditional distinction between the Amaryllidaceae and Liliaceae seemed less significant, and it began to appear that the traditional Amaryllidaceae were really several different groups which had independently become epigynous. For example, epigyny in *Alstroemeria* and its allies is claimed

---

[9] These studies, by Ulrich Hamann, have now been completed and are cited in the references for the order Liliales. Hamann concludes that the Philydraceae are most nearly related to the Pontederiaceae, and that these two families are probably best treated as a peripheral subgroup within the Liliales.

to be of different origin, and even different nature, from epigyny in other Amaryllidaceae. An attempt to substitute the form of the inflorescence for the position of the ovary as the critical distinction between the Liliaceae and Amaryllidaceae has also left students of the group unsatisfied. At least until these uncertainties are resolved, I prefer to submerge the Amaryllidaceae in the Liliaceae.

The question even arises as to whether the Agavaceae should be returned to the Liliaceae. The Agavaceae all have an unusual karyotype consisting of a few large chromosomes and numerous much smaller ones. Several genera, including *Yucca* and *Agave*, have five large and 25 small chromosomes. This same karyotype also occurs, however, in *Hosta*, a liliaceous genus which has not usually been considered to be closely allied to the Agavaceae. Recent embryological studies have emphasized the diversity rather than the unity of the Agavaceae, even though *Yucca* and *Agave* are themselves much alike. Further work is obviously in order.

The Smilacaceae have traditionally been retained in the Liliaceae, except by authors who subdivide that family into a number of smaller ones. However, they differ from the great bulk of the Liliaceae in having vessels distributed throughout the plant, as well as in the more obvious characters of habit. The Smilacaceae are in several respects parallel to the Dioscoreaceae and are probably equally worthy of family status.

The monotypic genus *Trichopus* is here maintained as an admittedly aberrant member of the Dioscoreaceae, in preference to establishing it as a trivial satellite family.

The families of the Liliales are marked by mixtures of obviously adaptive and seemingly nonadaptive characters. The Liliaceae and Iridaceae exploit the bulbous and cormose habit more fully than any other group of plants, but the adaptive significance of the characters which distinguish the two families from each other remains to be demonstrated. The Dioscoreaceae, Smilacaceae, and Stemonaceae are mostly broad-leaved climbers, but the floral characters which distinguish these families from each other are of doubtful ecological importance. The Taccaceae and Cyanastraceae are plants whose broad leaves may reflect adaptation to moist, shaded habitats. In this respect they resemble the otherwise very different orders Cyclanthales and Zingiberales, in other subclasses. The Avagaceae, Xanthorrhoeaceae, and Velloziaceae are another habital group, with firm, narrow leaves and often arborescent stem, occurring mostly in warm, dry places. Some of the taxonomic differences in this group (e.g., vessels in Velloziaceae, chaffy perianth in Xanthorrhoeaceae) may also be of considerable biological importance.

SYNOPTICAL ARRANGEMENT OF THE FAMILIES OF LILIALES[10]

1. Endosperm starchy (sometimes also with some fat); stomates with 2 subsidiary cells (4 or more in some Philydraceae); perianth more or less distinctly irregular; plants either aquatic, or with only a single stamen

    2. Tepals 4 (one member of the outer set derived by connation of 2, one member of the inner set missing), not united into a tube; plants terrestrial; sepal nectaries wanting; stamen solitary         **1. Philydraceae**

    2. Tepals (4) 6, usually connate below into a perianth tube; plants aquatic; septal nectaries present; stamens (including staminodia) 3 or 6         **2. Pontederiaceae**

1. Endosperm (when present) with reserve cellulose and/or fat or protein, sometimes also with some starch; stomates in most families usually without subsidiary cells; flowers with more than one stamen; plants seldom aquatic

    3. Leaves typically narrow and parallel-veined, sometimes broader and more net-veined, but only seldom with both a broad, net-veined blade and a distinct petiole

        4. Vessels chiefly or wholly confined to the roots; stomates typically without subsidiary cells; perianth and ovary seldom conspicuously hairy or glandular

            5. Habit liliaceous (mostly soft-leaved herbs from a bulb or rhizome); stamens 6 or 3

                6. Stamens (including staminodes) mostly 6, seldom more than 6 or only 4; ovary superior or less often inferior         **3. Liliaceae**

                6. Stamens 3; ovary inferior         **4. Iridaceae**

            5. Habit agavaceous (mostly coarse, often shrubby or arborescent plants with firm, perennial leaves); stamens 6

---

[10] Information on the distribution of vessels in the Cyanastraceae, Philydraceae, Stemonaceae, and Taccaceae provided by Vernon I. Cheadle, personal communication.

7. Perianth usually more or less corolloid, not chaffy; X = 16 or more, usually with a few large chromosomes and more numerous small ones **5. Agavaceae**

7. Perianth dry and chaffy; X = 11 **6. Xanthorrhoeaceae**

4. Vessels well distributed in all vegetative organs; stomates mostly with 2 subsidiary cells; perianth, or ovary, or both, commonly densely hairy or glandular

8. Shrubs or subshrubs with the leaves crowded at the top of a usually branching stem, the old leaf-bases persistent and clothing the stem; plants without branched hairs **7. Velloziaceae**

8. Scapose or subscapose herbs; inflorescence often with branched hairs **8. Haemodoraceae**

3. Leaves with a definite petiole and an expanded (often large), more or less net-veined blade

9. Plants acaulescent; vessels confined to the roots

10. Ovules numerous on parietal placentae; endosperm well developed, rich in protein; chalazosperm wanting; leaves often much cleft **9. Taccaceae**

10. Ovules 2 in each of the 3 locules (but usually only one locule fertile), on axile placentae; endosperm none; chalazosperm abundant, rich in starch and fat; leaves undivided **10. Cyanastraceae**

9. Plants leafy-stemmed, usually climbing

11. Flowers dimerous; vessels confined to the roots **11. Stemonaceae**

11. Flowers trimerous; vessels well distributed in all vegetative organs

12. Ovary superior; nectaries staminal or staminodial; plants not twining, mostly climbing by tendrils and without prominent tubers **12. Smilacaceae**

12. Ovary inferior; nectaries septal; plants mostly twining,
without tendrils, usually with large tubers
**13. Dioscoreaceae**

## SELECTED REFERENCES

Ayensu, E. S., Notes on the anatomy of the Dioscoreaceae. Ghana Jour. Sci.
5: 19–23. 1965.

————, Taxonomic status of *Trichopus*: anatomical evidence. Jour. Linn.
Soc. 59: 425–430. 1966.

Belval, H., A propos des idées de Hutchinson sur les Amaryllicacées. Bull. Soc.
Bot. Fr. 85: 486–489. 1938.

Burkill, I., The organography and the evolution of Dioscoreaceae, the family
of the yams. Jour. Linn. Soc. 56: 319–412. 1960.

Buxbaum, F., Morphologie der Blüte und Frucht von *Alstroemeria* und der
Anschluss der Alstroemerioideae bei den echten Liliaceae. Osterr. Bot. Zeits.
101: 337–352. 1954.

Cave, M. S., Cytological observations of some genera of the Agavaceae. Madrono 17: 163–170. 1964.

Clausen, R. T., A review of the Cyanastraceae. Gentes Herb. 4: 293–304.
1940.

Fahn, A., The anatomical structure of the Xanthorrhoeaceae Dumort. Jour.
Linn. Soc. 55: 158–184. 1953.

Hamann, U., Embryologische, morphologisch-anatomische und systematische
Untersuchungen an Philydraceen. Willdenowia Beih. 4: 1–178. 1966.

McKelvey, S. D., and K. Sax, Taxonomic and cytological relationships of
*Yucca* and *Agave*. Jour. Arn. Arb. 14: 76–81. 1933.

Nietsch, H., Zur systematischen Stellung von *Cyanastrum*. Osterr. Bot. Zeits.
90: 31–52. 1941.

Sato, D., Karyotype alteration and phylogeny in Liliaceae and allied families.
Jap. Jour. Bot. 12: 57–161. 1942.

Schlittler, J., Die systematische Stellung der Gattung *Petermannia* F. v. Muell.
und ihrer phylogenetischen Beziehungen zu den Luzuriagoideae Engl. und
den Dioscoreaceae Lindl. Vjschr. Naturf. Ges. Zürich 94, Beih. 1: 1–28.
1949.

Swamy, B. G. L., Observations on the floral morphology and embryology of
*Stemona tuberosa* Lour. Phytomorph. 14: 458–468. 1964.

Whitaker, T., Chromosome constitution in certain monocotyledons. Jour.
Arn. Arb. 15: 135–143. 1934.

Wunderlich, R., Die Agavaceae Hutchinsons im Lichte ihrer Embryologie,
ihres Gynözeum- Staubblatt- und Blattbaues. Osterr. Bot. Zeits. 94: 437–
502. 1950.

ORDER 2. **Orchidales**

The order Orchidales (Microspermae) as here defined consists of four families and about 20,000 species. The vast majority of the species belong to the single family Orchidaceae. The Burmanniaceae have about 130 species, the Corsiaceae about ten, and the Geosiridaceae only one. The total number of species in these three families is less than the probable error in estimating the number of species of Orchidaceae.

The Orchidales are mycotrophic, sometimes nongreen Liliidae with very numerous, tiny seeds that have an undifferentiated embryo and very little or no endosperm. The lack of differentiation of the embryo is doubtless at least in part a consequence of mycotrophy, but the number and size of seeds and the loss of endosperm are presumably due to other factors. At least there are many other mycotrophic angiosperms which have seeds of ordinary size with a well developed endosperm. The ovary in the Orchidales is always inferior and apparently does not have typical septal nectaries, although it is possible that some of the ovarian nectaries in both the Burmanniaceae and the Orchidaceae are derived from septal nectaries.

The combination of mycotrophy and numerous, tiny seeds offers certain evolutionary opportunities as well as imposing some limitations. The plants are physiologically dependent on their fungal symbionts, sometimes even for food, sometimes only for other factors as yet not fully understood, but in any case they can grow only where their symbiont finds the conditions suitable. The dustlike seeds of the Orchidales are admirably adapted to being carried by the wind and lodging in the bark of trees, and many of the Orchidaceae are epiphytes. The production of many ovules is of course of no value if the ovules do not get fertilized. One way to increase the likelihood of fertilization is to offer special attractions to a restricted set of potential pollinators, and to have the pollen stick together in masses so that many grains are transported by a single pollinator.

Only the Orchidaceae have efficiently exploited the evolutionary opportunities of the order. The Geosiridaceae, Corsiaceae, and Burmanniaceae may be regarded as groups which failed to seize the opportunity and have therefore remained relatively unsuccessful. The student should of course recognize that such phraseology is metaphorical, and that here as elsewhere it is the combination of mutation, selection, and the accidents of survival that determines the evolutionary result.

The floral characters which distinguish the Orchidaceae from the other families of the order clearly relate to their progressive specialization for insect pollination and massive transfer of pollen grains. The adaptations by different orchids to special pollinators are numerous, highly diverse, and often even comical. Such adaptations of course put the plants in thrall to their pollinators, and we have already noted that mycotrophy also imposes certain limitations. The result of the interaction of all the factors is that although the orchids are one of the largest (perhaps *the* largest) families of flowering plants, they are not among the most abundant plants in terms of number of individuals, and not many of the species have a wide geographical range or any great ecological amplitude.

The Orchidales are evidently derived from the Liliales. All of the characters by which the Orchidales differ from the Liliales represent evolutionary advances, i.e., they are phyletically acropetal. Within the Liliales, only that segment of the Liliaceae that has often been treated as the Amaryllidaceae has the characters from which those of the Orchidales might well have arisen. Hutchinson has suggested that *Apostasia* and *Neuwiedia*, here taken as primitive Orchidaceae, may be related to *Curculigo* (Liliaceae) and *Campylosiphon* (Burmanniaceae) as well as to more typical orchids.

## SYNOPTICAL ARRANGEMENT OF THE FAMILIES OF ORCHIDALES

1. Stamens 3 or more, free from the style; pollen grains not cohering in pollinia; flowers regular or less often irregular; terrestrial plants, most species with reduced leaves and without chlorophyll; nectaries, at least in the Burmanniaceae, ovarian

   2. Flowers regular or nearly so; stamens 3 or 6

      3. Stamens 3, opposite the sepals; sepals valvate
         **1. Geosiridaceae**

      3. Stamens 6, or more often 3 and opposite the petals; sepals imbricate　　**2. Burmanniaceae**

   2. Flowers highly irregular; stamens 6　　**3. Corsiaceae**

1. Stamens 1 or 2, rarely 3, mostly adnate to the style; pollen grains in most genera cohering in pollinia; flowers more or less strongly irregular; terrestrial or often epiphytic plants, with or less often without chlorophyll; nectaries diverse, but seldom ovarian　　**4. Orchidaceae**

## SELECTED REFERENCES

Ames, O., The evolution of the orchid flower. Am. Orchid Soc. Bull. 14: 355–360. 1946.

Dodson, C. H., The importance of pollination in the evolution of the orchids of tropical America. Am. Orchid Soc. Bull. 31: 525–534; 641–649; 731–735. 1962.

Dressler, R. L., and C. H. Dodson, Classification and phylogeny in the Orchidaceae. Ann. Mo. Bot. Gard. 47: 25–68. 1960.

Garay, L. A., On the origin of the Orchidaceae. Bot. Mus. Leafl. Harvard Univ. 19: 57–96. 1960.

Jonker, F. P., A monograph of the Burmanniaceae. Med. Bot. Mus. Rijksuniv. Utrecht 51: 1–279. 1938.

———, Les Géosiridacées, une nouvelle famille de Madagascar. Rec. Trav. Bot. Neerl. 36: 473–479. 1939.

Pai, R. M., Studies in the floral morphology of the Burmanniaceae. I. Vascular anatomy of the flower of *Burmannia pusilla* (Wall. ex Miers) Thw. Proc. Indian Acad. Sci. B63: 301–308. 1966.

Withner, C. L., ed., *The Orchids: A Scientific Survey*. New York: The Ronald Press, 1959 (Vol. 32 of Chronica Botanica).

# List of the classes, subclasses, orders, and families of Magnoliophyta

A. **Class Magnoliatae**

I. **Subclass Magnoliidae**

1. **Order Magnoliales**
   1. Austrobaileyaceae
   2. Lactoridaceae
   3. Magnoliaceae
   4. Winteraceae
   5. Degeneriaceae
   6. Himantandraceae
   7. Annonaceae
   8. Myristicaceae
   9. Canellaceae
   10. Illiciaceae
   11. Schisandraceae
   12. Eupomatiaceae
   13. Amborellaceae
   14. Trimeniaceae
   15. Monimiaceae
   16. Gomortegaceae
   17. Calycanthaceae
   18. Lauraceae (Cassythaceae)
   19. Hernandiaceae (Gyrocarpaceae)

2. **Order Piperales**
   1. Chloranthaceae
   2. Saururaceae
   3. Piperaceae

3. **Order Aristolochiales**
   1. Aristolochiaceae

4. **Order Nymphaeales**
   1. Nymphaeaceae (Barclayaceae, Cabombaceae)
   2. Nelumbonaceae
   3. Ceratophyllaceae

5. **Order Ranunculales**
   1. Ranunculaceae (Glaucidiaceae, Helleboraceae, Hydrastidaceae, Podophyllaceae)
   2. Circaeasteraceae (Kingdoniaceae)
   3. Berberidaceae (Nandinaceae)
   4. Lardizabalaceae (Sargentodoxaceae)
   5. Menispermaceae

**365**

   6. Coriariaceae
   7. Corynocarpaceae
   8. Sabiaceae
6. **Order Papaverales**
   1. Papaveraceae
   2. Fumariaceae (Hypecoaceae)

II. **Subclass Hamamelidae**
   1. **Order Trochodendrales**
      1. Tetracentraceae
      2. Trochodendraceae
   2. **Order Hamamelidales**
      1. Cercidiphyllaceae
      2. Eupteleaceae
      3. Platanaceae
      4. Didymelaceae
      5. Hamamelidaceae (Altingiaceae)
      6. Myrothamnaceae
   3. **Order Eucommiales**
      1. Eucommiaceae
   4. **Order Urticales**
      1. Ulmaceae
      2. Barbeyaceae
      3. Moraceae
      4. Cannabaceae
      5. Urticaceae
   5. **Order Leitneriales**
      1. Leitneriaceae
   6. **Order Juglandales**
      1. Rhoipteleaceae
      2. Picrodendraceae
      3. Juglandaceae
   7. **Order Myricales**
      1. Myricaceae
   8. **Order Fagales**
      1. Balanopaceae
      2. Fagaceae
      3. Betulaceae (Corylaceae)
   9. **Order Casuarinales**
      1. Casuarinaceae

III. **Subclass Caryophyllidae**
   1. **Order Caryophyllales**
      1. Phytolaccaceae (Achatocarpaceae, Agdestidaceae, Barbeuiaceae, Gyrostemo-naceae, Petiveriaceae, Stegonospermaceae)
      2. Nyctaginaceae
      3. Didiereaceae
      4. Cactaceae
      5. Aizoaceae (Ficoidaceae, Mesembryanthemaceae, Tetragoniaceae)
      6. Molluginaceae

7. Caryophyllaceae (Alsinaceae, Illecebraceae)
8. Portulacaceae
9. Basellaceae
10. Chenopodiaceae (Dysphaniaceae, Halophytaceae)
11. Amaranthaceae

**2. Order Batales**
1. Bataceae

**3. Order Polygonales**
1. Polygonaceae

**4. Order Plumbaginales**
1. Plumbaginaceae

**IV. Subclass Dilleniidae**

**1. Order Dilleniales**
1. Dilleniaceae
2. Paeoniaceae
3. Crossosomataceae

**2. Order Theales**
1. Ochnaceae (Lophiraceae, Strasburgeriaceae, Wallaceaceae)
2. Rhopalocarpaceae (Sphaerosepalaceae)
3. Sarcolaenaceae (Chlaenaceae)
4. Dipterocarpaceae
5. Caryocaraceae
6. Theaceae (Asteropeiaceae, Bonnetiaceae, Pellicieraceae, Pentaphylacaceae, Ternstroemiaceae, Tetrameristaceae)
7. Stachyuraceae
8. Actinidiaceae (Saurauiaceae)
9. Marcgraviaceae
10. Quiinaceae
11. Elatinaceae
12. Medusagynaceae
13. Guttiferae (Clusiaceae, Hypericaceae)

**3. Order Malvales**
1. Elaeocarpaceae
2. Scytopetalaceae
3. Tiliaceae
4. Sterculiaceae (Byttneriaceae)
5. Bombacaceae
6. Malvaceae

**4. Order Lecythidales**
1. Lecythidaceae (Asteranthaceae, Barringtoniaceae)

**5. Order Sarraceniales**
1. Sarraceniaceae
2. Nepenthaceae
3. Droseraceae

**6. Order Violales (Parietales, max. pars.)**
1. Flacourtiaceae (Lacistemaceae, Samydaceae)
2. Peridiscaceae
3. Scyphostegiaceae
4. Violaceae

    5. Turneraceae
    6. Passifloraceae
    7. Malesherbiaceae
    8. Bixaceae (Cochlospermaceae)
    9. Cistaceae
  10. Tamaricaceae
  11. Frankeniaceae
  12. Dioncophyllaceae
  13. Ancistrocladaceae
  14. Fouquieriaceae
  15. Hoplestigmataceae
  16. Achariaceae
  17. Caricaceae
  18. Loasaceae
  19. Begoniaceae
  20. Datiscaceae
  21. Cucurbitaceae

**7. Order Salicales**
    1. Salicaceae

**8. Order Capparales**
    1. Tovariaceae
    2. Capparaceae (Koeberliniaceae, Pentadiplandraceae)
    3. Cruciferae (Brassicaceae)
    4. Resedaceae
    5. Moringaceae

**9. Order Ericales**
    1. Cyrillaceae
    2. Clethraceae
    3. Ericaceae (Vacciniaceae)
    4. Epacridaceae
    5. Empetraceae
    6. Pyrolaceae
    7. Monotropaceae

**10. Order Diapensiales**
    1. Diapensiaceae

**11. Order Ebenales**
    1. Sapotaceae (Sarcospermataceae)
    2. Ebenaceae
    3. Styracaceae
    4. Lissocarpaceae
    5. Symplocaceae

**12. Order Primulales**
    1. Theophrastaceae
    2. Myrsinaceae (Aegicerataceae)
    3. Primulaceae (Coridaceae)

**V. Subclass Rosidae**

**1. Order Rosales**
    1. Eucryphiaceae
    2. Cunoniaceae (Baueraceae, Brunelliaceae)

3. Davidsoniaceae
4. Pittosporaceae
5. Byblidaceae (Roridulaceae)
6. Columelliaceae
7. Hydrangeaceae (Philadelphaceae)
8. Grossulariaceae (Brexiaceae, Escalloniaceae, Iteaceae, Montiniaceae, Phyllonomaceae, Pterostemonaceae, Tetracarpaeaceae)
9. Bruniaceae
10. Alseuosmiaceae
11. Crassulaceae
12. Cephalotaceae
13. Saxifragaceae (Eremosynaceae, Francoaceae, Parnassiaceae, Penthoraceae, Vahliaceae)
14. Rosaceae (Amygdalaceae, Drupaceae, Malaceae, Pomaceae)
15. Neuradaceae
16. Chrysobalanaceae
17. Leguminosae (Caesalpiniaceae, Fabaceae, Mimosaceae, Papilionaceae)

2. **Order Podostemales**
   1. Podostemaceae

3. **Order Haloragales**
   1. Haloragaceae
   2. Gunneraceae
   3. Hippuridaceae
   4. Theligonaceae (Cynocrambaceae)

4. **Order Myrtales**
   1. Sonneratiaceae
   2. Lythraceae
   3. Peneaceae
   4. Crypteroniaceae
   5. Thymelaeaceae (Aquilariaceae, Gonystylaceae)
   6. Trapaceae (Hydrocaryaceae)
   7. Dialypetalanthaceae
   8. Myrtaceae (Heteropyxidaceae)
   9. Punicaceae
   10. Onagraceae
   11. Oliniaceae
   12. Melastomataceae
   13. Combretaceae

5. **Order Proteales**
   1. Elaeagnaceae
   2. Proteaceae

6. **Order Cornales**
   1. Rhizophoraceae
   2. Davidiaceae
   3. Nyssaceae
   4. Alangiaceae
   5. Cornaceae (Mastixiaceae, Torricelliaceae)
   6. Garryaceae

7. **Order Santalales**
   1. Medusandraceae

  2. Dipentodontaceae
  3. Olacaceae (Aptandraceae, Erythropalaceae, Octoknemataceae, Schoepfiaceae)
  4. Opiliaceae
  5. Grubbiaceae
  6. Santalaceae
  7. Loranthaceae
  8. Misodendraceae
  9. Balanophoraceae
 10. Cynomoriaceae

**8. Order Rafflesiales**
  1. Mitrastemonaceae
  2. Hydnoraceae
  3. Rafflesiaceae (Cytinaceae)

**9. Order Celastrales**
  1. Geissolomataceae
  2. Hippocrateaceae (Capusiaceae)
  3. Celastraceae (Chingithamnaceae, Goupiaceae)
  4. Siphonodontaceae
  5. Stackhousiaceae
  6. Salvadoraceae
  7. Aquifoliaceae (Phellineaceae)
  8. Icacinaceae
  9. Cardiopteridaceae
 10. Dichapetelaceae (Chailletiaceae)

**10. Order Euphorbiales**
  1. Buxaceae (Simmondsiaceae)
  2. Euphorbiaceae
  3. Daphniphyllaceae
  4. Aextoxicaceae
  5. Pandaceae

**11. Order Rhamnales**
  1. Rhamnaceae
  2. Leeaceae
  3. Vitaceae (Ampelidaceae)

**12. Order Sapindales**
  1. Staphyleaceae
  2. Melianthaceae
  3. Greyiaceae
  4. Connaraceae
  5. Stylobasiaceae (Surianaceae)
  6. Akaniaceae
  7. Sapindaceae (Bretschneideriaceae, Dodonaeaceae)
  8. Hippocastanaceae
  9. Aceraceae
 10. Burseraceae
 11. Anacardiaceae (Podoaceae)
 12. Julianaceae
 13. Simaroubaceae (Balanitaceae, Irvingiaceae, Kirkiaceae)
 14. Cneoraceae

15. Rutaceae
16. Meliaceae
17. Zygophyllaceae (Nitrariaceae, Peganiaceae)

**13. Order Geraniales**
1. Oxalidaceae (Averrhoaceae, Lepidobotryaceae)
2. Geraniaceae (Biebersteiniaceae, Dirachmaceae, Ledocarpaceae, Vivianaceae)
3. Limnanthaceae
4. Tropaeolaceae
5. Balsaminaceae

**14. Order Linales**
1. Humiriaceae
2. Erythroxylaceae
3. Linaceae (Ctenolophonaceae, Hugoniaceae, Ixonanthaceae)

**15. Order Polygalales**
1. Malpighiaceae
2. Trigoniaceae
3. Vochysiaceae
4. Tremandraceae
5. Xanthophyllaceae
6. Polygalaceae (Diclidantheraceae)
7. Krameriaceae

**16. Order Umbellales**
1. Araliaceae
2. Umbelliferae (Apiaceae, Hydrocotylaceae)

**VI. Subclass Asteridae**

**1. Order Gentianales**
1. Loganiaceae (Antoniaceae, Potaliaceae, Spigeliaceae, Strychnaceae)
2. Gentianaceae
3. Apocynaceae (Plocospermaceae)
4. Asclepiadaceae (Periplocaceae)

**2. Order Polemoniales**
1. Nolanaceae
2. Solanaceae
3. Convolvulaceae (Dichondraceae, Humbertiaceae)
4. Cuscutaceae
5. Menyanthaceae
6. Polemoniaceae (Cobaeaceae)
7. Hydrophyllaceae
8. Lennoaceae

**3. Order Lamiales**
1. Boraginaceae (Ehretiaceae, Heliotropiaceae)
2. Callitrichaceae
3. Verbenaceae (Avicenniaceae, Chloanthaceae, Stilbeaceae, Symphoremaceae)
4. Phrymaceae
5. Labiatae (Lamiaceae, Menthaceae, Tetrachondraceae)

**4. Order Plantaginales**
1. Plantaginaceae

**5. Order Scrophulariales (Personales)**
1. Buddlejaceae

    2. Oleaceae
    3. Scrophulariaceae
    4. Myoporaceae
    5. Globulariaceae (Selaginaceae)
    6. Gesneriaceae
    7. Orobanchaceae
    8. Bignoniaceae
    9. Acanthaceae
  10. Pedaliaceae (Martyniaceae, Trapellaceae)
  11. Lentibulariaceae
  12. Hydrostachyaceae

**6. Order Campanulales**
    1. Pentaphragmataceae
    2. Sphenocleaceae
    3. Campanulaceae (Lobeliaceae)
    4. Stylidiceae (Donatiaceae)
    5. Brunoniaceae
    6. Goodeniaceae

**7. Order Rubiales**
    1. Rubiaceae

**8. Order Dipsacales**
    1. Caprifoliaceae (Sambucaceae)
    2. Adoxaceae
    3. Valerianaceae
    4. Dipsacaceae
    5. Calyceraceae

**9. Order Asterales**
    1. Compositae (Ambrosiaceae, Asteraceae, Carduaceae, Cichoriaceae)

**B. Class Liliatae**

**I. Subclass Alismatidae**

**1. Order Alismatales**
    1. Butomaceae
    2. Limnocharitaceae
    3. Alismataceae

**2. Order Hydrocharitales**
    1. Hydrocharitaceae

**3. Order Najadales**
    1. Aponogetonaceae
    2. Scheuchzeriaceae
    3. Juncaginaceae (Lilaeaceae)
    4. Najadaceae
    5. Potamogetonaceae
    6. Ruppiaceae
    7. Zannichelliaceae
    8. Zosteraceae (Cymodoceaceae, Posidoniaceae)

**4. Order Triuridales**
    1. Petrosaviaceae
    2. Triuridaceae

II. **Subclass Commelinidae**
1. **Order Commelinales**
1. Rapateaceae
2. Xyridaceae
3. Mayacaceae
4. Commelinaceae (Cartonemataceae)
2. **Order Eriocaulales**
1. Eriocaulaceae
3. **Order Restionales**
1. Flagellariaceae (Hanguanaceae)
2. Anarthriaceae
3. Ecdeiocoleaceae
4. Restionaceae
5. Centrolepidaceae
4. **Order Juncales**
1. Juncaceae
2. Thurniaceae
5. **Order Cyperales**
1. Cyperaceae
2. Gramineae (Poaceae)
6. **Order Typhales**
1. Sparganiaceae
2. Typhaceae
7. **Order Bromeliales**
1. Bromeliaceae
8. **Order Zingiberales**
1. Strelitziaceae
2. Lowiaceae
3. Heliconiaceae
4. Musaceae
5. Zingiberaceae
6. Costaceae
7. Cannaceae
8. Marantaceae

III. **Subclass Arecidae**
1. **Order Arecales**
1. Palmae (Arecaceae)
2. **Order Cyclanthales**
1. Cyclanthaceae
3. **Order Pandanales**
1. Pandanaceae
4. **Order Arales**
1. Araceae
2. Lemnaceae

IV. **Subclass Liliidae**
1. **Order Liliales**
1. Philydraceae
2. Pontederiaceae

   3. Liliaceae (Alliaceae, Alstroemeriaceae, Aphyllanthaceae, Amaryllidaceae, Asparagaceae, Colchicaceae, Hypoxidaceae, Melanthiaceae, Petermanniaceae, Philesiaceae, Ruscaceae, Trilliaceae)
   4. Iridaceae
   5. Agavaceae (Dracaenaceae)
   6. Xanthorrhoeaceae
   7. Velloziaceae
   8. Haemodoraceae (Tecophilaeaceae)
   9. Taccaceae
   10. Cyanastraceae
   11. Stemonaceae (Roxburghiaceae)
   12. Smilacaceae
   13. Dioscoreaceae (Stenomeridaceae, Trichopodaceae)
**2. Order Orchidales**
   1. Geosiridaceae
   2. Burmanniaceae (Thismiaceae)
   3. Corsiaceae
   4. Orchidaceae (Apostasiaceae)

# Glossary

The definitions in the glossary are principally for usage within the angiosperms. Many of the terms also have a broader meaning, or a different meaning in some other group.

**acotyledonous**   Without cotyledons.

**acropetal**   Near the tip or distal end, as opposed to near the base or proximal end; proceeding from the proximal to the distal end.

**adnate**   Grown together or attached; applied only to unlike organs, as stipules adnate to the petiole, or stamens adnate to the corolla.

**adventitious**   Originating from mature nonmeristematic tissues, especially if such a development would not ordinarily be expected.

**aestivation**   The arrangement of flower parts in the bud, especially the position of the petals (or sepals) with respect to each other.

**ament**   A dense, bracteate spike or raceme with a nonfleshy axis bearing many small, naked or apetalous flowers; a catkin.

**Amentiferae**   A group (now considered to be artificial) of dicotyledons characterized by the production of flowers in aments.

**amphitropous ovule**   An ovule with the body half-inverted, so that the funiculus is attached near the middle, and the micropyle points at right angles to the funiculus; hemianatropous.

**anatropous ovule**   An ovule with the body fully inverted, so that the micropyle is basal, adjoining the funiculus.

**androecium**   All of the stamens of a flower, considered collectively.

**androgynophore**   A common stalk, arising from the receptacle, on which both the androecium and the gynoecium are borne.

**anemophilous**   Pollinated by wind.

**angiosperm**   A member of the group of plants (Magnoliophyta) characterized by having the ovules enclosed in an ovary.

**anisomerous**   With a different number of parts; usually applied to parts of a small and definite number which is less than the number of some other kind of part. E.g., if there are five sepals and five petals but fewer than five stamens, the stamens are anisomerous.

**anthesis**   The period during which a flower is fully expanded and functional.

**anthocyanin**   A chemical class of flavonoid pigments, ranging in color from blue or violet to purple or red, often found in the central vacuole of a cell, especially in petals.

*anthoxanthin* A group of flavonoid pigments closely allied to anthocyanin, but ranging in color from yellow or orange to orange-red, differing chemically from the anthocyanins primarily in that the heterocyclic ring is more oxidized.

*antipetalous* In front of (on the same radius as) the petal(s).

*antisepalous* In front of (on the same radius as) the sepal(s).

*aperturate* Having one or more apertures or openings; as applied to pollen grains, having one or more thin spots in the wall or gaps in the exine.

*apetalous* Without petals.

*apocarpous* With the carpels free from each other (or with only one carpel).

*apotropous ovule* An ovule which (if erect) has the raphe ventral (between the partition and the body of the ovule), or which (if pendulous) has the raphe dorsal; a sapindaceous ovule.

*appendiculate* Having one or more appendages.

*archegonium* A specialized structure, composed of more than one cell, within which an egg is produced.

*aril* A specialized outgrowth from the funiculus which covers or is attached to the mature seed; more loosely, any appendage or fleshy thickening of the seed coat.

*autotrophic* Nutritionally independent, making its own food from raw materials obtained more or less directly from the substrate; usually interpreted to include mycorhizal as well as nonmycorhizal plants, so long as they are photosynthetic.

*axile placenta* A placenta along the central axis (or along the vertical midline of the septum) of an ovary with two or more locules.

*berry* The most general type of fleshy fruit, derived from a single ovary, and with the pericarp fleshy throughout.

*betacyanin* A chemical class of nitrogenous, water-soluble pigments, ranging in color from blue or violet to purple or red, found in the central vacuole of a cell, especially in petals, in some kinds of plants.

*betalain* A chemical class of nitrogenous, water-soluble pigments, consisting of betacyanins and betaxanthins.

*betaxanthin* A chemical class of pigments closely allied to betacyanins, but ranging in color from yellow or orange to orange-red.

*bisporangiate strobilus* A strobilus with both micro- and megasporangia.

*bisporic embryo sac* An embryo sac derived from two megaspores.

*bitegmic ovule* An ovule with two integuments.

*bract* Any more or less reduced or modified leaf associated with a flower or an inflorescence, but not part of the flower itself.

*calyx* All of the sepals of a flower collectively.

*campylotropous ovule* An ovule with the body distorted by unequal growth, so that the micropyle is brought near the funiculus and chalazal end.

*capsule* A dry, dehiscent fruit composed of more than one carpel.

*carpel* The fertile leaf (megasporophyll) of an angiosperm, which bears the ovules.

*carpophore*   The part of the receptacle which in some kinds of flowers is prolonged between the carpels as a central axis.

*caruncle*   An excrescence near the hilum of some seeds; a sort of aril.

*catkin*   An ament.

*cellular endosperm*   Endosperm in which the formation of partitions (cell
walls) follows immediately after mitosis, even in the earliest ontogeny, so
that there is no initial free-nuclear stage.

*chalaza*   The part of the ovule or seed which lies at the opposite end from the
micropyle, in line with the antipodal end of the embryo sac.

*colpate pollen grain*   A pollen grain with the aperture(s) small and nearly
isodiametric.

*compound ovary or pistil*   An ovary or pistil composed of more than one carpel.

*connate*   Grown together or attached; applied only to like organs, as filaments
connate into a tube, or leaves connate around the stem.

*connective*   The tissue which connects the pollen sacs of an anther.

*contorted (aestivation)*   Same as convolute.

*convolute*   Arranged so that each petal (or sepal) has one edge exposed (covering the adjoining petal) and the other edge covered, no one member being
wholly external or wholly internal to the others.

*corolla*   All of the petals of a flower collectively.

*corona*   A set of petal-like structures or appendages between the corolla and
the androecium, derived by modification of the corolla or of the androecium.

*cotyledon*   A leaf of the embryo of a seed.

*crassinucellate*   With the nucellus several cells thick, at least at the micropylar
end.

*cyanogenetic*   Producing cyanide.

*cyme*   A broad class of inflorescences characterized by having the terminal
flower bloom first, commonly also with the terminal flower of each branch
blooming before the others on that branch.

*dehiscent*   Opening at maturity, releasing or exposing the contents.

*deliquescent*   With the central axis melting away irregularly into a series of
smaller branches.

*dichotomous*   Forking more or less regularly into two branches of about equal
size.

*dioecious*   Producing male and female flowers on separate individuals.

*diplostemonous*   With two cycles of stamens.

*disk (of a flower)*   An outgrowth from the receptacle, surrounding the base
of the ovary; often derived by reduction of the innermost set of stamens.

*distichous*   Arranged in two vertical or longitudinal rows, as the leaves of
an iris.

*distinct*   Not connate with similar organs.

*drupe*   A fleshy fruit with a firm endocarp that permanently encloses the
usually solitary seed, or with a portion of the endocarp separately enclosing
each of two or more seeds.

*dyad*   A set or group of two.

*embryo sac*  The female gametophyte of an angiosperm, within which the embryo begins to develop.

*endosperm*  The food storage tissue of a seed that is derived from the triple fusion nucleus of the embryo sac.

*entomophagous*  Insect-eating; insectivorous.

*entomophilous*  Pollinated by insects.

*epigynous*  With the perianth and stamens attached near the top of the ovary, rather than beneath it, i.e., with the ovary inferior.

*epipetalous*  Attached to the petals or corolla.

*epiphyte*  A plant without connection to the soil, growing attached to another plant but not deriving its food or water from it.

*epitropous ovule*  An ovule which (if erect), has the raphe dorsal, or which (if pendulous) has the raphe ventral (between the partition and the body of the ovule).

*exine*  The outer wall layer of a pollen grain.

*exomorphic*  Pertaining to external form.

*extrastaminal*  Outside of (as opposed to within) the androecium.

*follicetum*  A group of lightly cohering follicles.

*follicle*  A fruit, derived from a single carpel, which dehisces along one suture (the seed-bearing one) at maturity.

*free-central placenta*  A placenta consisting of a free-standing column or projection from the base of a compound, unilocular ovary.

*free-nuclear*  With scattered nuclei that are not separated by partitions or cell walls.

*funiculus*  The stalk of an ovule.

*gametophyte*  The generation which has *n* chromosomes and produces gametes as reproductive bodies. In angiosperms the female gametophyte is the embryo sac and the male gametophyte is the pollen grain.

*gamophyllous*  With the leaves (or segments) connate.

*gamopetalous*  With the petals connate, at least toward the base; sympetalous.

*gamosepalous*  With the sepals connate, at least toward the base; synsepalous.

*geraniaceous ovule*  An epitropous ovule.

*gymnosperm*  A member of a group of plants characterized by having ovules that are not enclosed in an ovary.

*gynobasic style*  A style which is attached directly to the receptacle as well as to the base of the carpel(s).

*gynoecium*  All of the carpels of a flower, collectively.

*gynophore*  A central stalk in some flowers, bearing the gynoecium.

*halophyte*  A plant adapted to growth in salty soil.

*haplostemonous*  With a single cycle of stamens.

*helobial endosperm*  Endosperm in which the first division of the triple fusion nucleus is followed by the formation of a partition, after which one chamber develops along the nuclear pattern, and the other along the cellular pattern.

*heterotrophic*  Parasitic or saprobic, as opposed to autotrophic.

*hydrophyte*   A plant adapted to life in the water.

*hypanthium*   A ring or cup around the ovary, usually formed by the union of the lower parts of the calyx, corolla, and androecium; when the petals and stamens appear to arise from the calyx tube, that part of the apparent calyx tube which is below the attachment of the petals is the hypanthium.

*hypogynous*   With the perianth and stamens attached directly to the receptacle; more generally, beneath the gynoecium.

*idioblast*   An individual cell that is markedly different from those surrounding it, without being an essential part of an obviously integrated, functioning system.

*imbricate*   Arranged in a tight spiral, so that the outermost member has both edges exposed, and at least the innermost member has both edges covered; more loosely, a shingled arrangement.

*indusium*   A more or less cupulate outgrowth from the style, below the stigmas.

*inferior ovary*   An ovary with the other floral parts (calyx, corolla, and androecium) attached to its summit.

*integument*   One of the one or two layers which partly enclose the nucellus of the ovule; the forerunner of the seed coat.

*intercalary meristem*   A meristem, detached from the apical meristem, which produces primary tissues; e.g., the meristem at the base of the leaf blade of grasses.

*internal phloem*   Primary phloem which lies between the primary xylem and the pith; intraxylary phloem.

*intrastaminal*   Within (as opposed to outside of) the androecium.

*involucel*   Diminutive of involucre; an involucre of the second order.

*involucre*   Any structure which surrounds the base of another structure; in angiosperms usually applied to a set of bracts beneath an inflorescence.

*irregular flower*   A flower in which the petals (or less often the sepals) are dissimilar in form or orientation.

*isomerous*   With the same number of parts as something else; a flower that has isomerous stamens has the same number of stamens as sepals or petals.

*jaculator*   A modified funiculus which expels the seeds in the Acanthaceae.

*karyotype*   All of the chromosomes of a nucleus, with reference to relative size and shape as well as number.

*laminar*   Thin and flat, as in a leaf blade.

*laminar placentation*   An arrangement of the ovules on the ventral surface (rather than the margins) of the carpel.

*latex*   A colorless to more often white, yellow, or reddish liquid, produced by some plants, characterized by the presence of colloidal particles of terpenes dispersed in water.

*laticifer*   A tube containing latex.

*liana*   A climbing, woody vine.

*locule*   A seed cavity (chamber) in an ovary or fruit; a compartment in any container.

*loculicidal*   Dehiscing along the midrib or outer median line of each locule, i.e., "through" the locules.

*lycopsid*   A member of a group of plants characterized by alternate or opposite microphylls and axillary sporangia.

*megaphyll*   A leaf of the type associated with leaf gaps in the stele, typically with a branching vein system; opposite of microphyll.

*megaphyte*   A thick-stemmed, soft shrub that is anatomically much like an herb.

*megasporangium*   A sporangium which produces megaspores.

*megaspore*   A spore which may develop into a female gametophyte.

*megasporophyll*   A sporophyll which bears one or more megasporangia.

*mericarp*   An individual carpel of a schizocarp.

*mesophyte*   A plant adapted to an ordinary amount of water stress; intermediate between hydrophyte and xerophyte.

*microphyll*   A leaf, usually small, with an unbranched midvein whose departure from the stele does not leave a gap; opposite of megaphyll.

*micropyle*   The opening through the integuments of an ovule to the nucellus.

*microsporangium*   A sporangium which produces microspores.

*microspore*   A spore which develops into a male gametophyte.

*microsporophyll*   A sporophyll which bears one or more microsporangia.

*monad*   One of many individuals which are free from each other rather than attached in groups.

*monadelphous stamens*   Stamens with the filaments connate.

*monocarpic*   Blooming only once and then dying; usually applied to certain perennials, as the century plant, but technically applicable to annuals and biennials as well.

*monoecious*   With unisexual flowers, both types borne on the same individual.

*monophyletic*   With a unified evolutionary ancestry.

*monopodial*   With the branches or appendages arising from a simple axis. Compare sympodial.

*monosporic*   Derived from a single spore.

*multilacunar node*   A node with more than three gaps.

*mycorhiza*   A symbiotic association of a fungus and the root of a vascular plant; by extension, other associations of fungi with higher plants.

*mycotrophic*   Evidently modified as a result of mycorhizal association.

*myrosin*   An enzyme involved in the formation of mustard oil.

*nectary*   Any structure which produces nectar, usually in association with a flower.

*neoteny*   A synchronic disharmony in which a plant attains sexual maturity while remaining permanently juvenile in some vegetative features.

*node*   A place on a stem where a leaf is (or has been) attached.

*nonaperturate*   Without apertures.

nucellus   The megasporangial wall of a seed plant, which typically encloses the female gametophyte.

nuclear endosperm   Endosperm which has a free-nuclear stage in early ontogeny, the formation of partitions being delayed until several or many mitotic divisions have occurred.

ontogeny   The developmental history of an individual.

ornithophilous   Pollinated by birds.

orthotropous ovule   A straight (unbent) ovule with the micropyle at the opposite end from the stalk.

parietal placenta   A placenta along the walls or on the intruded partial partitions of a compound, unilocular ovary.

pentalacunar node   A node with five gaps.

pentamerous   With five parts of a kind.

perfect flower   A flower with both an androecium and a gynoecium.

perianth   All of the sepals and petals (or tepals) of a flower, collectively.

pericarp   The ovary wall of a fruit.

perigynous   Having the perianth and stamens united into a basal saucer or cup (the hypanthium) distinct from the ovary; more generally, around the base of the gynoecium, as a perigynous disk.

perisperm   Food storage tissue in the seed, derived from the nucellus.

petal   A member of the second set of floral leaves (i.e., the set just internal to the sepals), usually colored or white and serving to attract pollinators.

phenetic   Pertaining to the expressed characteristics of an individual, without regard to its genetic nature.

phylad   A natural group, of whatever rank, considered from the standpoint of its evolutionary history or prospects; an evolutionary line.

phyllosporous   With the ovules borne on specialized or modified leaves, rather than on the ends of telomes.

phylogeny   The evolutionary history of a group.

pistil   The female organ of a flower, composed of one or more carpels, and ordinarily differentiated into ovary, style, and stigma.

placenta   The tissue of the ovary to which the ovules are attached.

pollen   The mass of young male gametophytes (pollen grains) of a seed plant, at the stage when they are released from the anther.

pollinium   A coherent cluster of many pollen grains, transported as a unit during pollination.

polypetalous   With the petals separate from each other.

polyphyletic   Of more than one evolutionary origin.

polystemonous   With many stamens.

poricidal   Opening by pores.

pseudanthium   A cluster of small or reduced flowers, collectively simulating a single flower.

pseudomonomerous ovary   An ovary seemingly composed of a single carpel, but phyletically derived from a compound ovary.

pteridosperm   A seed fern; a member of the class Lyginopteridatae.

*pteropsid* A member of a group of vascular plants characterized by having megaphylls and leaf gaps.

*quincuncial* Consisting of five members, in a 2/5 phyllotaxy, so that two members have both margins exposed, two have both margins covered, and one has one margin exposed and the other covered.

*raceme* An inflorescence with pedicellate flowers arising in acropetal sequence from an unbranched central axis.

*racemose* Pertaining to a broad class of inflorescences characterized by flowering in acropetal sequence.

*rachis* A main axis, such as that of a compound leaf.

*raphe* The part of the funiculus which is permanently adnate to the integument of the ovule, commonly visible as a line or ridge on the mature seed coat.

*raphide* A needle-shaped crystal; raphides occur in certain cells of some kinds of plants.

*receptacle* The end of the stem, to which the other flower parts are attached.

*replum* A persistent, septumlike or framelike placenta which bears ovules on the margins, as in the Cruciferae.

*ruminate* Irregularly ridged and sulcate, looking as though chewed.

*samara* An indehiscent, winged fruit.

*sapindaceous ovule* An apotropous ovule.

*scalariform* Ladderlike, with cross-bars connecting vertical members; a scalariform vessel has the secondary wall in a ladderlike pattern.

*schizocarp* A fruit which splits into separate carpels at maturity.

*schizogenous* Originating by splitting or separation of tissue.

*sclerophyll* A firm leaf, with a relatively large amount of strengthening tissue, which retains its firmness (and often also its shape) even when physiologically wilted.

*sepal* A member of the outermost set of floral leaves, typically green or greenish and more or less leafy in texture.

*septicidal* Splitting through the septa, so that the carpels are separated.

*septum* A partition; in an ovary, a partition formed by the connate walls of adjacent carpels.

*spadix* A spike with small, crowded flowers on a thickened, fleshy axis.

*spathe* A large, usually solitary bract subtending and often enclosing an inflorescence; the term is used only in the monocotyledons.

*sphenopsid* A member of a group of vascular plants characterized by whorled leaves and terminal strobili.

*sporophyll* A leaf (often more or less modified) which bears one or more sporangia.

*stachyosporous* With the ovules borne terminally on telomes, rather than on specialized or modified leaves.

*stamen* The microsporophyll of an angiosperm.

*staminode* A modified, infertile stamen.

*stele*   The primary vascular structure of a stem or root, together with any tissues (such as the pith) which may be enclosed.

*stigma*   The part of the pistil which is receptive to pollen.

*strobilus*   A cluster of sporophylls on an axis; a cone.

*subsidiary cells (of the stomatal apparatus)*   The modified epidermal cells immediately adjacent to the guard cells.

*succulent*   A plant which accumulates reserves of water in the fleshy stems or leaves, due largely to the high proportion of hydrophilic colloids in the cell sap.

*superior ovary*   An ovary which is attached to the receptacle above the level of attachment of the other flower parts.

*Sympetalae*   A group (now considered to be artificial) of dicotyledons characterized by having the petals connate at least toward the base.

*sympetalous*   With the petals connate, at least toward the base.

*sympodial*   With the apparent main axis consisting of a series of usually short axillary branches.

*syncarpous*   With the carpels united to form a compound pistil.

*synsepalous*   With the sepals connate, at least toward the base; gamosepalous.

*tapetum*   A nutritive tissue of the anther which degenerates during the development of the pollen.

*taxon (pl. taxa)*   Any taxonomic entity, or whatever rank.

*taxonomy*   A study aimed at producing a system of classification of organisms which best reflects the totality of their similarities and differences.

*telome*   An ultimate branch of a dichotomously branching stem.

*tenuinucellate*   With the nucellus consisting of a single layer of cells.

*tepal*   A sepal or petal, or a member of an undifferentiated perianth.

*tetrad*   A group of four; especially a group of four spores or pollen grains.

*tetradinous*   Coherent in tetrads.

*tetrasporic embryo sac*   An embryo sac derived from four megaspores.

*thalloid*   Resembling or consisting of a thallus.

*thallus*   A plant body which is not clearly differentiated into roots, stems, and leaves.

*tracheid*   The most characteristic cell type in xylem, being long, slender, tapered at the ends, with a lignified secondary wall and without living contents at maturity.

*translator*   A structure connecting the pollinia of adjacent anthers in the Asclepiadaceae.

*triaperturate*   With three apertures.

*trilacunar node*   A node with three gaps.

*trilocular*   With three locules.

*trimerous*   With parts in sets of three.

*umbel*   A racemose inflorescence with greatly abbreviated axis and elongate pedicels; in a compound umbel the primary branches are again umbellately branched at the tip.

*uniaperturate*   With a single aperture.

unilacunar node   A node with a single gap.

unilocular   With a single locule.

unisexual   With androecium but not gynoecium, or with gynoecium but not androecium.

unitegmic ovule   An ovule with a single integument.

valvate (aestivation)    Arranged with the margins of the petals (or sepals) adjacent throughout their lengths, without overlapping.

vasicentric   Concentrated around the vessels.

vessel   A xylem tube formed from several vessel segments (modified tracheids with imperfect or no end-walls) set end to end.

xeromorphic   With a form suggesting adaptation to dry conditions.

xerophyte   A plant adapted to life in dry places.

# INDEX

# Index

Names in italics are those of taxonomic or nomenclatural synonyms, not here admitted as accepted names of accepted taxa. Page numbers in bold face indicate illustrations.

# T